Chemomechanical Instabilities
in Responsive Materials

NATO Science for Peace and Security Series

This Series presents the results of scientific meetings supported under the NATO Programme: Science for Peace and Security (SPS).

The NATO SPS Programme supports meetings in the following Key Priority areas: (1) Defence Against Terrorism; (2) Countering other Threats to Security and (3) NATO, Partner and Mediterranean Dialogue Country Priorities. The types of meeting supported are generally "Advanced Study Institutes" and "Advanced Research Workshops". The NATO SPS Series collects together the results of these meetings. The meetings are co-organized by scientists from NATO countries and scientists from NATO's "Partner" or "Mediterranean Dialogue" countries. The observations and recommendations made at the meetings, as well as the contents of the volumes in the Series, reflect those of participants and contributors only; they should not necessarily be regarded as reflecting NATO views or policy.

Advanced Study Institutes (ASI) are high-level tutorial courses intended to convey the latest developments in a subject to an advanced-level audience

Advanced Research Workshops (ARW) are expert meetings where an intense but informal exchange of views at the frontiers of a subject aims at identifying directions for future action

Following a transformation of the programme in 2006 the Series has been re-named and re-organised. Recent volumes on topics not related to security, which result from meetings supported under the programme earlier, may be found in the NATO Science Series.

The Series is published by IOS Press, Amsterdam, and Springer, Dordrecht, in conjunction with the NATO Public Diplomacy Division.

Sub-Series

A.	Chemistry and Biology	Springer
B.	Physics and Biophysics	Springer
C.	Environmental Security	Springer
D.	Information and Communication Security	IOS Press
E.	Human and Societal Dynamics	IOS Press

http://www.nato.int/science
http://www.springer.com
http://www.iospress.nl

Series A: Chemistry and Biology

Chemomechanical Instabilities in Responsive Materials

edited by

P. Borckmans
Service de Chimie Physique et Biologie Théorique
Université Libre de Bruxelles, Belgium

P. De Kepper
Centre de Recherche Paul Pascal
Université de Bordeaux, France

A. R. Khokhlov
Chair of Physics of Polymers and Crystals
Moscow State University, Russia

and

S. Métens
Matière et Systèmes Complexes
Université Paris 7 - Denis Diderot, France

Springer

Published in cooperation with NATO Public Diplomacy Division

Proceedings of the NATO Advanced Study Institute on
Morphogenesis Through the Interplay of Nonlinear Chemical Instabilities
and Elastic Active Media
Cargese, Corsica, France
2–14 July 2007

Library of Congress Control Number: 2009929396

ISBN 978-90-481-2992-8
ISBN 978-90-481-2991-1
ISBN 978-90-481-2993-5 (eBook)

Published by Springer,
P.O. Box 17, 3300 AA Dordrecht, The Netherlands.

www.springer.com

Printed on acid-free paper

CONTENTS

FOREWORD

The present volume includes most of the material of the invited lectures delivered at the NATO Advanced Study Institute "Morphogenesis through the interplay of nonlinear chemical instabilities and elastic active media" held from 2th to 14th July 2007 at the Institut d'Etudes Scientifiques de Cargèse (http://www.iesc.univ-corse.fr/), in Corsica (France). This traditional place to organize Summer Schools and Workshops in a well equipped secluded location at the border of the Mediterranean sea has, over many years now, earned an increasing deserved reputation.

Non-linear dynamics of non equilibrium systems has worked its way into a great number of fields and plays a key role in the understanding of self-organization and emergence phenomena in domains as diverse as chemical reactors, laser physics, fluid dynamics, electronic devices and biological morphogenesis. In the latter case, the viscoelastic properties of tissues are also known to play a key role.

The control and formulation of soft responsive or "smart" materials has been a fast growing field of material science, specially in the area of polymer networks, due to their growing applications in bio-science, chemical sensors, intelligent microfluidic devices, ... Nature is an important provider of active materials whether at the level of tissues or at that of sub-cellular structures. As a consequence, the fundamental understanding of the physical mechanisms at play in responsive materials also shines light in the understanding of biological artefacts.

Such wide program is overly too ambitious, and the organizers approached it by selecting topics, to be made to interact quite naturally, with aspects of their own research interest.

On the one hand, nonlinear chemistry deals with autocatalytic chemical reactions that give rise to temporal oscillations, and propagating or stationary periodical waves of concentration (Turing patterns) through their interplay with matter diffusion. Such organization of reaction-diffusion system can, however, only show up in far from equilibrium condition. To sustain and control such dissipative structures, reactors have lately been built around a core of hydrogel for protection against the perturbative action of hydrodynamic currents, that arise mainly from the feeding of the reactants and excretions of the reactive products.

On the other hand, the swelling-shrinking properties of gels have lately also been well advertised as a source for a host of possible applications Among the first, soft contact lenses were developed and they were also used in the pharmaceutical and cosmetic fields. Their super absorbent properties also lead to application to sanitary items, disposable diapers. Therefore their study has become a field of research in itself. The discovery of their "volume phase transition" in response to external stimuli, around 1978, may be considered as a turning point. It triggered research to use gels as functional materials for artificial muscle, actuators, stimuli-responsive drug delivery systems, separation or purification means, valves, etc.

One of the current means to obtain large volume changes is to design gels sensitive to the chemical composition of the solvent impregnating their network. Therefore the mode of interaction of the two fields appeared obvious and became the theme of the School.

Its primary objective was to provide students and researchers, which are specialists of one or either of the two fields, with a sample of courses going from bottom ground to advanced ideas. Emphasis was brought on potential developments and opportunities that arises from the cross-fertilization of the two fields. The courses were provided by world wide distinguished experts, actively engaged in their respective domain, and covered both fundamental theoretical aspects and experiments.

The first, obvious set of lectures, treated nonlinear chemical reactions and the associate development of reaction-diffusion patterns which are the most widely used chemical engines in autonomous chemomechanical devices (De Kepper).

The theoretical fundamentals to deal with dynamical self-organization phenomena in nonlinear systems operating far from thermodynamic equilibrium are a must to understand and go beyond simple observation (Borckmans).

The specificity of polymer reactions was also addressed, as in some cases, these reactions can undergo dynamical instabilities of their own (Pojman).

The design concept of functional materials based on molecular synchronization has just started in the field of materials science and engineering, and polymer gels are becoming more important as a material which realizes the new concept. The advanced chemical formulation of novel functional gels with self-oscillating function has been reviewed (Yoshida).

The theoretical progresses in the description of systems combining nonlinear chemical kinetics and the swelling properties of visco-elastic materials were also presented (Boissonade, Métens). Emphasis was brought on emergent dynamical properties coming from the interplay of reaction and these materials/gels (Boissonade, Siegel). It was also pointed out that further developments are necessary to describe the actual properties of gels out of equilibrium. The internal stress, defined as the stress that is maintained within a system by itself, without the aid of external supports or constraints plays an important role (Sekimoto).

It is often asserted as a drawback that gel machines working on the transduction of energy between chemistry and mechanical property would age quite rapidly and become fragile when cycling due to the development of cracks. It was shown, in recent formulations of double network gels that this challenge could be strikingly overcome (Osada).

Some applications of advanced gel materials sensitive to chemistry and autonomous drug delivery devices were described (Yoshida, Siegel).

Chemical light-sensitive media were considered on a new interesting perspective in the context of information processing. A novel class of image evolution modes has been designed on materials exhibiting oscillating swelling-shrinking cycles in the presence of an oscillatory chemical reaction (Khokhlov).

Lectures were also devoted to the presentation of some passive and active motion in biology. Vesicles systems are studied under non-equilibrium conditions through shear flow. Some rheological properties as stress average and effective viscosity are presented. Actin gels, that play an important role in living systems were also analyzed through a model that involves symmetry breaking transitions (Misbah).

The school has provided optimal conditions to stimulate contacts between young and senior scientists. All of the young scientists have also received the opportunity to present their works and to discuss them with the lecturers during two posters sessions that spanned the duration of the School. Additional general discussions took place during several round tables. More than 60 lecturers and students from 15 countries have participated in the ASI.

We are grateful to the North Atlantic Treaty Organization for their valuable support of the Advanced Study Institute, that lies at the origin of the publication of this volume. The meeting was an opportunity for a warm interactive atmosphere besides the scientific exchanges. We want to warmly thank those who, locally, contributed to its success: the director, Professor Elisabeth Dubois-Violette, and the staff of the Institute d'Etudes Scientifiques de Cargèse, Brigitte, Dominique, Nathalie, and Pierre-Eric. Financial support from the Collectivités Territoriale Corse is also acknowledged.

P. Borckmans, P. De Kepper, A. Khokhlov, and S. Métens

FROM SUSTAINED OSCILLATIONS TO STATIONARY REACTION-DIFFUSION PATTERNS

P. De Kepper (dekepper@crpp-bordeaux.cnrs.fr),
J. Boissonade (boisson@crpp-bordeaux.cnrs.fr)
*Université de Bordeaux and CNRS, Centre de recherche Paul Pascal,
115,av. Schweitzer, F-33600 Pessac, France*

I. Szalai (pisti@chem.elte.hu)
*Institute of Chemistry, L. Eötvös University,
P.O. Box 32, H-1518 Budapest 112, Hungary*

Abstract. A brief overview of the developments of oscillating chemical reactions and sustained reaction diffusion patterns is presented. Focus is made on experimental tools and know-hows to study and create these nonequilibrium time and space chemical structures. Different specific examples are provided.

Keywords: oscillations, bistability, reaction-diffusion, chemical patterns, excitability waves, Turing patterns

1. Oscillatory chemical systems

1.1. INTRODUCTION

Chemically driven oscillating systems have been reported since at least the end of the 19th century. Among these early observations, the most noticeable are the periodic dissolution of metals [1], the periodic oxidation of phosphorous vapor in bottles, and the periodic decomposition of hydrogen peroxide on metallic mercury [2]. For an early history of chemical oscillatory systems, see the review by Hedges and Meyer [3].

Until the first decade of the 20th century, all oscillatory chemical system involved an heterogeneous process and the overall phenomenon was not clearly understood. When the first homogeneous kinetic oscillatory reactions were discovered by Bray [4] in 1921 and by Belousov [5] in 1951, the concept of homogeneous oscillator was discarded by most chemists of the time. The paradigm of an oscillator was the pendulum where the oscillations of the position overshoot the equilibrium position due to the mechanical inertia of the system. Thermodynamic principles clearly establish that chemical

P. Borckmans et al. (eds.), *Chemomechanical Instabilities in Responsive Materials*,
© Springer Science + Business Media B.V. 2009

1

reactions cannot overshoot their equilibrium composition, so that no homogeneous reaction process can move back and forth across the equilibrium value. Hence, it was thought that any observed oscillatory chemical system had to be associated to an "extra physical mechanism" to make apparent periodic overshooting of the equilibrium possible. However, electrochemical oscillators [6] were known since 1828, but were seldom considered as valid dynamical analogs of what was observed in the pioneering works of Bray and Belousov. It took Bray 3 years to convince himself, through careful patient experimentation, that the oscillatory phenomenon he was observing, was the result of a homogeneous kinetic process. Even so, the homogeneous hypothesis was rejected by most other chemists. Thirty years later, it took seven years to Belousov to have his observations of another oscillatory reaction to be accepted for publication in an obscure Russian medical journal without peer review [5]. The observation of chemical oscillations in isothermal single phase systems became more widely accepted only after the theoretical developments of thermodynamic of irreversible process by the School of Brussels, lead by the 1977 Nobel price of chemistry, Ilya Prigogine [7]. These formal theoretical studies showed that homogeneous chemical kinetic systems could admit periodic dynamic solutions if they involved appropriate nonlinear mechanisms and if the system evolved at finite distance from thermodynamic equilibrium. The conceptual break was further admitted when, in the early 1970s, Noyes and his collaborators provided a kinetic explanation [8,9] for the oscillatory behavior of the reaction discovered by Belousov, in terms of admitted halogen chemistry kinetics. At that time more detailed dynamic studies of this oscillatory reaction was already taken over by a young Russian physiologist Anatol Zhabotinsky and the reaction is now known as the Belousov-Zhabotinsky or BZ reaction.

1.2.   THE KINETIC BASIS

Chemical oscillators are unlike mechanical oscillators such as pendulum or springs. Chemical reactions have no momentum. Oscillations are the result of competing kinetic processes involving nonlinear feedbacks. When a feedback accelerates a process, it is said to be a positive feedback. When it slows down the process, it is said to be a negative feedback.

A simple formal example of an oscillatory mechanism including positive and negative feedback steps is the Gray-Scott model [10]:

$$\text{input process} \xrightarrow{k_0} A \quad \text{(feed)} \tag{1}$$

$$A \xrightarrow{k_u} X \quad \text{(initiation)} \tag{2}$$

$$A + 2X \xrightarrow{k_1} 3X \quad \text{(autocatalysis)} \tag{3}$$

$$X \xrightarrow{k_2} Q \quad \text{(scavenging)} \tag{4}$$

The autocatalytic reaction (3) exerts a positive feedback on component X while reaction (4) which independently opposes to the increase of X is a negative feedback process. The rate law for X is given by

$$\frac{dX}{dt} = k_1 A X^2 + k_u A - k_2 X.$$

The consumption of A in reaction (3) also limits the rate of increase of X and can be considered as a negative feedback. However, this contribution to the rate term, directly linked to the production of X, cannot lead to an oscillatory instability in an homogenenous system. In the absence of the autocatalytic reaction (3), this would be a simple linear cascade of reactions that could not produce oscillations.

In real chemical systems, autocatalysis usually appears in a multistep process, like:

$$A + mX \longrightarrow nY \quad \text{with} \quad n > m \tag{5}$$

$$Y \longrightarrow \dots \longrightarrow X \tag{6}$$

where the sequence of reactions (6) is fast, so that the rate of reaction (5)

$$v_2 = \frac{dX}{dt} = (n - m)k_2 A^\alpha X^\beta$$

is rate determinant for this process. Multistep autocatalysis can be found in the BZ reaction, where the metal ion $M^{n+}/M^{(n+1)+}$ catalyzed production of bromous acid (Br(III)) goes through bromine-dioxide (Br(IV)) [9]:

$$Br(V) + Br(III) \longrightarrow 2Br(IV) \xrightarrow{M^{n+}/M^{(n+1)+}} 2Br(III). \tag{7}$$

In biochemical systems gorverned by Michaelis-Menten type mechanisms, positive feedbacks often appear in the form of substrate inhibitions [11], where as a result of some fast equilibria among some intermediate species in a multistep process, the effective rate of an overall reaction

$$A + Y \longrightarrow P \tag{8}$$

increases when the concentration of the substrate Y decreases and A is in excess. In inoganic systems, such a behavior was found in the oxidation of iodide ions by chlorite ions [12], where the rate law of reaction (8) takes the form:

$$\frac{dY}{dt} = -\frac{k_a A Y}{k_b + Y^2}$$

that remains valid as long as the concentration of Y is not too small in regard to $\sqrt{k_b}$.

Positive feedbacks are always destabilizing processes, while the antagonist negative feedbacks are stabilizing. Whatever the complexity of the kinetic mechanism, oscillations develop only when the positive feedback evolves on a shorter time scale τ_{act} than the negative feedback τ_{inh}. In a crude sense, there must be a "delay" between the positive and negative feedbacks actions. A more detailed classification of positive and negative feedbacks are provided, in abstract form, by Tyson [13], and in real chemistry models, by Luo and Epstein [14]. A theoretical discussion on the development of oscillatory instabilities in nonlinear dynamical systems is provided in Chapter 3.

Since, in closed systems, all chemical reactions spontaneously evolve to their thermodynamic equilibrium, it results that in such conditions oscillations could only be observed for a brief while, when the chemical composition is still at finite distance to equilibrium. Actually, batch oscillatory reactions are very few. They require that the initial reagents be sparingly consumed at each oscillatory cycle. The pioneer batch oscillatory reactions, discovered by Bray [4] and Belousov [5], which exhibit long lived oscillatory behaviors, are quite exceptional reactions.

1.3. FROM STEADY STATE BISTABILITY TO SUSTAINED OSCILLATIONS IN OPEN CHEMICAL REACTORS

The simplest open chemical reactor is the so called Continuous Stirred Tank Reactor (CSTR). It is a vigorously stirred mixture of chemical in a tank of fixed volume, permanently refreshed by constant flows of reagents. In the ideal case, the input flows are supposed to instantaneously and uniformly blend into the contents of the tank with a conservation of the input volume at any time. Though this idealization is commonly used in model calculations there are well documented observations showing that, in nonlinear chemical systems, even finely dispersed residual feed mixing inhomogeneities in the reacting solution can dramatically affect the overall dynamics [15, 16]. Some careful control of the stirring effect is always highly recommended to evaluate the role of mixing imperfections in the observed CSTR dynamics [15].

In the ideal case, the asymptotic state of the CSTR contents is entirely determined by the initial composition and the following control parameters: $[C_i]_0$ the concentration that species C_i would have in the mixed input flow prior to any reaction, the average residence time τ_0 that chemicals have in the reactor – ideally taken as the filling time of the reactor – and the temperature T_0 of the thermostatting bath. In the case of significant heat production one has also to take into account the heat exchanges of the reactor.

In open reactors, a stationary state is the result of a balance between the matter exchange fluxes of the reactor with the environment (input and output fluxes) and the chemical processes inside the reactor (reactant consumption and product formation). All real open chemical systems exhibit at least one stationary state. These states may be stable or unstable. Let us illustrate this in the ideal CSTR approximation. Consider a reaction process which can be fully characterized by one parameter, the extent of reaction ξ. The reaction rate $v_r(\xi)$ is always a positive scalar function of ξ that drops to zero at thermodynamic equilibrium. In the presence of direct matter exchanges with the environment, one can define a transfer rate $v_t(\xi)$ which is proportional to the reactants concentration differences between the reactor contents and the reactor environment. Thus $v_t(\xi)$ is a linearly increasing function of ξ. A steady state ξ_s is naturally reached when the reaction rate $v_r(\xi)$ equals the transfer rate $v_t(\xi)$. This can be solved graphically by the intersection of the $v_r(\xi)$ and $v_t(\xi)$ curves (Figure 1). The stability of the solution can be deduced from the local relative slopes at the intersection point. The general stability condition writes [17]:

$$\frac{\left(\frac{dv_r}{d\xi}\right)_{\xi_s}}{\left(\frac{dv_t}{d\xi}\right)_{\xi_s}} < 1 \tag{9}$$

In most reactions $v_r(\xi)$ is a monotonically decreasing function of ξ; hence the system has a unique stable stationary state (Figure 1a). However, reactions with positive feedback mechanisms may exhibit non-monotonic evolution of the reaction rate on the way to equilibrium. Thus, the $v_r(\xi)$ curve may have several intersections with the $v_t(\xi)$ curve, as sketched in Figure 1b where the two curves intersect three times corresponding to three stationary

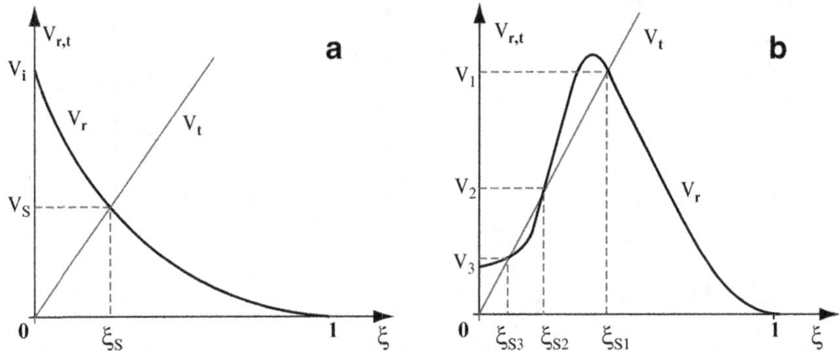

Figure 1. Graphic resolution of steady state solutions in a CSTR. Reaction v_r and reactor transfer v_t rates as a function of the extent of reaction ξ: (a) reaction with simple decay to equilibrium; (b) autoactivated reaction.

state solutions. According to Eq. (9), the two outer solutions are stable while the intermediate one is unstable. This corresponds to the standard bistability phenomenon. The low extent steady state branch is commonly referred to as the flow branch (F), since the flow terms dominate this state. The steady state branch with a high extent of reaction is referred to as the thermodynamic branch (T). The composition of this latter is the direct continuation of that found at the thermodynamic equilibrium. Bistability is typically observed when a clock "reaction" – i.e., a reaction that, in batch, suddenly accelerates to the equilibrium composition after a more or less long slow rate induction period – is operated in a CSTR.

Tristability has been observed in an number of cases [18, 19]. There is no theoretical limit to the number of possible stationary state in an arbitrarily complex reaction system. Figure 2a is a schematic representation of the stationary values that a species X, produced by the reaction, can take as a function of the feed concentration U_0 of a species U, in a bistable system. When U_0 is increased from a low value, the stationary concentration X is high and slowly decreases, following branch 1, until a critical value X_{A1} is reached at $U_0 = U_A$. Beyond this feed value the steady X value drops to X_{A3} and, as U_0 is changed, the X value now follows branch 3. However, if the feed value is decreased beyond U_B, X reaches X_{B3} another critical value, at which X suddenly jumps to X_{B1}. For $U_B < U_0 < U_A$, there are two stable steady values of X and transitions between branch 1 and 3 occur with hysteresis. In this range of parameters, the state which is actually observed depends on the initial conditions and on the history of parameter changes. Branch 2 corresponds to unstable steady states.

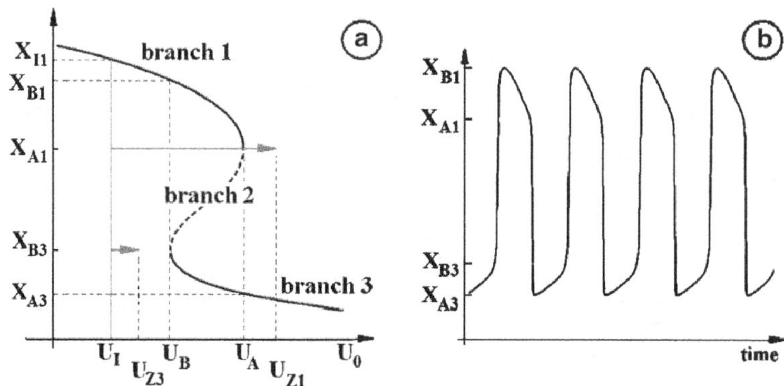

Figure 2.   (a) Bistability in a constraint U_0 response X plot: *Thick dark solid curves* correspond to stable steady states; *thick dash curve correspond* to unstable steady states. *Arrows* correspond to the feedback amplitude on the U_0 value exerted by Z_0. For detail see text. (b) schematic relaxation oscillations resulting from the periodic cycling and transition from one branch to the other at X_{A1} and X_{B3}.

It has been theoretically shown that such a bistable system can be readily changed into an oscillatory system by addition of an appropriate feedback reaction [20]. To set up minds, let us expand the previous simply bistable system by adding to the feed a species Z which would react with X to produce additional amounts of U. The original bistable subsystem then experience an effective greater value of U that would depend on the actual value of Z and X. Let $U' = U_0 + f(X, Z_0)$ be the effective value of U. The effective constraint U' is now a function of X and, at fixed value of X, its deviation from U_0 is controlled by Z_0. The effect of Z will be naturally bigger when the system is on branch 1, than on branch 3 of the bistable system. Suppose that, on the initial bistable subsystem, the reactor is fed at $U_0 = U_I$. Suppose, in addition that Z_0 is large enough for the effective concentration of U to appear equal to U_{Z1} when the system state belongs to branch 1 and equal to U_{Z3} when it belongs to branch 3. In the absence of Z, U' is equal to U_0 which corresponds to a steady state on branch 1 with a value X_{I1}. However, if now Z_0 is introduced, the reaction between X and Z cause the value of U' to increase. Concentration U tends to reach the value U_{Z1} but when crossing U_A, the high value of X is no longer sustainable so that it rapidly drops to X_{A3}. This jump makes U' decrease. It now tends to reach the apparent U_{Z3} value. The representative composition point of the system shifts to the left following the low X branch of the initial bistable system. When U drifts down beyond U_B, the low value of X cannot be kept anymore and it rapidly switches to X_{B1} on the top branch. Concentration X being high again, the effective U' value increases again and the cycle is repeated indefinitely. The periodic switches between the two branches generate periodic oscillations (Figure 2b). The amplitude of the feedback on U_0 depends on Z_0 and one can intuitively understand that, as Z_0 is increased, the width of the bistability domain along U_0 axis decreases gradually until it is transformed into a domain of sustained oscillations. This mechanism leads to a typical cross-shape phase diagram where, as shown in Figure 3, a domain of bistability exchanges with a domain of large amplitude oscillations at a critical point. Such a cross-shape diagram is obtained only if the switching time between the two attracting branches are significantly shorter than the time the system drifts on them. When this switching time increases, the relaxational character of the oscillations softens and they eventually vanish whereas in the phase diagram the bistability phenomenon vanishes through a standard cusp point with increasing Z_0 [20].

Detailed analytical unfolding of models based on the above dynamic properties shows that the cross-shape diagram topology sketched in on Figure 3 is a caricature of the actual bifurcation diagrams. In the vicinity of the "crossing" point a wealth of subregions with the coexistence of

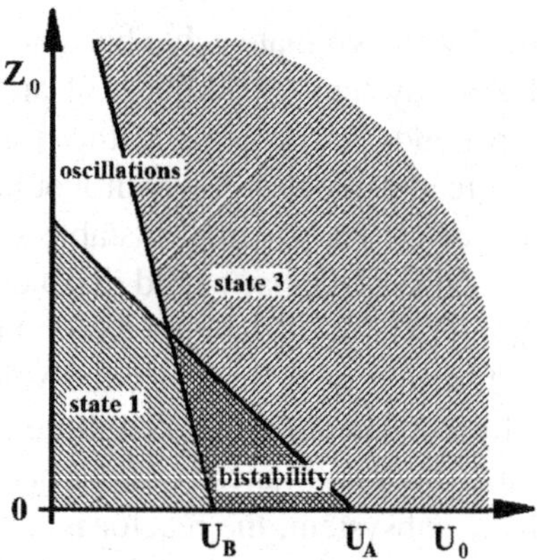

Figure 3.   Typical cross-shape phase diagram obtained when a feedback with increasing amplitude controlled a Z_0 acts on the bistable control variable U_0.

oscillations and steady state solutions are expected but in most experimental observations, the extension of these subregions lies below the usual experimental accuracy.

This hand-waving route to relaxation oscillations was turned into a generic way to produce chemical oscillations [21]. The method consists in operating an autoactivated reaction in a CSTR to find the appropriate conditions for steady state bistability and then in introducing a feedback species which would operate on an appropriate time scale. The large majority of oscillatory reactions discovered since the mid-1980s result more or less directly from this method. Correlatively, most oscillatory systems observed in a CSTR also show bistability. In the following section, we illustrate the above general presentation of oscillatory reactions by a few specific examples.

1.4.   THE CASE OF *PH* DRIVEN OSCILLATORS

One of the most interesting class of the oscillatory chemical reactions is that of pH-oscillators [22]. These systems are driven by a proton (or hydroxide-ion [23]) autoactivated reaction which, when coupled to an appropriate inhibitory process (negative feedback) operating on a slower time-scale generate autonomous oscillations. A simple model proposed by Rábai [24] summarizes a typical chemical mechanism for pH-oscillators:

$$A^- + H^+ \;\rightleftharpoons\; AH \tag{10}$$

$$AH + H^+ + B \;\longrightarrow\; 2\,H^+ + P^- \tag{11}$$

$$H^+ + C^- \;\longrightarrow\; Q \tag{12}$$

This model contains a fast protonation equilibrium (10), an autocatalytic production of protons (11) and a proton consuming reaction (12)[25].

In some pH-oscillators water can directly contribute to the autocatalytic proton production [25]. The autocatalytic reaction (11), is usually a redox process where A is a reducing agent and B is an oxidant. The proton consuming process (12) can be a separated reaction or just a competing kinetic pathway to a side product of the overall reaction between A and B. The above generic model exhibits bistability and sustained oscillations in CSTR conditions [24]. We illustrate this approach with the bromate–sulfite reaction, a prototypical pH-oscillator, for which there is a nearly quantitative kinetic model [24, 26].

The bromate–sulfite reaction is a clock (or Landolt-type [27]) reaction, that shows after an initial slow induction period a sharp transition from pH ~7.5 to pH ~2 under batch conditions [28]. In a CSTR the bromate–sulfite reaction shows bistability and large amplitude oscillations. A phase diagram in the ($[\mathrm{BrO_3^-}]_0$, $[\mathrm{H_2SO_4}]_0$) plane is presented in Figure 4. In the range of bistability, depending on the history of the system, the CSTR contents can be in the "Flow" (F) state, characterized by a high redox potential and a high pH (~7) or in the "Thermodynamic" (T) state characterized by a low redox potential and low pH (~3). The oscillations are only observed at very long residence times [29] or at very high bromate/sulfite feed ratios [26]. Domains of bistability and oscillation are connected through a typical cross-shape phase diagram topology. The bistability phenomenon can be accounted for by the following four steps mechanism.

$$\mathrm{H_2O} \;\rightleftharpoons\; \mathrm{H^+ + HO^-} \tag{13}$$

$$\mathrm{SO_3^{2-} + H^+} \;\rightleftharpoons\; \mathrm{HSO_3^-} \tag{14}$$

$$\mathrm{HSO_3^- + H^+} \;\rightleftharpoons\; \mathrm{H_2SO_3} \tag{15}$$

$$\mathrm{BrO_3^- + 3HSO_3^-} \;\longrightarrow\; \mathrm{3SO_4^{2-} + Br^- + 3H^+} \tag{16}$$

$$\mathrm{BrO_3^- + 3H_2SO_3} \;\longrightarrow\; \mathrm{3SO_4^{2-} + Br^- + 6H^+} \tag{17}$$

These steps lead to an autoactivated production of protons through the oxidation of protonated sulfur(IV) compounds to sulfate ions, a sulfur(VI) compound which is not protonated under moderately low pH conditions. To account for the temporal oscillations, Rábai [29] proposed that the oxidation of sulfite proceeds through a second reaction channel to produce dithionate, an other relatively stable sulfur compound and where protons are now trapped in less dissociated water molecules:

$$\mathrm{BrO_3^- + 6\,HSO_3^-} \;\longrightarrow\; \mathrm{3\,S_2O_6^{2-} + Br^- + 3\,H_2O} \tag{18}$$

Associated to the protonation equilibrium of the sulfite ions, this channel is proton consuming. It provides a negative feedback with appropriate slow evolution time scale. Model calculations are [26] in quasi-quantitative agreement with the experimental observations, as illustrated in Figure 4.

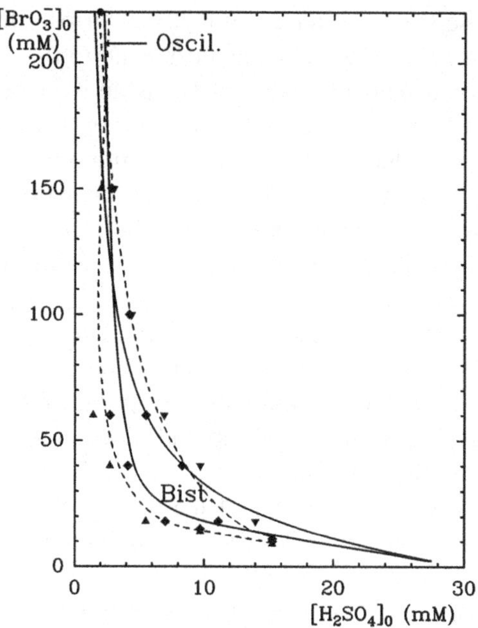

Figure 4. Nonequilibrium phase diagram of the bromate–sulfite reaction: full curves, numerical calculations; symbols and dash curves, experimental observations (▲, F state (high pH); ▼, T state (low pH); ♦, bistability; ●, oscillations).

The oscillatory capacities of the bromate–sulfite auto activated reaction can be considerably amplified by introducing other species, to drive the proton consuming process. This can be done in several ways:

– Adding an additional reducing agent [28]:

$$BrO_3^- + 6\,Fe(CN)_6^{4-} + 6\,H^+ \longrightarrow Br^- + 6\,Fe(CN)_6^{3-} + 6\,H_2O$$

– Adding an additional oxidizing agent [30]:

$$2\,MnO_4^- + 10\,HSO_3^- + 11\,H^+ \longrightarrow 2\,Mn^{2+} + 5\,HS_2O_6^- + 8\,H_2O$$

– Addition of a heterogeneous reaction on a piece of marble [31]:

$$CaCO_3 + H^+ \longrightarrow Ca^{2+} + HCO_3^-$$

The properties of pH-oscillators have been used in many different ways in the past decade: (i) Orbán and coworkers [32] have proposed a method to produce different targeted ion oscillators (Ca^{2+}, Al^{3+} or F^-) by coupling complexation and precipitation reactions sensitive to the pH of the solution.

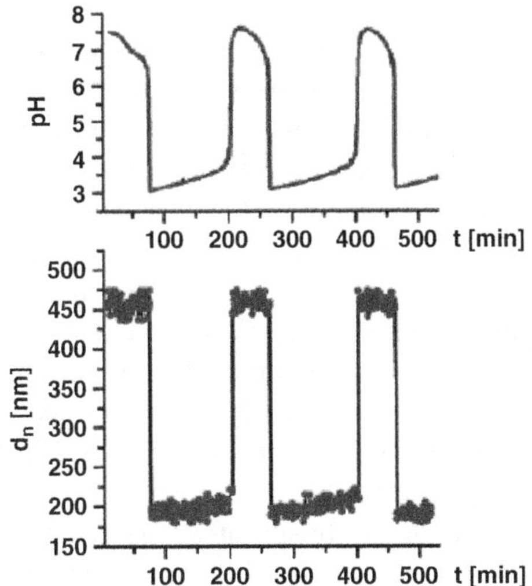

Figure 5. Periodic pH and associated size changes of nanogel beads (measured by light scattering) driven by the bromate–sulfite pH-oscillator.

In particular, the $BrO_3^- - SO_3^{2-} - Fe(CN)_6^{4-}$ oscillator is made to drive large amplitude pulses of free Ca^{2+} ions by introducing ethylene-diamine-tetraacetic acid (EDTA), a pH sensitive calcium ion complexing agent [33].

(ii) The pH-driven bistable and oscillatory reactions are also used to produce different types of autonomous motion devices converting chemical energy to mechanical work at a molecular scale. These experiments generally use the stretch-coil transition of natural or synthetic organic polymers driven by the pH changes of the reaction solution. Cyclic conformational changes of DNA [34] as well as different poly-n-isopropylacrylamide-*co*-acrylate gel networks were used to create periodic mechanical [35–37] or light transmitting [38] devices. The latter effect is illustrated on Figure 5, with a suspension of chemosensitive nanogels. Interestingly, such volume transition can be used as a feedback process to make a pH-bistable device to oscillate even in the absence of appropriate kinetic feedback. Such chemomechanical instability is detailed in Chapter 4.

2. Chemical reaction-diffusion patterns

2.1. INTRODUCTION

In the above section we have considered the development of self-organization phenomena in well mixed (homogeneous) reacting solutions where space and transport of species played no role. In unstirred reacting solutions, beside the above temporal instabilities (oscillations or multistabilities), new sources of instabilities and new spatio-temporal self-organization phenomena can emerge from the interplay of nonlinear reaction processes and transport processes. In the following, we shall focus on chemical patterns that result from the sole competition between reaction and molecular diffusion, the so called reaction-diffusion patterns. However, unstirred reacting chemical solutions can readily develop convective fluid patterns [39], due to local differences in buoyancy or in surface tension. In the presence of such convective instabilities even simple linear reactions can initiate transient concentration patterns in an initially uniform solution. In a number of cases, these convective patterns were initially mistaken with reaction-diffusion patterns [40–42]. These spurious effect can be avoided by operating in finely porous media such as hydrogels.

For simplicity, among the diffusion driven systems, we shall further distinguish the single phase systems from those for which the the pattern development mechanism is associated to a phase separation. An archetype of the latter is the periodic precipitation banding patterns observed when two different solutions of electrolytes which react to form a sparingly soluble salt are set in diffusive contact. The first report of such patterns traces back to Liesegang [43] and is the first example of well defined chemical patterns. This spatially periodic precipitation phenomenon is often evoked to account for patterns (e.g. stripes or concentric rings) found in rocks, such as malachite and agate [44]. Their mechanism of formation is quite involved. Several highly nonlinear processes seem to play a role: colloidal crystal nucleation, growth and aggregation, delayed nucleation due super-saturation, Oswald ripening [45], ... Depending on which nonlinear mechanism is put forward, different approaches have been proposed to account for the wealth of observations [46, 47]. In all cases, the observed patterns are the result of transient dynamics in reactive media with finite chemical resource, i.e. on their way to equilibrium. Interestingly, the phenomenon has been recently used to produce patterned materials for technological applications [48]. Development of spatial structures in systems involving phase transitions are not included in this chapter.

We elaborate on systems where the patterning mechanism only relies on homogeneous kinetics and molecular diffusion. The simplest form of such reaction-diffusion phenomena is the propagation of an undamped chemical activity front in an unstirred solution of a "clock reaction" and the first observation goes back to Luther [49]. However, the main impulse was due to the observation of periodic traveling wave patterns in solutions of BZ reagents. These were first reported by Zaikin and Zhabotinsky [50] and later by Winfree [51]. The initial studies were performed in closed reactors so that only the most robust properties of these systems (phase wave and triggered wave dynamics, spiral wave core dynamics [52], wave train and curvature dispersion relations [53], etc.) could be studied. Since, in batch reactor, the BZ reaction keeps far equilibrium properties during a much longer times than any other studied system, the large majority of spatial studies were performed on different versions of this reaction until the late 1990s. Although new waves properties are still discovered in variants of this reaction [54], the modern breakthrough of spatial studies was made by the development and systematic use of open spatial reactors. In the early 1990s, they enable the first observations [55] of the spontaneous development of stationary concentration patterns, predicted by the British mathematician Turing [56], and soon after that of observation stationary labyrinthine patterns [57] resulting repulsive interactions of chemical activity fronts.

2.2. OPEN SPATIAL REACTOR DESIGN

In the following, we describe the basic principles of open spatial reactors and the major reactor geometries that were developed for specific studies. For mathematical tractability, many theoretical studies on chemical reaction-diffusion systems use elementary conditions to maintain the reaction at controlled distance to thermodynamic equilibrium: either the "pool-chemical" approximation or the "continuously-fed- unstirred-reactor" (CFUR) [58]. In the former, one assumes that the concentrations of initial reagents are fixed in space and time and that end products are irreversibly produced. These conditions can be only briefly approximated by the few oscillatory reactions which exhibit longlived batch oscillations, like the BZ or the CDIMA reactions [59]. In the CFUR approximation, the system is thought as a uniform array of small diffusively coupled CSTRs where the feed parameters are generally taken to be independent from the diffusivity of species. It has been used to account for pattern development in thin open spatial reactors operated with "clock reaction" reactions. However, this approximation is impossible to meet in real devices, since every space point must be evenly refreshed and the spatial dynamic of species must be unconstrained by this feed process [58].

The challenge of open spatial reactors is to feed all space points in order to maintain the system far from equilibrium, without introducing macroscopic fluid mixing or convective transport. To account for the above requirements, the feed can only be made by diffusive exchanges of matter with the environment at the boundaries of the system, and not in the bulk. To quench any spurious fluid flow, the core of an open spatial reactor is made of a piece of finely porous material, generally a soft hydrogel (e.g. polyacrylamide, agarose, polyvinyl alcohol), with one or two opposites faces in contact with the contents of permanently refreshed, rigorously stirred, tanks of reagents (Figure 6). Chemicals diffuse and react inside the gel matrix. In experiments

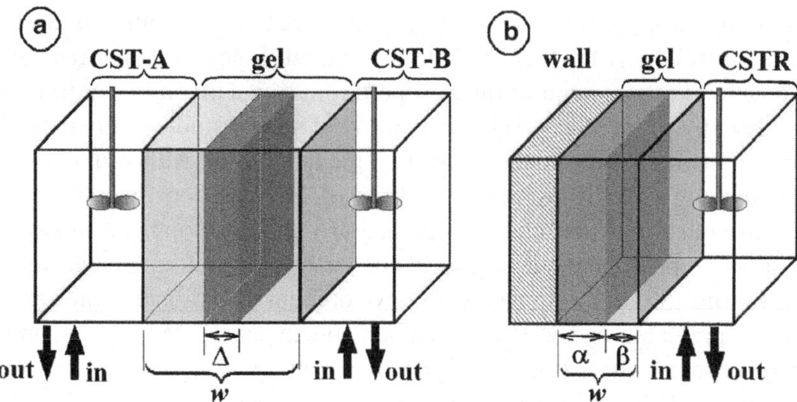

Figure 6.   Schematic principles of open spatial reactors. (a) Two-side-fed-reactor (TSFR), a gel slab (in gray) is in contact by two opposite faces with the contents of two continuous stirred tanks (CST); (b) One-side-fed-reactor (OSFR), a thin gel slab is in contact with the contents of a continuous stirred tank reactor (CSTR). Δ and α are the actual part where reaction-diffusion patterns can develop. For details see text.

described in this chapter, the volume fraction of water to the gel network is large enough for the diffusivities to remain close to those in pure water.

In two side fed reactors (TSFR), complementary A and B sets of chemicals are provided at opposite faces, separated by a distance w (Figure 6a). This naturally induces cross composition ramps of A and B between the feed boundaries. Concentration conditions for pattern development are only met in a more or less thick substratum Δ of the gel parallel to the feed surfaces.

The main advantage of this asymmetric feeding mode is that it avoids the development of the temporal composition instabilities (the oscillatory and multistability phenomena described above) in the feed tanks. However, the composition ramps associated to this feeding mode make possible the simultaneous development of different spatial instabilities at different adjacent positions. This makes difficult the interpretation of the observed patterns [60] and the development of theoretical models [61, 62].

The main advantage of using asymetric two-side-fed reactors is that the separated boundary feed compositions can be made non-reactive. This way, the temporal composition instabilities (the oscillatory and multistability phenomena) described above are avoided in the feed tanks. This feed mode is also little sensitive to small boundary "imperfections" and extended systems with effective uniform boundary values are easily obtained. However, the composition ramps associated to this feeding mode make possible the simultaneous development of different spatial instabilities at different adjacent positions. Furthermore, the width and the composition of the patterning stratum cannot be independently controlled.

For these reasons the one-side-fed-reactors (OSFR) have become more popular (Figure 6b). All reagents are fed through a unique CSTR. The feed composition at the CSTR/gel boundary relies on the chemical state of the CSTR which is controlled by the concentration C_{i0} of species in the input flow, the residence time τ and, the initial composition in the case of mutistability phenomenon in the CSTR. It is possible to keep the CSTR in a stationary state close to the input composition by a short enough residence time. To avoid that the gel contents significantly feedback onto the CSTR contents, the volume of the tank should be approximatively two orders of magnitude greater than the volume of the gel. In typical experiments, the stability of the contents of the CSTRs is monitored by Pt or pH electrodes and the concentration patterns of colored species in the gel are monitored by cameras connected to a computer with an image grabber board. For technical reasons, one-side-fed-reactors are about one order of magnitude thinner than their two-side-fed counterparts. This strong confinement in the direction orthogonal to the feed surface makes them to better approximate one or two dimensional systems. However, except in special cases, the gel thickness cannot be neglected. Actually, due to the concentration continuity at the contact with the uniform concentration in the CSTR contents, no pattern can develop in a vicinity β of the feed boundary ('chemical boundary layer') (Figure 6b). Patterns display some "structure" in the feed direction. The value of β also sets a lower limit to the thickness w of the OSFR for the development of patterns while the overall reaction time sets a maximum size to the thickness of an OSFR.

Based on the above considerations, different reactor geometries have been developed, the most typical are illustrated in Figure 7. One and two side feed modes have been used in the annular (Figure 7a) and in the disc (Figure 7b) shaped reactors. Annular reactors enable to see how concentration patterns

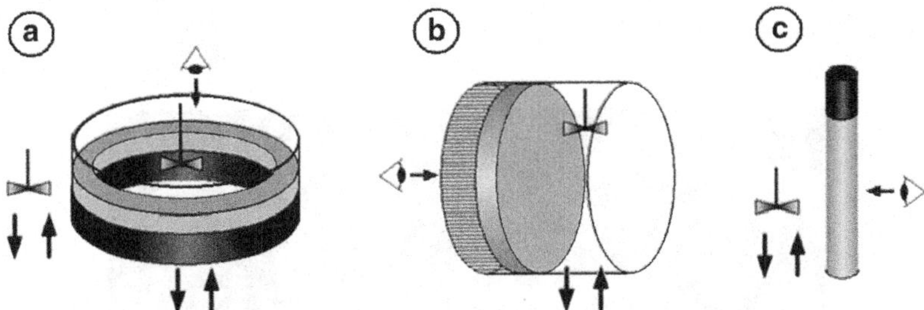

Figure 7. Schematic representations of different open spatial reactor geometries: up and down arrows indicate the relative position of chemical feed exchanges of the stirred tanks. (a) annular reactor in the TSFR feed mode; (b) disc reactor in the OSFR feed mode; (c) thin cylindrical OSFR a geometry used in chemomechanical experiments.

develop in the width of the gel. This geometry of reactor has periodic boundary conditions along the circumference that mimic aspects of infinite media in this direction. Disc shaped reactors are well suited to the observation of patterns that develop in planes parallel to the feed surface. If w is small enough, the observed patterns are reminiscent of two-dimensional patterns obtained in numerical model studies. These two geometries of reactors provide complementary views of the actual pattern concentration profiles. The last geometry of reactor (Figure 7c) is a thin cylinder that can be considered as a confined one-side-fed reactor with mechanically unconstrained boundary along the axis. It has been used for the studies of chemomechanical patterns. In this case the cylindrical reactors are made of chemoresponsive gels such as the pH-responsive networks (see Chapter 4).

In the following, different aspects of the development of reaction-diffusion patterns in open spatial reactors are illustrated.

2.3.   TWO DIMENSIONAL WAVE PATTERNS IN THE BELOUSOV-ZHABOTINSKY REACTION

2.3.1.   *Traditional batch observations, the notion of excitability waves*
Before we consider pattern development in open spatial reactors, let us briefly recall the most typical pattern observations made in traditional unstirred thin layers of the Belousov-Zhabotinsky reaction. This reaction is nowadays a well known classroom demonstration system for chemical oscillations and reaction-diffusion patterns and still probably the most studied oscillatory reaction. As mentioned above, this reaction has the peculiar property to be able to exhibit several hundreds of oscillation cycles in a closed (batch) reactor. It has made this reaction an ideal playground for the study of reaction-diffusion patterns in quiescent thin layers of solution in Petri dishes. These patterns appear as periodic traveling concentric rings of chemical activity or spirals (Figure 8). These undamped waves are based on the "excitability" property of the reaction mixture [63].

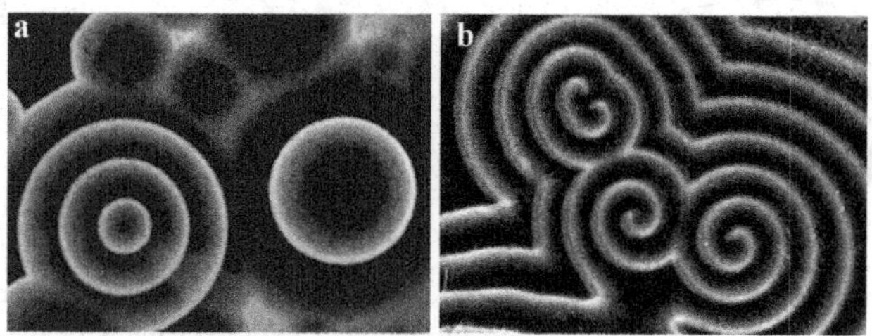

Figure 8.   Typical (a) target wave and (b) spiral wave patterns observed in a thin layer of BZ reaction solution.

In excitable systems, a small but supercritical (chemical) stimulus is amplified by the local reaction dynamics. This amplification induces a superthreshold contamination of the surroundings by diffusion which further spreads the activity without damping. Similar to a dry prairie fire, the local burning grass heats the neighboring dry grass which sets on fire and spreads leaving behind a burned inactive area. Like prairie flame circles colliding front mutually annihilate because of the refractoriness of the composition behind the front. Like in wild prairies where the grass can grow again after a time, the chemically "burned" area eventually recovers and an oxidation activity fronts can start again. Target patterns are commonly the result of periodic local excitations due to the presence of different "catalytic" impurities (e.g. dusts or scratches on the dish) that create local oscillations with different periods. Spiral waves result from the rupture of a circular wave. This induces a singularity in the wave front propagation and due to the refractory zone behind the chemical activity front the free end of the wave rotates around this singularity producing an expanding spiral wave. The core of a spiral can be thought as a punctual defect of a planar wave-train. The frequency generated by the rotation of the wave around this singularity is the maximum possible for a given excitable medium [51]. The most standard versions of the BZ reaction consist in the oxidation of an organic species (e.g. malonic acid) by an acidic bromate solution, in the presence of a metal ion catalyst, such as Ce(III)/Ce(IV) or $[Fe(II)phen)]_3^{2+}/[Fe(III)phen)]_3^{3+}$. The latter is preferred in demonstrations because of the dramatic color change from red, in the reduced state, to blue in the oxidized state [51].

2.3.2. *Sustained wave trains in a ring reactor: "excyclon" dynamics*

When the BZ reaction is operated in a two-side-fed gel reactor, such as illustrated in Figure 7a, other waveform patterns are observed and new properties can be studied. In the following experimental example, the whole reactor remains in the reduced red (dark color) state in the absence of any perturbation. No spontaneous oscillation or wave develop in the annulus. Nonetheless, trigger waves can be initiated, in the annular gel, by a superthreshold light pulse. Since the medium is circularly isotropic, the perturbation triggers a pair of counter propagating waves. Due to the cross composition gradients that naturally develop between the two feed boundaries, the waves are localized in the central part of the annulus and take the shape of blue crescents (light color), with a leading head and two side-tails. After traveling through half of the circle, the pair of waves meet and annihilate and no other waves forms. Using the "vulnerability" [64] property of excitability waves one can trigger, in the vicinity of an existing one, an excess number of waves traveling in one direction. The waves in excess in that direction then endlessly travel

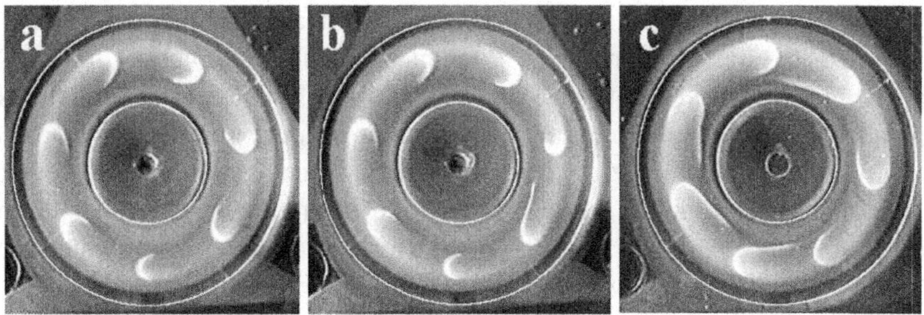

Figure 9.   Pinwheel patterns of seven excyclons on an annular TSFR. The head of all U-shape waves (excyclons) are equally spaced and move clockwise. (a) Pattern with identical shape excyclons; (b) Pattern with one excyclon exhibiting a stable outgrowth; (c) Pattern made of unstable irregular shaped excyclons. Long tails grow and die at "chaotically" distributed positions. Bromate, sulfuric acid and ferroin are fed at the inner rim (diameter = 20 mm) while malonic acid, sulfuric acid and ferroin are fed at the outer rim (diameter = 40 mm).

along the annular shape reactor. They have been called "excyclons" [65] and the global structure a "pinwheel" [66]. In the particular experimental conditions of Figure 9 [65], one to a maximum of twelve waves could be fitted into the annulus. In the asymptotic state the heads of the excyclons form an equally spaced wavetrain (Figure 9a). Consistently, the dispersion curve of these wavetrains is monotonous [65]. However, above a critical number which depends on experimental conditions, excyclons may exhibit fixed or dynamic "topological defects". In patterns made of n waves, all the excyclons may not have the same U-shape. Some excyclons may develop an outgrowth in the radial direction with a stretch along the inner rim of the gel, as shown in Figure 9b. Generally, in the asymptotic state the "elongated" excyclons lock in relative position; the whole pinwheel pattern still rotates as a solid body. When there are two or more elongated excyclons on the wavetrain, they may exhibit multiple stable relative distributions for the same total number of waves. Under other conditions the shape of the excyclons become unstable: elongated shapes constantly grow and disappear at different wave positions without ever settling onto a fixed pattern. A snapshot of such irregular wave-shape pattern is seen in Figure 9c. The above wave pattern snapshots illustrate the pattern confinement and changes in the chemical medium properties induced by the cross gradients of chemical feeds in a TSFR. This gives a flavor of the complexity and of the possibility for new dynamical behaviors that emerges in such spatial reactors.

Many other types of wavetrains instabilities yet not experimentally explored in such open systems can be envisioned like oscillatory spacing or fixed irregular spacing when the dispersion relations are no more monotonous. Such a non-monotonous dispersion relation has been observed in a recent variant of BZ reaction [54].

2.4. STATIONARY PATTERNS

2.4.1. *An empirical approach*

Before illustrating the diversity of stationary patterns, we provide an empirical explanation on how molecular diffusion, a dispersing mechanism that ordinarily smears out any concentration inhomogeneity in a solution, can induce and stabilize concentration patterns. The existence of such stationary structures was first analytically demonstrated by Turing [56]. In the proposed mechanism, patterns develop spontaneously in an initially uniform system. It is presented as a possible model for the spontaneous development of shapes and patterns observed during biological morphogenesis. His "reaction-diffusion" theory of morphogenesis mostly escaped notice until the development of nonlinear concepts in chemical dynamics [7] and, more recently, their experimental evidence. We shall present them in an heuristic way. More rigorous developments can be found in Chapter 3 and in references there in.

Let us consider, as Meinhardt [67], an activator-inhibitor system where the activator produces its own inhibitor. Suppose a stable uniform state where the activatory and inhibitory processes balance each other and operate on the same time scale so that the system would not undergo an oscillatory instability. Consider a local excess of the activator, due to a perturbation or a concentration fluctuation (Figure 10a). This increased concentration of the activator induces, over the same time scale, an associated increase of inhibitor, as schematized on Figure 10b). If the diffusion coefficients of the activator and inhibitor are identical, activator and inhibitor equally spread and the further local development of the activator is hindered. The levels of the activator and inhibitor turn back to their initial stable uniform state composition. However, if in similar conditions, the inhibitor diffuses faster than the activator, it escapes faster from the center and spreads over a wider area than the activator. As a result there will be shortage in the inhibitor at the center and the concentration of the activator will grow (Figure 10c). while around this activatory peak, the inhibitor is in excess and impedes the further invasion of the activator in the surroundings The localized activatory peak grows until saturation of the process. A stable activatory peak forms and settles for as long as the reagents are supplied. In this heuristic description, it is understandable that an other activatory peak can only develop at a finite minimal distance from an existing one (Figure 10d). In the special case of the Turing bifurcation the initial homogeneous state is unstable to spatial perturbations in a finite range of wavelength. Fluctuations are random. The peaks of activator compete for space. From this competition a well ordered pattern ruled by the nonlinear interactions emerges Figure 10e). In uniform two-dimensional systems, near to onset, the pattern selection process leads to two ordered

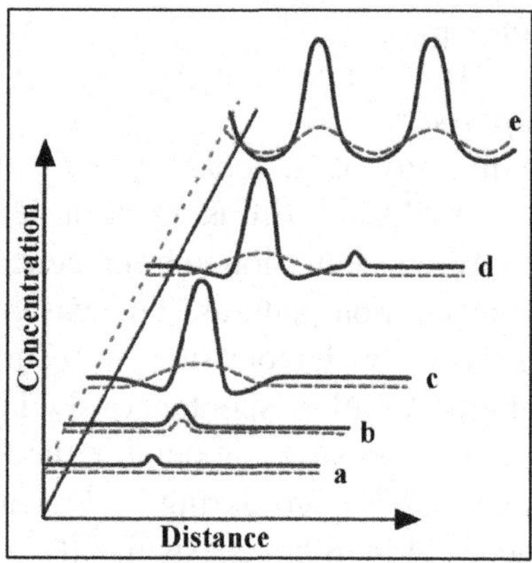

Figure 10.   Schematic growth mechanism of a Turing pattern in an activator (*full black curves*) inhibitor (*dashed gray curves*) system: (a) fluctuation in the concentration of the activator; (b) followed by the associated increase of the inhibitor; (c) the inhibitor spreads faster than the activator and stabilizes the peak of activator; (d) another fluctuation of the activator can grow only beyond a critical distance from a well developped peak; (e) the competition between growing activator peaks settles into a periodic pattern.

stable pattern planforms: hexagonal arrays of maxima or minima or parallel stripes [68–70]. In uniformly fed three-dimensional systems the spatial concentration modulations organize in body-centered-cubic, hexagonal arrays of columns, and lamellea structures [71]. More "exotic" three dimensional patterns are also possible far from onset: double gyroid [72] or arrays of Scherk minimal surfaces [73]. However, none of these more exotic patterns have been experimentally identified, so far.

There is an important property toTuring structures that distinguish them from other nonlinear patterns found in other field of physics. Their wavelength depend only on intrinsic parameters, namely, concentrations, kinetic constants, and diffusion coefficients, and not on the size and geometry of the system, provided of course that this system is larger than this wavelength. Geometric parameters only play a minor role in the selection or orientation of patterns.

The fundamentals of this spontaneous spatial instability of a steady uniform state relies on the existence of an appropriate difference in the diffusion coefficient of species. In two variable systems the conditions are $D_{act} < D_{inh}$. In many variables systems, the difference in the diffusion coefficients can be transferred to species, in connection with the activatory kinetic loop, other than the main activator.

In multistable and excitable systems other routes to the development of stationary patterns are possible but the long range inhibition and short range activation together with a time scale separation $\tau_{inh} \geq \tau_{act}$ is the royal way to these developments. Note that, for multi-time scale systems with non uniform initial conditions there is no strict demonstration that differences in the diffusion coefficient are absolutely required. There are no known necessary and sufficient conditions for the onset of stationary patterns in these more complex cases [74].

2.4.2. *The control of effective time and space scale of a reaction process*

The spontaneous symmetry breaking properties of Turing patterns make them popular in theoretical approaches of different aspects of biological morphogenesis. Full demonstration of the implication of such a straightforward reaction-diffusion instability in a real living system, has not been provided so far, though a number of "in favor" hints are found in the literature (e.g., teeth growth of alligators ([75], stripe patterns on the skin of the zebra fish [76]). As seen above, the major routes to the development of stationary patterns require appropriate space and time scale separation between activatory and inhibitory processes. Large differences in the transport of species are readily found in cells and biological tissues, because of membranes, complex active structures, and the large spectrum of molecule sizes involved in these systems. *A contrario*, the aqueous chemical reactions on which we focus in this chapter involve relatively small size molecules with diffusion coefficients typically within a factor of two of $10^{-5} \mathrm{cm}^2 \mathrm{s}^{-1}$, except for protons and hydroxyl ions for which these values are about one order of magnitude higher.

Fortunately, as mentioned above, the short range activation requirement may not necessarily be connected to the the actual slow diffusivity of the main activatory species, but this property can be indirectly induced by an other low mobility species. The simplest and most effective mechanism was proposed by Lengyel and Epstein [77] to account for the pioneer observation of Turing patterns in the CIMA reaction.

Let us consider a chemical system involving species X and Y with respective concentrations x and y, diffusion coefficients D_x and D_y, and ruled by the rate equations below:

$$\frac{\partial x}{\partial t} = f(x, y, \ldots) + D_x \nabla^2 x \tag{19}$$

$$\frac{\partial y}{\partial t} = g(x, y, \ldots) + D_y \nabla^2 y \tag{20}$$

...

Suppose one adds a fast reversible complexing agent S of species X governed by the fast equilibrium

$$S + X \overset{k_+, k_-}{\rightleftharpoons} C \tag{21}$$

with the equilibrium constant $K_{eq} = k_+/k_-$. The complexing agent S is assumed to be a large enough molecule so that both the diffusion of S and the diffusion of C, the complexed form of X, are negligible and can be considered as null. For simplicity, C is assumed to be non reactive,

  The set of dynamical equations of the system are modified and taking into account the conservation of the quantity $S_0 = S + C$, one gets

$$\frac{\partial x}{\partial t} = f(x, y, \ldots) + k_+ S.x - k_- C + D_x \nabla^2 x \tag{22}$$

$$\frac{\partial C}{\partial t} = -k_+ S.x + k_- C + D_C \nabla^2 C \tag{23}$$

whereas Eq. (20) is unchanged. The complex is in rapid equilibrium with X, so that one can eliminate Eq. (23) by summing Eqs. (22) and (23):

$$\frac{\partial(x + C)}{\partial t} = (1 + S.K_{eq})\frac{\partial x}{\partial t} = f(x, y, \ldots) + D_x \nabla^2 x \tag{24}$$

Assuming S is in large excess, so that $S \approx S_0 \gg C$, and setting $\sigma = 1/(1 + S_0 \cdot K_{eq})$ a renormalized form of Eq. (22) is recovered.

$$\frac{\partial x}{\partial t} = \sigma[f(x, y, \ldots) + D_x \nabla^2 x] \tag{25}$$

Introducing the slow diffusing complexing agent does not change the steady state solution of the system but rescales the space and time scales of species X by a factor σ and thus can change the time and space stability of this steady state. This derivation can be made more rigorous but the result remains valid in the frame of our approximations. If X is an activatory species, it can have an apparent diffusion coefficient lower than the that of the inhibitory species, say Y, for which the diffusion and time scale of evolution remain unchanged. Appropriate conditions for the development of stationary patterns are shown to be reached this way. This approach was further generalized by Pearson and Bruno [78] and by Strier and Ponce-Dawson [79] who show that, whatever the number of variables and the number of complexing agents, or if the inertness of the complex is partly relaxed, a Turing bifurcation can be obtained. Furthermore, the Turing bifurcation is obtained for parameter values where, in the absence of the complex, the steady uniform system is temporally unstable going from an unstable focus to an unstable node. In other words, it is "deep" into an oscillatory state domain. The quenching of the oscillations is associated to the absence of reactivity of the hypothetic complex, a requirement that is very commonly fulfilled by macromolecular complexes. Other route to stationary patterns, e.g. those resulting from front interactions in a bistable system, do not unnecessarily need to start with oscillatory conditions in the absence of complexing agent. The addition of a low mobility complexing agent has effectively been shown to be a corner stone in a systematic method to produce stationary patterns in open spatial reactors, whether they originate from a Turing instability or from the interaction of traveling fronts, as we shall see in Section 2.7.

2.5. TURING PATTERNS IN THE CIMA REACTION FAMILY

2.5.1. *Experimental observations*

The most remarkable fall out from the early development of open spatial reactors is the first observation of sustained stationary chemical patterns resulting from a Turing instability [55]. These were obtained by operating with the Chlorite-Iodide-Malonic Acid (CIMA) reaction [59], a newly discovered oscillatory reaction at that time. The reaction is one of the very few that is able to exhibit transient oscillations in batch conditions. It was shown that during the oscillatory behavior, the actual main reagents are chlorine dioxide, iodine and malonic acid [80], while the activatory and inhibitory processes were essentially driven by the iodide and chlorite ions, respectively. It was suggested [77] and experimentally corroborated [81] that, in this reaction, the necessary slower diffusion of the activator, relative to the inhibitor, is mediated by the formation of a reversible complex C between the iodide, iodine and starch S, a large very low mobility molecule, initially used as a color indicator.

$$I^- + n\, I_2 + S \rightleftharpoons C \text{ (colored complex)}$$

Consistently, it was shown that the increased concentration of S – starch or Polyvinyl-alcohol (PVA) an other polyiodide macromolecular complexing agent – could control the transition from oscillations and traveling waves to stationary patterns [81]. These two type of patterns, observed in a two-side-fed strip reactor (a reactor geometry similar to the one in Figure 7a) are illustrated in Figure 11a and b where the dark and clear regions correspond to high and low concentrations of the colored polyiodide complex C. At low concentration of starch, localized trains of traveling wave are observed, while, at higher starch concentration, a row of equally spaced standing clear spots is obtained. Both type of patterns develop in a subregion Δ parallel to the feed boundaries that are located at the top and bottom of the figures. The width Δ of the pattern area and the number of rows of spots depend on feed parameters, as illustrated Figure 11c, but also on the gradients of reagents governed by geometric parameters such as the distance w between the two feed boundaries [60].

This latter property is used in the experiment presented in Figure 12 where patterns are made to develop in a beveled disc reactor. One witnesses,

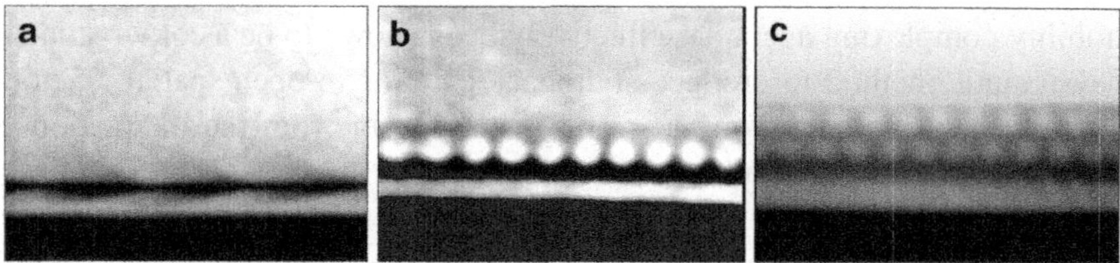

Figure 11.   Traveling and stationary patterns localized in a subregion Δ of a TSFR: (a) Train of arrow-shaped waves traveling to the left; (b) Single row of clear spots (Turing pattern), wavelength $\lambda = 0.19$ mm; (c) Double row of clear spots, wavelength $\lambda = 0.16$ mm. Chlorite is fed at the top while iodide and malonic acid are fed at the bottom, the width of the agarose gel reactor is $w = 3.0$ mm; the dark color correspond to high concentrations of iodide and iodine while the clear patterns correspond to areas with very low concentrations of iodide. Starch concentration: (a) 15g/liter; (b) and (c) 45g/liter. Potassium iodide: (a) and (b) 1.5 mM; (c)3.0 mM. All other parameters are fixed.

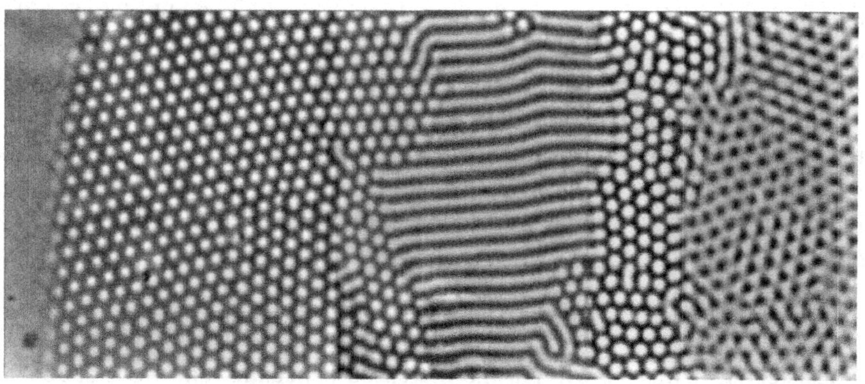

Figure 12.   Turing pattern planforms developed in disc reactor in the TSFR mode with a ramp of thickness and opposedly fed by a chlorite and an iodide solution as in Figure 11. The distance between the circular faces varies from 1.7 mm (*left*) to 3.2 mm (*right*) over a distance of 21 mm. About 70% of the disc diameter is represented.

from left to right: a featureless medium gray region, a domain of hexagonal array of clear spots (low iodide locations), a domain of parallel clear and darker stripes, a new domain of very contrasted hexagonal array of spots and finally patterns with more unusual symmetries. The first two pattern planforms are the two standard stable spatial pattern symmetries predicted for two-dimensional Turing patterns (see Chapter 3). The experimental patterns actually correspond to "monolayers", that is to conditions where the patterns in a cross-section through the thickness of the gel would have the aspect of Figure 11b. The observations are consistent with the theoretical prediction that when, in a three-dimensional system, the patterning region is confined to a monolayer by ramps of parameters, the stable Turing patterns modes are similar to those of a genuine two-dimensional system. However, the onset of patterns and their relative stability can be significantly modified [62]. Note that these monolayer patterns are actually three-dimensional structures: spots are spheroids and stripes are, here, clear cylindrical structures laying parallel to the feed surface. The re-entrant hexagonal pattern observed past half way in the figure would correspond to arrays of short cylinders with their axis now orthogonal to the feed surface. Hexagonal arrays of columnar structures are expected in three-dimensional systems [71] (see Chapter 3). As we move to the right, the thickness Δ over which pattern develops in the third direction increases and the seemingly non standard planforms result from "Moiré" effects due to different phase combinations of superposed arrays of beaded or cylindrical structures.

This illustrates, the already mentioned difficulties associated to the use of TSFRs. Modeling, the observations in this geometry is a great challenge which would require advanced kinetic models to properly handle the large range of reactant concentrations scanned through the ramps. Computational attractive, reduced kinetic models of reactions with nonlinear kinetic mechanisms are usually suited only over relatively restricted ranges of concentrations. However, tentative tries have been made in this direction to estimate the width and position of the patterning stratum in the CIMA reaction [82].

2.5.2. *Modeling experimental results in one-side-fed reactors*

To avoid the above problems, most recent studies are performed in thin OS-FRs. In such spatial reactor feed mode, one can distinguish two cases: First, the few reactions that can exhibit transient batch oscillations where the main reagents are only partly consumed during one oscillatory period. Second, the batch "clock" reactions where, after the "tick" of the clock, at least one of the major reactants is nearly totally consumed.

Let us illustrate the first case with the CDIMA reaction operated in a thin disc OSFR (see Figure 7). According to the work of Lengyel and Epstein on this reaction, if no external gradients are imposed, the variations of the major reactant concentrations are small on the distance of a wavelength or over a period of oscillation [80]. If all the input reactants are fed onto one side of a thin enough film of gels, i.e., with a thickness of about one wavelength or less, we expect the concentrations of these reactants to be almost constant in space and time all through the gel. Accordingly, the thin disc reactor should approximate a uniformly constrained two-dimensional system. Experiments were performed in a 0.2 mm thick agarose-gel disc OSFR separated from the contents of a CSTR by a rigid membrane, 0.02 mm thick, with a lower diffusivity than the gel. We shall see that this membrane enables to approximate an effective two-dimensional pool-chemical system. The OSFR was

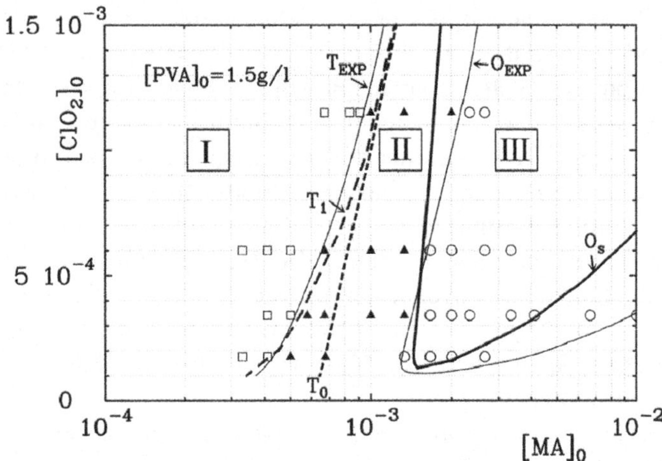

Figure 13. Spatial pattern phase diagram of the chlorine dioxide-iodine-malonic acid (CDIMA) reaction operated in a disc OSFR: Thin curves and symbols (□, stationary states; ▲, Turing patterns; ○, oscillatory states), experimental observations; thick curves, computed stability limits (T_0 subcritical Turing bifurcation, T_i limit of the subcritical region; O_{EXP} and O_S limit of the domain of oscillations. More details in the text.

operated in conditions where the CSTR contents are always in a monostable steady state composition. The resulting phase diagram of Figure 13 was established step by step, using the malonic acid concentration ($[MA]_0$) as the bifurcation parameter for different fixed values of chlorine dioxide ($[ClO_2]_0$) [81]. For an appropriate feed of the CSTR, the standard sequence of two-dimensional hexagonal and striped patterns arrays are observed, as in the case of monolayers in the TSFR, but the dimensionality of the patterns is now directly controlled by the geometric parameters of the reactor. From left to right (Figure 13), one can distinguish three regions: (i) a region of uniform stationary state, (ii) a region of stationary hexagonal and striped Turing patterns, (iii) a region where the CSTR contents oscillate and for which the dynamics in the gel was disregarded. A remarkably faithful modeling could be obtained [81] using using a two variable skeleton kinetic model of the reaction and a two-dimensional "pool-chemical" approximation of the main reactants. In the model approach a seven-variable kinetic model [83] is first used to determine the CSTR composition in the absence of the OSFR. Taking advantage that the initial reactants (chlorine dioxide, iodine and malonic acid) can be in stoichiometric excess relative to dynamical variables such as $[I^-]$ and $[ClO_2^-]$ the resulting concentrations of the reactants were fixed in the gel which was taken as two-dimensional. In doing this, the kinetic model can be reduced to only two variables. Computed results are in excellent agreement with the experimental observations, for the Turing and Hopf limits (Figure 13), the different pattern planform domain distributions, and the dependence of the wavelength on chemical feed parameters [81].

In spite of this agreement, one can be astonished that this approximation works so well and even that patterns are experimentally observed in such a thin gel. In Section 2.3, we had argued that the CSTR defines uniform fixed boundary conditions at the contact area with the OSFR and that in the vicinity of such boundary conditions no pattern should develop. The typical depth of the boundary layer β and the pattern wavelength rely on the same competition between reaction and diffusion and should be of the same order. Here the wavelength and the thickness of the gel are similar (i.e., ~0.2mm). The homogeneity onto the feed face should force uniformity across the whole thin gel. In fact the pattern developments are made possible because the presence of a thin membrane with lower diffusive properties than the gel part of the OSFR. This introduces mixed boundary conditions at the feed interface with different consequences for two classes of species: For major reactants, which are in excess, these mixed boundary properties show little differences with the previous fixed concentration boundary conditions, whereas for the intermediate species directly involved in the formation of Turing patterns, they are found to practically define no-flux boundary conditions at the membrane, discarding the uniformity constraints. The patterns are thus free to develop in planes parallel to the faces [81].

2.6. SPATIAL BISTABILITY IN THE CDI REACTION

When "clock reactions" are used, the decoupling between the feed and dynamic variables cannot be insured any more, since at least one main feed species is directly involved in the pattern development. In this case, a well defined chemical boundary layer develops. It gives rise to a new property of autoactivated systems: spatial bistability [84]. Spatial bistability is an OSFR direct analog of the "temporal" bistability described in section 1.3 and is readily observed in a CSTR operated with a "clock reaction". A clock reaction is characterized by a slow induction period τ_{ind} during which the concentration of initial species are slowly consumed, while the autoactivatory species built up to a critical value beyond which the reaction dramatically accelerates and reaches a composition close to that of equilibrium in a time $\tau_{switch} \ll \tau_{ind}$. When such reactions are operated in a CSTR, depending on how the residence time τ_0 compares to τ_{ind}, the contents are either in a steady state with a composition close to that of the feed composition or in a steady state close to thermodynamic equilibrium. In the case of the OSFR, the characteristic diffusion time τ_w of chemicals from the feed boundary to the deep core of the spatial reactor plays the same role as the residence time τ_0 of a CSTR. Obviously, for non-equilibrium instabilities and patterns to develop in the gel, the contents of the CSTR must not belong to the thermodynamic branch. OSFR dynamics are always studied with the CSTR contents kept on the unreacted steady state branch (low extent of reaction). If the diffusion coefficients are of order D, and w is the distance of the boundary to the deep core of the gel, the distribution of concentrations is controlled by the characteristic time $\tau_w = w^2/D$, over which species in the core are exchanged with the boundary. If $\tau_w \ll \tau_{ind}$, the composition in the OSFR remains in a state with a relatively low extent of reaction that we shall refer to as the F state (for feed value). If, on the contrary, $\tau_w \gg \tau_{ind}$ the composition in the deepest part of the OSFR switches to a reacted state. Under the combined actions of the diffusion and of the autocatalytic process, the reacted composition extends over a large portion of the width w of the OSFR. It has been shown that if the major limiting feed species X controlling the autoactivatory process is almost totally converted during the switch, as usual in clock reactions, the spread of the reacted composition stops at a distance $\beta \approx DS|\Delta X|/\dot{Q}$, where ΔX is the concentration drop of X during the switch, S is the surface through which the exchanges are performed at the CSTR/gel boundary, and $\dot{Q}$ is the total amount of X converted by time unit in the gel [85]. As a matter of fact, in most realistic cases, the chemical consumption/production flux is essentially located at the switching position β and this position does not depend on w for as long as the OSFR size is larger. This new concentration profile which mimics in space the batch clock behavior in time was been dubbed the FT state or "mixed state". When $w \leq \beta$, the FT state cannot be maintained and the only possible stable state is the F state. β acts as a finite boundary layer that insures the continuity of the reacted composition in the core of the OSFR and the unreacted state at the contact boundary with the CSTR contents. In a fictitious experiments where one would continuously change w, a hysteresis phenomenon would be observed as a function of w (Figure 14). When w increases from a low value, the system remains in F state until $w > w^*$ where τ_w is large enough for the composition in the deep core of the gel to switch to the reacted state and the whole system remains in a FT state. Now, it is necessary to decrease w down to $w = \beta$ to switch back to F, so that F and FT states are stable for $\beta \leq w \leq w^*$ as schematically illustrated on Figure 14.

This is spatial bistability. As for temporal bistability, the range of spatial bistability can be controlled by the chemical feed parameters or the temperature of the system. A typical example of F and FT state profiles, obtained by numerical simulations of a detailed kinetic model of the CDI reaction is shown in Figure 15. At the phase diagram level, these numerical calculations semi-quantitatively account for the experimental observations (Figure 16) [85] .

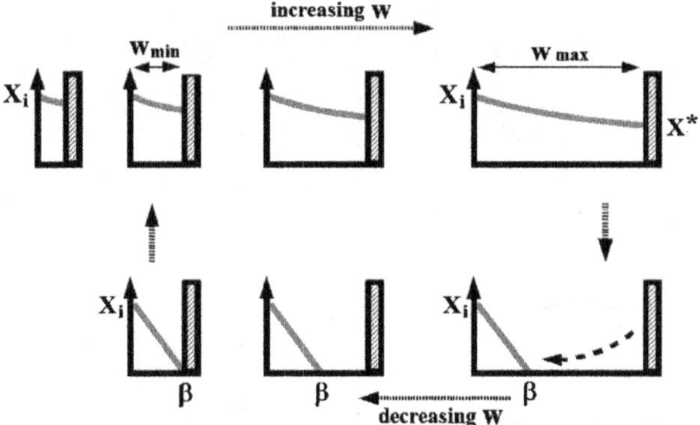

Figure 14.   Schematic spatial bistability. Concentration profile of species X fed with the initial concentration X_i at the CSTR/gel boundary on the left of a gel with a width w limited by an impermeable boundary (hatched bloc) on the right. Representations with increasing w, top row, and decreasing w, *bottom row*. X^* is the critical concentration at which the reaction suddenly accelerates. Switching from one type of profile to the other occurs with hysteresis at w_{min} and w_{max}.

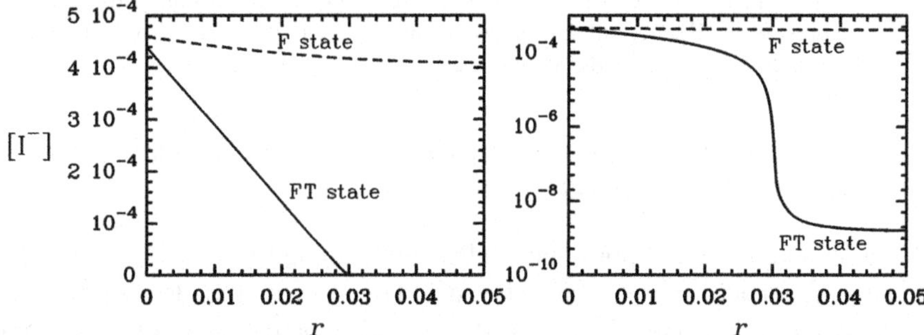

Figure 15.   Computed F and FT state iodide [I^-] concentration profiles as as function of r the distance from the feed boundary ($r = 0$)to the opposite impermeable wall at $r = 0.05$; (*left*) linear scale, note the quasi-linear decrease of the concentration in the FT state between $r = 0$ and $r = 0.03$; (*right*) log scale, note the sudden concentration drop at $r = 0.05$, this reaction front acts as a sink [85].

In systems extended in directions parallel to the feed boundary, in the spatial bistability domain, one can create interfaces between the two spatial states. Depending on the relative stability of the two states, one or the other states would expand at the expense of the other [85]. The control of the direction of propagation and the interaction of these interfaces on head-on collision can plays an important role in the development of stationary pulse patterns. However, such stationary structures require an expanded version of the CDI reaction [86], as explained in the next section.

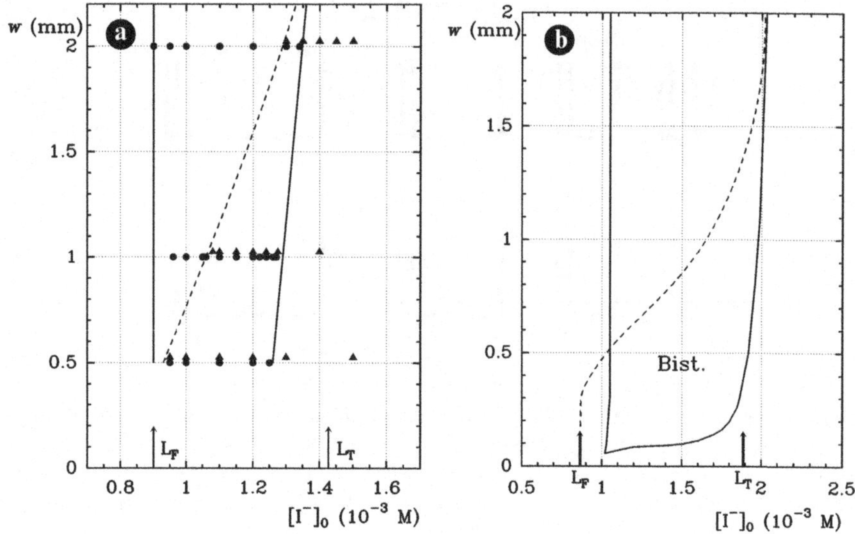

Figure 16.   CDI reaction phase diagrams when operated in a thin ($w = 1$mm) OSFR. Section in the ($[I^-]_0$, w) plane: (a) experimental observations; (b) computed results. Broken curves limit of stability of the F state; full curves low and high $[I^-]_0$ limits of the FT state. Below L_F the CSTR switches to the thermodynamic branch.

2.7.   A SYSTEMATIC DESIGN METHOD FOR STATIONARY PATTERNS AND ITS APPLICATIONS

The association of temporal bistability with an independently controlled negative feedback process has played a key role in the development of oscillatory reactions (see Section 1.3). Similarly, spatial bistability is playing a central role in the recent development of methods to discover stationary reaction-diffusion patterns in chemically different systems. In this case, beside adding and controlling the amplitude of a negative feedback process, the method must be supplemented by the development of appropriate space scale differences between the positive and negative feedback processes (see Section 2.4.2).

The presently most effective method, starts with the selection of a batch clock reaction which (i) can exhibit temporal bistability in a CSTR; (ii) can be made to oscillate or show enhanced oscillatory capacities when an *independent* species inducing a negative feedback process is added; (iii) where the main autoactivatory species can be reversibly and selectively binded to a non-reactive macromolecular complex.

The method was initially tested on the ferrocyanide-iodate-sulfite (FIS) reaction [87]. In the early 1990s, this reaction was found to produce stationary "labyrinthine" patterns and other remarkable dynamic behaviors such as "replicating spots" [57]. The pattern developments were reminiscent of front pairing interactions and of an Ising-Bloch front bifurcation [88–90] in bistable or excitable active media. However, until recently [87] no other group could reproduce these observations for unidentified reasons.

The reaction is a two-substrate pH bistable and oscillatory reaction [91]. Each of these substrates, oxidized by iodate ions (IO_3^-), play an antagonist role: The sulfite drives the proton (H^+) autoactivated process through its protonated form HSO_3^-),

$$IO_3^- + 3\,HSO_3^- \longrightarrow 3\,SO_4^{2-} + I^- + 3\,H^+ \qquad (26)$$

and the ferrocyanide ion ($Fe(CN)_6^{4-}$) oxidation drives the negative feedback process by consuming protons, through an independent kinetic process,

$$IO_3^- + 6\,Fe(CN)_6^{4-} + 6\,H^+ \longrightarrow 6\,Fe(CN)_6^{-3} + I^- + 3\,H_2O. \qquad (27)$$

The iodate-sulfite reaction is a long known batch clock reaction [27], where, after an induction time of the order of 10 min, the pH of the solution switches from ~6 to ~3. In this range of pH, the protons can be reversibly binded by carboxilate functions. Thus, macromolecular polyacids such as polyacrylate chains can act as suitable "complexing" agents for protons and are basically inert to oxihalogen oxidation. In the absence of macromolecular proton-binding species, the reaction develops spatial bistability when operated in an agarose gel OSFR. When amounts of $Fe(CN)_6^{4-}$ are progressively increased, the domain of spatial bistability shrinks and beyond a critical value it exchanges with a domain of spatio-temporal oscillations (Figure 17). Note that a cross-shape diagram is also recovered in the OSFR. In a disc

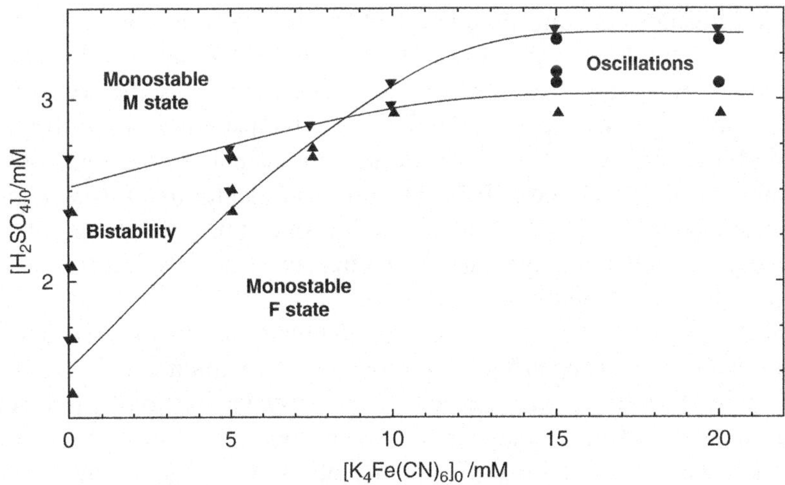

Figure 17.   Phase diagram of the FIS reaction in an annular agarose gel OSFR: Section in ([K$_4$Fe(CN)$_6$]$_0$, [H$_2$SO$_4$]$_0$) plane, in the absence of polyacrylate.

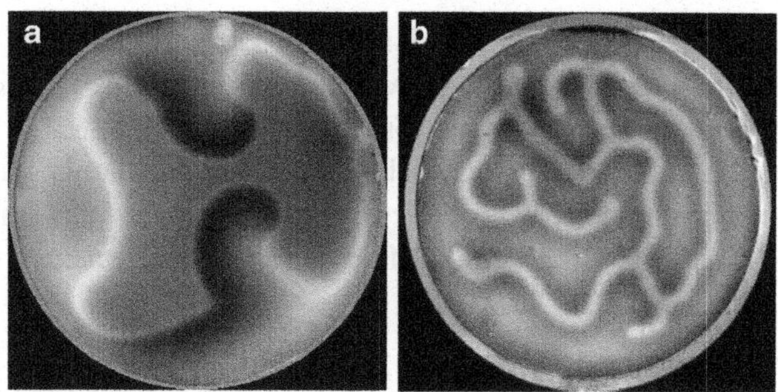

Figure 18. Patterns observed in an agarose gel disc OSFR with the FIS reaction: (a) Spiraling traveling domain of FT state (central structure) in the F state ('peripheral') domain, in the absence of polyacrylic acid; (b) stationary labinrythine pattern of FT state (clear gray structure) in the F state obtained when polyacrylic acid is introduced at a concentration of 4×10^{-3} M of carboxilate function units. Diameter of the disc, 18mm; thickness, $w = 0.75$mm.

OSFR (see Figure 7b), these oscillations can take the form of traveling low pH (FT-type state) waves in a high pH (F-type state) (Figure 18a).

If in the latter domain of the phase diagram a sufficient concentration of polyacrylate is added to the feed, stationary lamellar – "labyrinthine" like – low pH patterns are observed (Figure 18b), a result revisited nearly 15 years after the original discovery. The success of the original experiments probably resulted from an unintentional and uncontrolled slight hydrolysis of the amide functions of the polyacrylamide gel network used at that time. Supercritical concentrations of low mobility carboxilate functions could be reached by the hydrolysis of only a few tenths of percent of the amide functions of the gel network [92]. In the recent experiments the concentrations of these low mobility functions are directly controlled by the feed composition.

As a test for the method, it was shown that a sister chemical system, the thiourea-iodate-sulfite (TuIS), also a double substrate pH oscillatory reaction, could lead to stationary patterns. In this case, the patterns emerge through a Turing bifurcation when supercritical amounts of polyacrylic functions are fed in the agarose gel OSFR [93]. The different planforms of these patterns are illustrated on Figure 19. Note that the wavelength, $\lambda = 1.9$ mm, of these patterns is one order of magnidute greater than those observed with the CIMA reaction (Figures 11 and 12).

In the proposed design method, the amplitude of positive and negative feedbacks must be independently controlled for two different reasons: It provides more flexibility for exploring different combinations of the two antagonist processes. More fundamentally, the independent control of a negative feedback species with a diffusion coefficient not depending on the low mobility complexing agent enables to compensate for the accumulation of the activatory species in the core of the OSFR due to slower exchanges with the

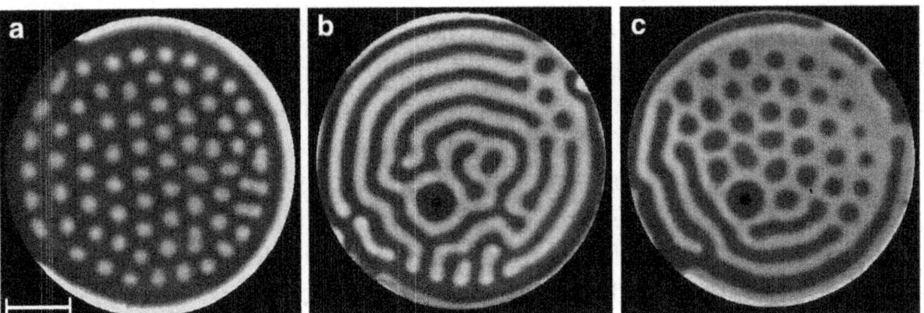

Figure 19.   Turing pattern planforms observed in the TuIS reaction operated in an agarose gel disc OSFR (thickness, $w = 0.75$ mm; diameter, 18 mm) fed. The different planform are obtained for slightly different acid feed concentrations.

CSTR. If not compensated, this accumulation would oppose to the stabilization of a pattern. Ferrocyanide and thiourea play this role in the above two systems and malonic acid plays a similar role in the stabilization of stationary pulses in the spatial bistable domain of an extended version of the CIMA reaction [86].

It is noteworthy that, beside appropriate space scale separation, the development of stationary patterns also requires appropriate time scales separation, specially when they develop through front interaction mechanisms. The activatory process should evolve on a longer time scale than the inhibitory process. In oscillatory systems, this is just the opposite. Yet, experience show that stationary patterns are more easily obtained in the domain or in the close vicinity of the domain of parameters for which the system oscillates in the absence of low mobility complexing agent for the activator. As seen in paragraph 2.4.2, if the complexed form of the activator is much less reactive than the free form, a quenching of the oscillatory dynamics follows along this same way. In fact, as stated above, conditions for reaching a Turing bifurcation by adding a complexing agent always correspond to an initially oscillatory domain of the phase diagram [78]. However, this prerequisite is, *a priori*, not necessary in the case of stationary patterns resulting from front pairing interactions. In the case of the FIS reaction, labyrinthine patterns can be observed for compositions corresponding to the spatial bistability domain just below the cross-point in Figure 17, but never in the absence of ferrocyanide.

3. Conclusion

In these lecture notes, we have provided a brief and somewhat truncated overview of the experimental developments in the field of oscillatory reactions and reaction-diffusion patterns. Many other studies including reaction-diffusion turbulence [94], pattern development under temporal and spatial forcing [95, 96] or in more complex microemulsion systems [97] have not been addressed here to keep things simple. Only examples of the most basic temporal and spatial phenomena are reported with emphasis on the method and hints. A good understanding of the basic features of oscillatory reactions is necessary since they are widely used as the driving chemical engines for chemomechanical devices described in this collection lectures.

Furthermore, many of the ideas which made successful the development chemical oscillations and patterns can be and are already used to produce emerging chemomechanical structures. The change in size of a gel supporting a bistable chemical reaction can act as a negative feedback to generate oscillatory size pulsations even when the reaction has no oscillatory property in itself. Similarly using the hysteretic gate properties of a responsive gel membrane in a compartmentalized reactor a chemical feedback reaction can induce concentration oscillation in one compartment. Detailed descriptions of those systems are found in this book.

Until, now there have been no studies of the coupling of stationary Turing-like reaction-diffusion patterns with property changes of responsive gels. The recent gain of control and development of those chemical patterns make such studies possible now. For example, the gel swelling-deswelling properties could introduce, in the game, a negative feedback operating on a different time scale than the activator and inhibitory processes and therefore new classes of patterning instabilities would be expected in such chemomechanical systems.

References

1. W. Ostwald, Z. *Phys. Chem.* **35**, 204 (1900).
2. G. Bredig, J. Weinmayer, Z. *Phys. Chem.* **42**, 601 (1903);
   G. Lippman, *Ann. Phys.* **149**, 546 (1873).
3. E.S. Hedges, J.E. Myers, *The Problem of Physico-chemical Periodicity*, (London, Edwards Arnold & Co, 1926), p. 38.
4. W.C. Bray, *J. Am. Chem. Soc.* **43**, 1262, (1921).
5. P. Belousov, *Sbornik Referatov po Radiatsionni Medditsine* p. 145 (1958).
6. M.G.T. Fechner, Schweigger, *J. Chem. Phys.* **53**, 129 (1828).
7. P. Glansdorff, I. Prigogine, *Thermodynamic Theory of Structure, Stability and Fluctuations*, (Wiley, New York 1971).
8. R.J. Field, E. Körös, R.M. Noyes, *J. Am. Chem. Soc.* **94**, 8648 (1972).
9. R.J. Field, R.M. Noyes, *J. Chem. Phys.* **60**, 1877 (1974).
10. P. Gray, S.K. Scott, *Chem. Eng. Sci.* **38**, 29 (1983).
11. P. Shen, R. Larter, *Biophys.* **67**, 1414 (1994).
12. D.M. Kern, C.-H. Kim, *J. Am. Chem. Soc.* **87**, 5309 (1965).
13. J.J. Tyson, *J. Chem. Phys.* **62**, 1010 (1975).

14. Y. Luo, I.R. Epstein, *Adv. Chem. Phys.*, **79**, 269 (1990).

15. J.C. Roux, P. De Kepper, J. Boissonade, *Phys. Lett.* **A 97**, 168 (1983);
M. Boukalouch, J. Boissonade, P. De Kepper, *J. Chim. Phys. (Paris)* **84**, 1353 (1987);
Y. Luo, I.R. Epstein, *J. Chem. Phys.* **85**, 5733 (1986).

16. A.K. Dutt, M. Menzinger, *J. Phys. Chem.* **94**, 4867 (1990);
M. Menzinger, A. Giraudi, *J. Phys. Chem.* **91**, 4391 (1987).

17. U. Franck, *Angew. Chem. Int. Ed.* **17**, 1 (1978).

18. P. De Kepper, A. Pacault, *Compt. Rend. Acad. Sci.* **C286**, 437 (1987).

19. M. Orbán, C. Dateo, P. De Kepper, I.R. Epstein, *J. Am. Chem. Soc.* **104**, 5911 (1982).

20. J. Boissonade, P. De Kepper, *J. Phys. Chem.* **84**, 501 (1980).

21. P. De Kepper, I.R. Epstein, K. Kustin, *J. Am. Chem. Soc.* **103**, 2133 (1981).

22. Gy. Rábai, M. Orbán, I.R. Epstein, *Acc. Chem. Res.* **23**, 258 (1990).

23. K. Kovacs, R.E. McIlwaine, S.K. Scott, A.F. Taylor, *Phys. Chem. Chem. Phys.* **9**, 3711 (2007).

24. Gy. Rábai, *ACH Mod. Chem.* **135**, 381 (1998).

25. Gy. Rábai, M.T. Beck, K. Kustin, I.R. Epstein, *J. Phys. Chem.* **93**, 2853 (1989).

26. Zs. Virányi, I. Szalai, J. Boissonade, P. De Kepper, *J. Phys. Chem.* **A 111**, 8090 (2007).

27. H. Landolt, *Ber. Dtsch. Chem. Ges.* **19**, 1316 (1886).

28. E.C. Edblom, Y. Luo, M. Orbán, K. Kustin, I.R. Epstein, *J. Phys. Chem.* **93**, 2722 (1989).

29. T.G. Szántó, Gy. Rábai, *J. Phys. Chem.* **A109**, 5398 (2005).

30. N. Okazaki, Gy. Rábai, I. Hanazaki, *J. Phys. Chem.* **A103**, 10915 (1999).

31. Gy. Rábai, I. Hanazaki, *J. Phys. Chem.* **100**, 10615 (1996).

32. K. Kurin-Csörgei, I.R. Epstein, M. Orbán, *Nature* **433**, 139 (2005).

33. K. Kurin-Csörgei, M. Orbán, I.R. Epstein, *J. Phys. Chem.* **A110**, 7588 (2006).

34. T. Liedl, M. Olapinski, F.C. Simmel, *Angew. Chem. Int. Ed.* **45**, 5007 (2006).

35. R. Yoshida, H. Ichijo, T. Hakuta, T. Yamaguchi, *Macromol. Rapid Commun.* **16**, 305 (1995).

36. V. Labrot, P. De Kepper, J. Boissonade, I. Szalai, F. Gauffre, *J. Phys. Chem.* **B109**, 21476 (2005).

37. C.J. Crook, A. Smith, R.A.L. Jones, A.J. Ryan, *Phys. Chem. Chem. Phys.* **4**, 1367 (2002).

38. I. Varga, I. Szalai, R. Meszaros, T. Gilányi, *J. Phys. Chem.* **B110**, 20297 (2006).

39. J. D'Hernoncourt, A. Zebib, A. De Wit, *Chaos* **17**, 013109 (2007).

40. A. Boiteux, B. Hess, *Ber. Bunsenges. Phys. Chem.* **84**, 392 (1980).

41. K. Showalter, *J. Chem. Phys.* **73**, 3735 (1980).

42. M. Orbán, *J. Am. Chem. Soc.* **102**, 4311 (1980).

43. R.E. Liesegang, *Naturwiss. Wochenschr.* **11**, 353 (1896).

44. H.K. Henisch, *Periodic Precipitation* (Pergamon Press, New York, 1991).

45. P.W. Voorhees, *J. Stat. Phys.* **38**, 231 (1985).

46. S.C. Müller, J. Ross, *J. Phys. Chem.* **A107**, 7997 (2003).

47. J.B. Keller, S.I. Rubinow, *J. Chem. Phys.* **74**, 5000 (1981).

48. I.T. Bensemann, M. Fialkowski, B.A. Grzybowski, *J. Phys. Chem.* **B109**, 2774 (2005).

49. R. Luther, *Z. Elektrochem.* **12**, 596 (1906);
K. Showalter, J.J. Tyson, *J. Chem. Educ.* **64**, 742 (1987).

50. A.N. Zaikin, A.M. Zhabotinsky, *Nature* **225**, 535 (1970).

51. A.T. Winfree, *Science* **175**, 634 (1972);
Scientific Am. **230**, 82 (1974);
Chaos **1**, 303 (1991).

52. Z. Nagy-Ungvarai, J. Ungvarai, S.C. Müller *Chaos* **3**, 15 (1993).

53. P. Foester, S.C. Müller, B. Hess, *Science* **241**, 686 (1988).

54. C.T. Hamik, N. Manz, O. Steinbock, *J. Phys. Chem.* **A105**, 6144 (2001).
N. Manz, S.C. Müller, O. Steinbock, *J. Phys. Chem. A* **A104**, 5895 (2000).

55. V. Castets, E. Dulos, J. Boissonade, P. De Kepper, *Phys. Rev. Lett.* **64**, 2953 (1990).

56. A. Turing, *Philos. Trans. R. S. London* **B327**, 37 (1952).

57. K.J. Lee, W.D. McCormick, Q. Ouyang, H.L. Swinney, *Science* **261**, 192 (1993);
K.J. Lee, W.D. McCormick, J.E. Pearson, H.L. Swinney, *Nature* **369**, 6477 (1994).

58. J.A. Vastano, J.E. Pearson, W. Horsthemke, H.L. Swinney, *J. Chem. Phys.* **88**, 6175 (1988).

59. P. De Kepper, I.R. Epstein, K. Kustin, M. Orbán, *J. Phys. Chem.* **86**, 170 (1982).

60. E. Dulos, P.W. Davies, B. Rudovics, P. De Kepper, *Physica* **D98**, 53 (1996);
B. Rudovics, E. Dulos, P. De Kepper, *Physica Scripta* **T67**, 43 (1996).

61. P. Borckmans, G. Dewel, A. De Wit, D. Walgraef, in *Chemical Waves and Patterns*, Eds. R. Kapral and K. Showalter (Kluwer, Dordrecht, 1995), p. 323.

62. V. Dufiet, J. Boissonade, *Phys. Rev.* **E53**, 4883 (1996).

63. A.S. Mikhailov, V.S. Zykov, *Physica* **D52**, 379 (1991).

64. M. Gomez-Gesteira, G. Fernandez-Garcia, A.P. Munuzuri, V. Perez-Munuzuri, V.I. Krinsky, C.F. Starmer, V. Perez-Villar, *Physica* **D76**, 359 (1994).

65. E. Dulos, J. Boissonade, P. De Kepper, in *Nonlinear Waves Processes in Excitable Media*, Eds. M. Markus and H.G. Othmer eds., p. 423 (Plenum Press, New York, 1991).

66. N. Kreisberg, W.D. Mc Cormic, H.L. Swinney, *J. Chem. Phys.* **91**, 6532 (1989).

67. H. Meinhardt, *Models of Biological Pattern Formation* (Academic Press, New York, 1982).

68. D. Walgraef, G. Dewel, P. Borckmans, *Phys. Rev.* **A21**, 337 (1980); P. Borckmans, A. De Wit, G. Dewel, *Physica* **A188**, 137 (1992).

69. B.A. Malomed, M.I. Tribeslskii, *Sov. Phys. JETP* **65**, 305 (1987);
L.M. Pismen, in *Dynamics of Nonlinear Systems*, Ed. V. Hlavacec (Gordon and Break, 1986).

70. V. Dufiet, J. Boissonade, *Physica* **A188**, 158 (1992).

71. A. De Wit, G. Dewel, P. Borckmans, D. Walgraef, *Physica* **D61**, 289 (1992).

72. H. Shoji, K. Yamada, D. Ueyama, T. Otha, *Phys. Rev.* **E75**, 046212 (2007).

73. A. De Wit, P. Borckmans, G. Dewel, *Proc Natl Acad Sci USA* **94**, 12765 (1997).

74. G. Nicolis I. Prigogine, *Self Organization in Nonequilibrium Systems* (Wiley, New York, 1977).

75. P.M. Kulesa, G.C. Cruywagen, S.R. Lubkin, P.K. Maini, J.S. Sneyd, J.D. Murray, *J. Biol. Syst.* **3**, 975 (1995).

76. S. Kondo, *Genes Cells* **7**, 535 (2002).

77. I. Lengyel, I.R. Epstein, *Science,* **251**, 650 (1990);
Proc. Natl. Acad. Sci. USA, **89**, 3977 (1992).

78. J.E. Pearson, W. Bruno *Chaos* **2**, 513 (1992).

79. D.E. Strier, S. Ponce Dawson, *PLoS ONE* **2**, 1053 (2007).

80. I. Lengyel, Gy. Rábai, I.R. Epstein, *J. Am. Chem. Soc.* **112**, 4606 (1990).

81. R. Rudovics, E. Barillot, P.W. Davies, E. Dulos, J. Boissonade, P. De Kepper, *J. Phys. Chem.* **A103**, 1790 (1999).

82. S. Kádár, I. Lengyel, I.R. Epstein, *J. Phys. Chem.* **99**, 4054 (1995).

83. I. Lengyel, Gy. Rábai, I.R. Epstein, *J. Am. Chem. Soc.* **112**, 4606 (1990).

84. P. Blanchedeau, J. Boissonade, *Phys. Rev. Lett.* **81**, 5007 (1998).

85. P. Blanchedeau, J. Boissonade, P. De Kepper, *Physica* **D147**, 283 (2000).

86. I. Szalai, P. De Kepper, *J. Phys. Chem.* **A108**, 5315 (2004);
D.E. Strier, P. De Kepper, J. Boissonade, *J. Phys. Chem.* **A109**, 1357 (2005).

87. I. Szalai, P. De Kepper, *J. Phys. Chem.* **A112**, 783 (2008);
I. Szalai, P. De Kepper, *Chaos*, **18**, 026105 (2008).

88. A. Hagberg, E. Meron, *Chaos* **4**, 477 (1994);
C. Elphick, A. Hagberg, E. Meron, *Phys. Rev.* **E51**, 3052 (1995).

89. H. Ikeda, M. Mimura, Y. Nishiura, *Nonlinear Anal. Theory Methods Appl.* **13**, 507 (1989).

90. S. Ponce Dawson, M.V. DAngelo, J.E. Pearson, *Phys. Lett.* **A265**, 346 (2000).

91. E.C. Edblom, M. Orbán, I.R. Epstein, *J. Am. Chem. Soc.* **108**, 2826 (1986);
E.C. Edblom, L. Gyrgyi, M. Orbn, I.R. Epstein, *J. Am. Chem. Soc.* **109**, 4876 (1987).

92. T. Tanaka, D. Fillmore, S.T. Sun, I. Nishio, G. Swislaw, A. Shah, *Phys. Rev. Lett.* **45**, 1636 (1980);
S.R. Sandler, W. Karo, *Polymer Synthesis*, **II** (Academic, New York, 1977).

93. J. Horváth, I. Szalai, P. De Kepper, *Science* **324**, 772 (2009).

94. Q. Ouyang, J.-M. Flesselles, *Nature* **379**, 143 (1996);
L.Q. Zhou, Q. Ouyang, *J. Phys. Chem.* **A105**, 112 (2001).

95. A.L. Lin, A. Hagberg, A. Ardelea, M. Bertram, H.L. Swinney, E. Meron, *Phys. Rev.*, **E62**, 3790 (2000);
K. Martinez, A.L. Lin, R. Kharrazian, X. Sailer, H.L. Swinney, *Physica*, **D 168–169**, 1 (2002).

96. I. Berenstein, L. Yang, M. Dolnik, A.M. Zhabotinsky, I.R. Epstein, *J. Phys. Chem.* **A 109**, 5387 (2005).

97. I.R. Epstein, V. Vanag, *Chaos*, **15**, 047510 (2005);

MECHANOCHEMICAL INSTABILITIES IN ACTIVE GELS

Ryo Yoshida[*]

Department of Materials Engineering, Graduate School of Engineering, The University of Tokyo, Japan

Abstract. Stimuli-responsive polymers and their application to smart materials have been widely studied. On the other hand, as a novel bio-mimetic polymer, we have been studying a polymer with an autonomous self-oscillating function by utilizing oscillating chemical reactions. The self-oscillating polymer is composed of poly(N-isopropylacrylamide) (PNIPAAm), in which Ru(bpy)$_3$ is incorporated as a catalyst for the BZ reaction. Under the coexistence of the BZ reactants (malonic acid, sodium bromate, and nitric acid), the polymer undergoes spontaneous cyclic soluble-insoluble changes or swelling-deswelling changes (in the case of gel) without any on-off switching of external stimuli. In this chapter, our recent studies on the self-oscillating polymer and the design of functional material systems using the polymer are summarized.

Keywords: the BZ reaction, oscillation, polymer, gels, swelling, actuator

1. Introduction

Polymer gels is a research field of polymer science which has seen rapid progress during the past 20–30 years. Gel can be widely defined as a cross-linked polymer network which is swollen by absorbing large amounts of solvent such as water. Theoretical study of the characteristics of gel had already proceeded in the 1940s, and the principle of swelling by water absorption based on thermodynamics had been clarified by Flory [1]. As an application of gel research, soft contact lenses were developed in the 1960s, and subsequently gels have been widely used in medical and pharmaceutical fields. Since a polymer which can absorb about 1,000 times as much water as its own weight was developed in the US in the 1970s, gels have been

---

[*] To whom correspondence should be addressed. e-mail: ryo@cross.t.u-tokyo.ac.jp

P. Borckmans et al. (eds.), *Chemomechanical Instabilities in Responsive Materials,*
© Springer Science + Business Media B.V. 2009

applied as super absorbent polymers in several industrial fields, mainly application to sanitary items, disposable diapers, etc. Further, in 1978, it was discovered by Tanaka [2] that gels change volume reversibly and discontinuously in response to environmental changes such as solvent composition, temperature, pH change, etc. (called "volume phase transition" phenomena) With this discovery as a turning point, research to use gels as functional materials for artificial muscle, robot hands (actuator), stimuli-responsive drug delivery systems (DDS), separation or purification, cell culture, biosensors, shape memory materials, etc. was activated [3–8].

Until now, fundamental and applied research which include many different fields such as elucidation of gelation mechanisms, analysis of physical properties and structure, functional control by molecular design, etc. have been done. Especially, from the early 1990s, new functional gels which include the following three functions in themselves; sensing an external signal (sensor function), judging it (processor function), and taking action (actuator function), have been developed by many researchers as "intelligent gels" or "smart gels".

Further, in recent years, the usefulness of gels has also been shown in the field of micromachines and nanotechnology. In addition to new synthetic methods to give unique functions by molecular design in nano-order scale including supramolecular design, the design and construction of micro or nano material systems with the biomimetic functions of motion, mass transport, transformation and transmission of information, molecular recognition, etc. have been attempted.

So far, many researchers have developed stimuli-responsive polymer gels that change volume abruptly in response to a change in their surroundings such as solvent composition, temperature, pH, and supply of electric field, etc. Their ability to swell and deswell according to conditions makes them an interesting proposition for use in new intelligent materials. In particular, applications for biomedical fields are extensively studied. One of the strategies of these applications is to develop biomimetic material systems with stimuli-responding function; i.e., systems in which the materials sense environmental changes by themselves and go into action. For these systems, the on-off switching of external stimuli is essential to instigate the action of the gel. Upon switching, the gels provide only one unique action, either swelling or deswelling.

This stimuli-responding behavior is temporary action toward an equilibrium state. In contrast, there are many physiological phenomena in our body that continue their own native cyclic changes. These phenomena exist over a wide range from cell to body level, as represented by the cell cycle, cyclic reaction in glycolysis, pulsatile secretion of hormones, pulsatile potential of nerve cells, brain waves, heartbeat, peristaltic motion in the digestive tract, and human biorhythms, etc. If such self-oscillation could be achieved for gels, possibilities would emerge for new biomimetic intelligent materials that exhibit autonomous rhythmical motion.

In this paper, a new design concept for polymer gels which exhibit spontaneous and autonomous periodic swelling-deswelling changes under constant conditions without on-off switching of external stimuli will be introduced. In the materials design, nonlinear dynamics of chemical reactions and characteristics of gels as open systems play an important role.

2. Design of self-oscillating gel

The mechanical oscillation of the gel is produced via an oscillating chemical reaction, called the Belousov-Zhabotinsky (BZ) reaction [9–11]. We attempted to convert the chemical oscillation of the BZ reaction to the mechanical changes of gels and generate an autonomic swelling-deswelling oscillation under nonoscillatory outer conditions. A copolymer gel which consists of N-isopropylacrylamide (NIPAAm) and $Ru(bpy)_3^{2+}$ was prepared. $Ru(bpy)_3^{2+}$, acting as a catalyst for the BZ reaction, is pendent to the polymer chains of NIPAAm (Figure 1). It is well known that homopolymer gels of NIPAAm have thermosensitivity and undergo an abrupt volume-collapse (phase transition) when heated at around 32°C. The poly(NIPAAm-co-$Ru(bpy)_3^{2+}$) gel has a phase transition temperature because of themosensitive constituent NIPAAm. The oxidation of the $Ru(bpy)_3^{2+}$ moiety caused not only an increase in the swelling degree of the gel, but also a rise in the transition temperature. These characteristics may be interpreted by considering an increase in hydrophilicity of the polymer chains due to the oxidation of Ru(II) to Ru(III) in the $Ru(bpy)_3$ moiety. As a result, it is expected that the gel undergoes a cyclic swelling-deswelling alteration when the $Ru(bpy)_3$ moiety is periodically oxidized and reduced under constant temperature. When the gel is immersed in an aqueous solution containing the substrate of

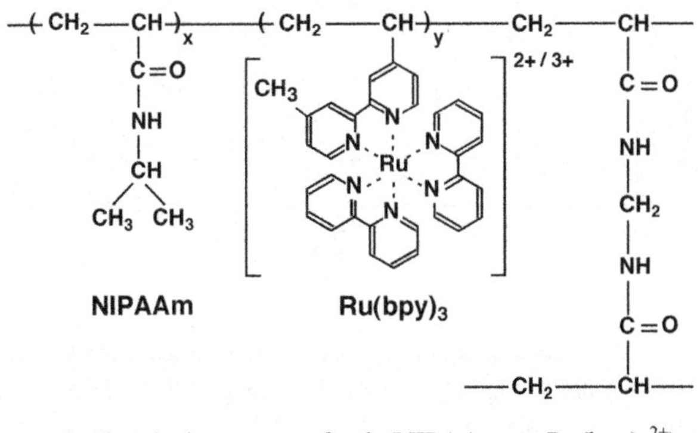

NIPAAm   Ru(bpy)3

Figure 1. Chemical structure of poly(NIPAAm-co-$Ru(bpy)_3^{2+}$) gel.

the BZ reaction except for the catalyst, the substrates penetrates into the polymer network and the BZ reaction occurs in the gel. Consequently, periodical redox changes induced by the BZ reaction produce periodical swelling-deswelling changes of the gel [12–14]. The gel has the cyclic reaction network in itself to generate periodic mechanical energy from the chemical energy of the BZ reaction.

3. Swelling-deswelling oscillation of the gel with periodic redox changes

3.1. SELF-OSCILLATION OF BULK GEL SMALLER THAN CHEMICAL WAVELENGTH

The miniature cubic poly(NIPAAm-co-Ru(bpy)$_3{}^{2+}$) gel (each length of about 0.5 mm) was immersed into an aqueous solution containing MA, sodium NaBrO$_3$, and HNO$_3$ at constant temperature (20°C). This outer solution comprised the reactants of the BZ reaction, with the exception of the catalyst. Therefore the redox oscillation does not take place in this solution. However, as it penetrates into the gel, the BZ reaction is induced within the gel by the Ru(bpy)$_3{}^{2+}$ copolymerized as a catalyst on the polymer chains. Under the reaction, the Ru(bpy)$_3{}^{2+}$ in the gel network periodically changes between reduced and oxidized states. In miniature gels sufficiently smaller than the wavelength of the chemical wave (typically several mm), the redox change of ruthenium catalyst can be regarded to occur homogeneously without pattern formation. Figure 2 shows the observed oscillating behavior under a microscope. Color changes of the gel accompanied by redox oscillations

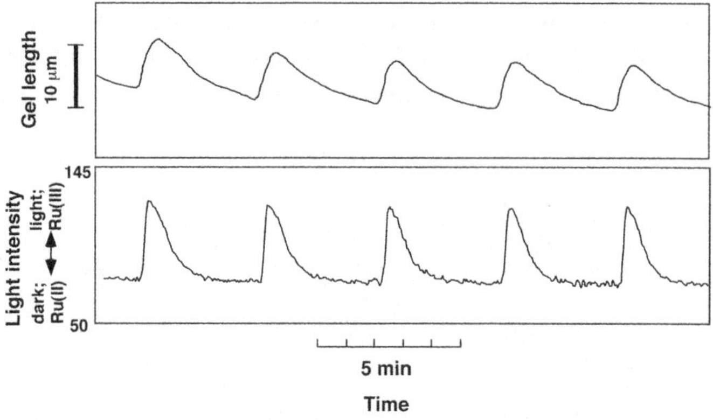

Figure 2. Periodic redox changes of the miniature cubic poly(NIPAAm-co-Ru(bpy)$_3{}^{2+}$) gel (lower) and the swelling-deswelling oscillation (upper) at 20°C [15]. Transmitted light intensity is expressed as an 8-bit grayscale value.

(orange: reduced state, light green: the oxidized state) were converted to 8-bit grayscale changes (dark: reduced, light: oxidized) by image processing. Due to the redox oscillation of the immobilized $Ru(bpy)_3^{2+}$, mechanical swelling-deswelling oscillation of the gel autonomously occurs with the same period as for the redox oscillation. The volume change is isotropic and the gel beats as a whole, like a heart muscle cell. The chemical and mechanical oscillations are synchronized without a phase difference (i.e., the gel exhibits swelling during the oxidized state and deswelling during the reduced state).

3.2. CONTROL OF OSCILLATION PERIOD AND AMPLITUDE

In order to enhance the amplitude of swelling-deswelling oscillations of the gel, control of the period and amplitude of the redox oscillation was attempted by varying the initial concentration of substrates. It is a general tendency that the oscillation period increases with a decrease in concentration of substrates. The variation in chemical oscillation leads to a change in the swelling-deswelling oscillation: i.e., the swelling-deswelling amplitude (the change in gel length, Δd) increases with an increase in the period and amplitude of the redox changes. Empirically, the relation between Δd [μm] and the substrate concentrations was expressed as:

$$\Delta d = 2.38[\mathrm{MA}]^{0.392}[\mathrm{NaBrO_3}]^{0.059}[\mathrm{HNO_3}]^{0.764}.$$

As a result, it is apparent that the swelling-deswelling amplitude of the gel is controllable by changing the initial concentration of substrates. So far, a swelling-deswelling amplitude with ca. 20% of the initial gel size has been obtained as a maximum value. When the amplitude of swelling-deswelling oscillation increased, the waveform of redox changes deformed to a rectangular shape with a plateau period [15]. From this result, it is supposed that not only energy transformation from chemical to mechanical change, but also a feedback mechanism from mechanical to chemical change acts in the synchronization process.

As an inherent behavior of the BZ reaction, the abrupt transition from steady state (non-oscillating state) to oscillating state occurs with a change in controlling parameter such as chemical composition, etc. This change is termed "bifurcation". Considering this characteristic, it is expected that the rhythmical motion of the gel can be controlled by changing substrate concentration during the oscillation. For example, if the [MA] is switched between the concentration regions of steady state and oscillating state, on-off control of the beating would be possible [16]. And also, as the gel has thermosensitivity due to the NIPAAm component, the beating rhythm can be also controlled by temperature [17].

3.3. PERISTALTIC MOTION OF GELS

When a one-dimensional rectangular piece of gel is immersed in an aqueous solution containing the three reactants of the BZ reaction. The chemical waves propagate in the gel at a constant speed in the direction of the gel length [18–21]. Considering the Ru(II) and Ru(III) zones represent simply the shrunken and swollen parts respectively, the locally swollen and shrunken parts move with the chemical wave, like the peristaltic motion of living worms. The propagation of the chemical wave makes the free end of the gel move back and forth at a rate corresponding to the wave propagation speed. As a result, the total length of the gel periodically changes. It was demonstrated by mathematical model simulations that the change in the overall gel length is equivalent to that in the remainder of gel length divided by the wavelength, because the swelling and the deswelling cancel each other per one period of oscillations under steady oscillating conditions [22,23].

We succeeded in measuring the oscillating force of cylindrical poly(NIPAAm-co-Ru(bpy)$_3^{2+}$) gel accompanied by the BZ reaction [24,25]. The measurements were made for three gels with different diameters, 0.65, 1.17, and 1.88 mm, at 15°C. It was found that the amplitude of oscillatory tensile stress of the reacting cylindrical gel with a diameter of 0.65 mm is much higher than the stress theoretically expected from the Donnan osmotic pressure. The amplitude is explained by an oscillatory change of the interaction parameter, χ, which might be induced by the hydration and dehydration of the chain due to the oscillatory charge density on the polymer chain. The oscillation behavior of the BZ reaction is found to be strongly dependent on the diameter of the gel. The staying period in the oxidative state and the period of the oscillation increase with a decrease in the diameter of the cylindrical gel.

And also, it was reported that the structural color behavior of a periodic ordered mesoporous gel synchronized with the BZ reaction [26]. We prepared a periodically ordered mesoporous gel which reveals "structural color" depending on its swelling ratio. To obtain the gel, we used as a template the closest-packing colloidal crystal composed of silica sphere particles 210 nm in diameter. The structural colored concentric rings which were spatiotemporally spread out on the porous gel were observed during the BZ reaction. The color tone of the structural color, which is determined by the swelling ratio of the gel, periodically changed. This is the first evidence that a self-sustaining peristaltic motion occurs on the surface of a gel.

4. Design of biomimetic microactuator using self-oscillating gel

4.1. CILIARY MOTION ACTUATOR USING SELF-OSCILLATING GEL

Recently, microfabrication technologies such as photolithography have also been attempted for preparation of microgels. Since any shape of gel can be created by these methods, application as a new manufacturing method for soft microactuators, microgel valves, gel displays, etc. is a possibility. Micro-fabrication of self-oscillating gel has also been attempted by photolithography for application to such micro-devices [27,28].

One of the promising fields of the MEMS is micro actuator array or distributed actuator systems. The actuators, which have a very simple actuation motion such as up and down motion, are arranged in an array form. If their motions are random, no work is extracted from this array. However, by controlling them to operate in a certain order, they can generate work as a system. A typical example of this kind of actuation array is a ciliary motion micro actuator array. There have been many reports of this. Although various actuation principles have been proposed, all the previous work is based on the concept that the motion of actuators is controlled by external signals. If a self-oscillating gel plate with a micro projection structure array on top were realized, it would be expected that the chemical wave propagation would create dynamic rhythmic motion of the structure array. This proposed structure could exhibit spontaneous dynamic propagating oscillation producing a ciliary motion array .

A gel plate with micro projection array was fabricated by molding [29–31]. First, moving mask deep-X-ray lithography was utilized to fabricate a PMMA plate with a truncated conical shape microstructure array. This step was followed by evaporation of a Au seed layer and subsequent electroplating of nickel to form the metal mold structure. Then, a PDMS mold structure was duplicated from the Ni structure and utilized for gel molding. The formation of gel was carried out by vacuum injection molding. A structure with a height of 300 μm and bottom diameter of 100 μm was successfully fabricated by the described process. The propagation of chemical reaction wave and dynamic rhythmic motion of the micro projection array were confirmed by chemical wave observation and displacement measurements. Figure 3 shows the measured lateral and vertical movements and the motion trajectory of the projection top. Motion of the top with 5 μm range in both lateral and vertical directions, and elliptical motion of the projection top were observed. The feasibility of the new concept of the ciliary motion actuator made of self-oscillating polymer gel was successfully confirmed. The actuator may serve as a micro-conveyer to transport micro- or nano-particles on the surface.

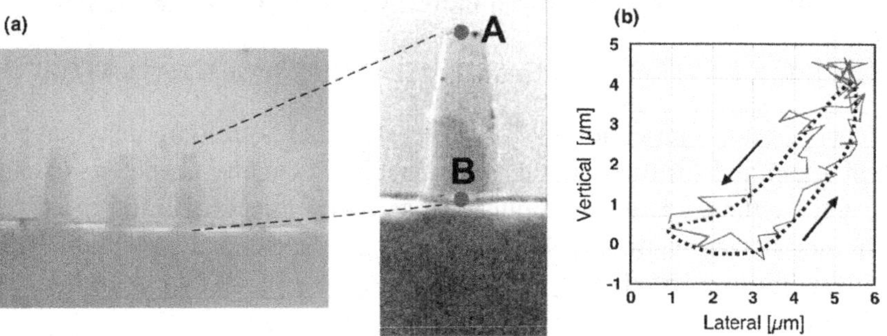

Figure 3. (a) Cross-sectional view of the micro projection and (b) measured motion of the projection top.

4.2. SELF-WALKING GEL

We are trying to develop a chemical robot, which is unlike a conventional electrically powered robot, by coupling with a PDMS membrane [32,33]. Further, we successfully developed a novel biomimetic walking-gel actuator made of self-oscillating gel [34]. To produce directional movement of gel, asymmetrical swelling-deswelling is desired. For these purposes, as a third component, hydrophilic 2-acrylamido-2-methylpropanesulfonic acid (AMPS) was copolymerized into the polymer to lubricate the gel and to cause anisotropic contraction. During polymerization, the monomer solution faces two different surfaces of plates; a hydrophilic glass surface and a hydrophobic Teflon surface. Since $Ru(bpy)_3^{2+}$ monomer is hydrophobic, it easily migrates to the Teflon surface side. As a result, a non-uniform distribution along the height is formed by the components, and the resulting gel has gradient distribution for the content of each component in the polymer network. At the surface side where the content of hydrophilic AMPS is higher, the swelling ratio of the gel membrane in water becomes larger than that at the opposite side in the same gel where the content of hydrophobic $Ru(bpy)_3^{2+}$ is higher. Consequently, in water, the gel strip always bends in the direction of the surface which was facing the Teflon plate during polymerization.

In order to convert the bending and stretching changes to one-directional motion, we employed a ratchet mechanism. A ratchet base with an asymmetrical surface structure was fabricated from an acrylic sheet. On the ratchet base, the gel repeatedly bends and stretches autonomously resulting in the forward motion of the gel, while sliding backwards is prevented by the teeth of the ratchet. Figure 4 shows successive profiles of the "self-walking" motion of the gel like a looper in the BZ substrate solution under constant temperature. The period of chemical oscillation was approximately

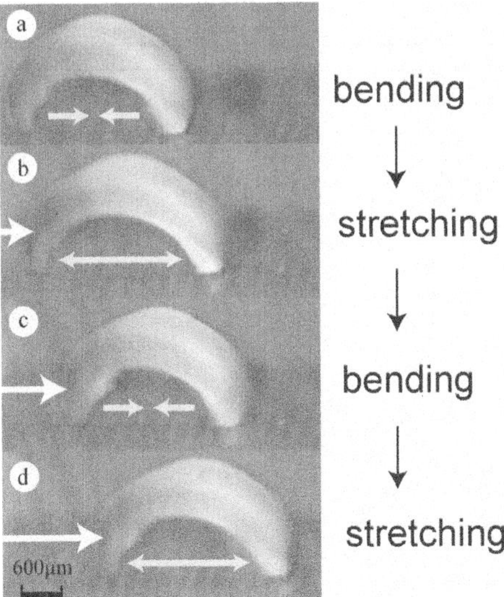

Figure 4. Time course of self-walking motion of the gel actuator. During stretching, the front edge can slide forward on the base, but the rear edge is prevented from sliding backwards. Oppositely, during bending, the front edge is prevented from sliding backwards while the rear edge can slide forward. This action is repeated, and as a result, the gel walks forward [34].

112 s, and the walking velocity of the gel actuator was approximately 170 μm/min. Since the oscillating period and the propagating velocity of chemical wave change with concentration of substrates in the outer solution, the walking velocity of the gel can be controlled.

5. Self-oscillating polymer chains and gel particles as "nano-oscillator"

5.1. SELF-OSCILLATION OF POLYMER CHAINS WITH RHYTHMICAL SOLUBLE-INSOLUBLE CHANGES

In self-oscillating gel, redox changes of $Ru(bpy)_3^{2+}$ catalyst are converted to conformational changes of polymer chain by polymerization. The conformational changes are amplified to macroscopic swelling-deswelling changes of the polymer network by crosslinking. Further, when the gel size is larger than chemical wavelength, the chemical wave propagates in the gel by coupling with diffusion. Then peristaltic motion of the gel is created. In this manner, a hierarchical synchronization process exists in the self-oscillating gel (Figure 5).

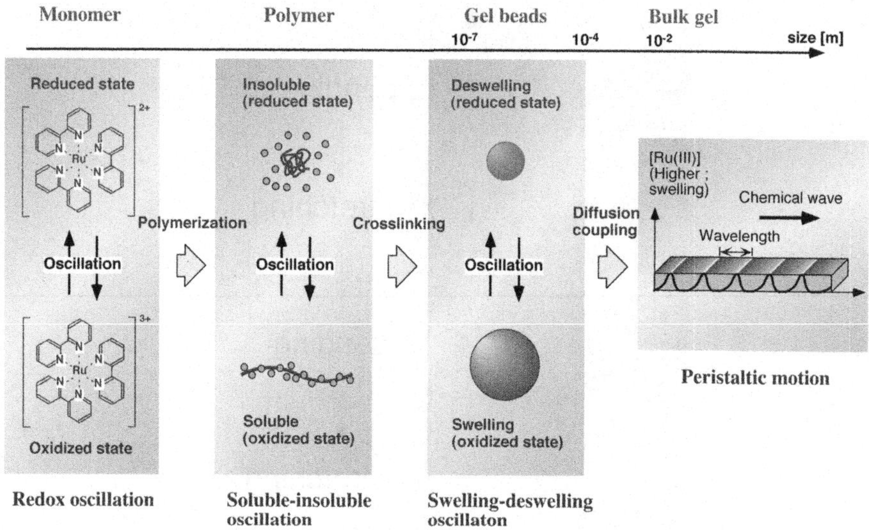

Figure 5. Hierarchical synchronization in self-oscillating gel.

These periodic changes of linear and uncrosslinked polymer chains can be easily observed as cyclic transparent and opaque changes for the polymer solution with color changes due to the redox oscillation of the catalyst [35]. Figure 6 shows the oscillation profiles of transmittance for a polymer solution which consists of linear poly(NIPAAm-co-Ru(bpy)$_3{}^{2+}$), MA, NaBrO$_3$ and HNO$_3$ at constant temperatures. The wavelength (570 nm) at the isosbestic point of reduced and oxidized states was used to detect the optical transmittance changes based on soluble-insoluble changes of the polymer, not on the redox changes of the Ru(bpy)$_3$ moiety. Synchronized with the periodical changes between Ru(II) and Ru(III) states of the Ru(bpy)$_3{}^{2+}$ site, the polymer becomes hydrophobic and hydrophilic, and exhibits cyclic soluble-insoluble changes.

5.2. SELF-FLOCCULATING/DISPERSING OSCILLATION OF MICROGELS

We then prepared submicron-sized poly(NIPAAm-co-Ru(bpy)$_3{}^{2+}$) gel beads by surfactant-free aqueous precipitation polymerization, and analyzed the oscillating behaviors [36–38]. In both cases of the reduced Ru(II) and the oxidized Ru(III) states, the microgels were also flocculated when they were heated above the volume phase transition temperature (VPTT) because the interparticle electrostatic repulsion is extremely low at a high salt concentration (~0.3 M). The microgels in the oxidized Ru(III) state became flocculated at slightly higher temperature (34°C) than those in the reduced

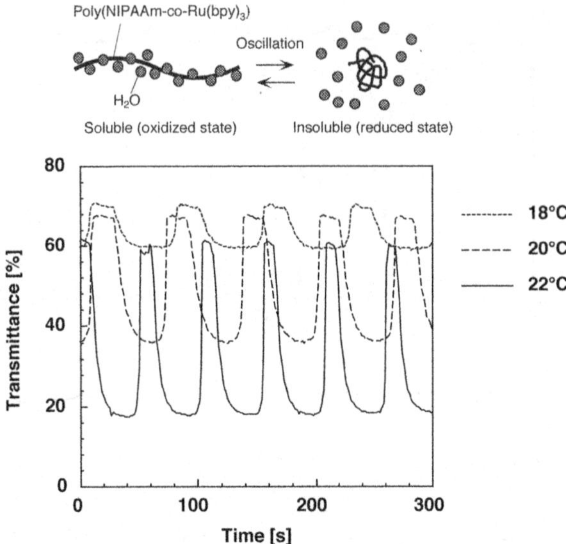

Figure 6. Oscillating profiles of optical transmittance for poly(NIPAAm-co-Ru(bpy)$_3^{2+}$) (Ru(bpy)$_3^{2+}$=5 wt %) solution at constant temperatures [35].

Ru(II) state (32°C). This temperature shift is due to an increase in hydrophilicity of the polymer by the increased charge density on the copolymer chains. As a result, microgels in the oxidized Ru(III) state show larger hydrodynamic diameters at each temperature because PNIPAAm-based microgels show continuous change of diameter below the VPTT. These deviations of the hydrodynamic diameters and differences of the colloidal stabilities at the same temperature should lead to self-oscillation of the microgels.

Figure 7 shows the oscillation profiles of transmittance for the microgel dispersions. First, temperature dependence of the oscillation was checked (Figure 7a). At low temperatures (20–26.5°C), on raising the temperature, the amplitude of the oscillation became bigger. The increase in the amplitude is due to the increased deviation of the hydrodynamic diameters between the Ru(II) and Ru(III) states. Furthermore, a remarkable change in the waveforms was observed between 26.5°C and 27°C. Then the amplitude of the oscillation dramatically decreased at higher temperature (27.5°C), and finally periodical transmittance change could not be observed at 28°C. The sudden change in the oscillation's waveforms should be related to the colloidal stability change between the Ru(II) and Ru(III) states. Here, the microgels should be flocculated due to lack of electrostatic repulsion at a high salt concentration when the microgels were deswollen. The temperature at

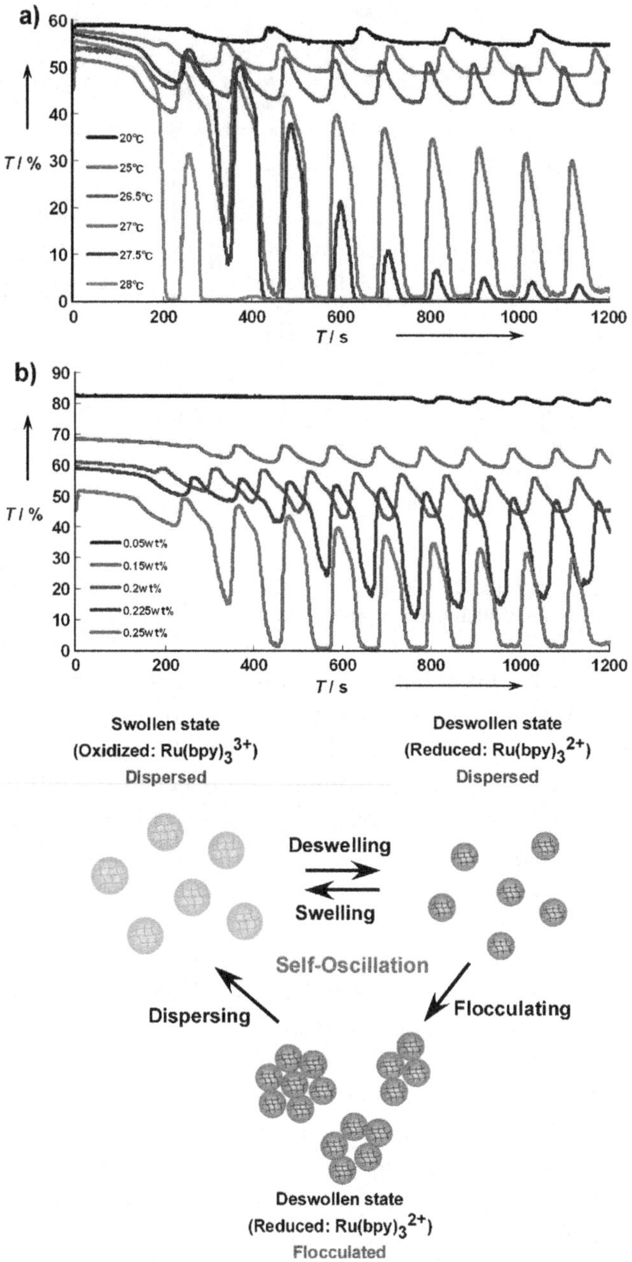

Figure 7. Self-oscillating profiles of optical transmittance for microgel dispersions. The microgels were dispersed in aqueous solutions containing MA, NaBrO$_3$, and HNO$_3$. Microgel concentration was 0.25 wt %. (a) Profiles measured at different temperatures. (b) Profiles measured at different microgel dispersion concentrations at 27°C. (c) Self-oscillation of microgels around the phase-transition temperature [38].

which the colloidal stability change is different from that of the remarkable change in the waveforms was observed, because colloidal stability is sensitive to ionic strength and dispersion concentration (being easier at higher concentrations and ionic strengths). In this system, because precise adjustment of ionic strength is very difficult, we checked a dependence of dispersion concentrations on oscillation at 27°C where the remarkable change in the waveform was observed in Figure 7a. As can be seen clearly from Figure 7b, the waveforms of the oscillation are related to the dispersion concentrations: remarkable change in the waveforms was only observed at higher dispersion concentrations (>0.225 wt %). The self-oscillating property makes microgels attractive for future development as microgel assemblies, drug/gene controlled release, and optical and rheological applications.

6. Design of nano-actuating systems

6.1. AFM OBSERVATION OF IMMOBILIZED SELF-OSCILLATING POLYMER

Further, by grafting the polymers or arraying the gel beads on the surface of substrates, we have attempted to design self-oscillating surfaces as nano-conveyers to transport cells, etc. with the spontaneous propagation of chemical waves. The polymer was covalently immobilized on a surface and self-oscillation was directly observed at a molecular level by scanning probe microscopy [39]. A self-oscillating polymer was synthesized using $Ru(bpy)_3$ monomer, NIPAAm and an N-succinimidyl group (NAS, a component for linking to the substrate). The synthesized copolymer was immobilized on an aminosilane-coupled glass plate. The immobilized copolymer was measured by SPM operating in tapping mode. While no oscillation was observed in pure water, nano-scale oscillation was observed in an aqueous solution of the BZ reaction which consisted of malonic acid, $NaBrO_3$, and HNO_3 (Figure 8). The amplitude was about 10–15 nm and the period was about 70 s, although some irregular behaviour was observed. Although no stirring could lead to the observed irregularity, the oscillation was reproducibly observed.

The amplitude of oscillation of the immobilized polymer (about 10–15 nm) was less than that in solution, as observed by DLS (23.9 and 59.6 nm). This smaller amplitude may be because the structure of the immobilized polymer was a loop-train-tail: the moving regions were shorter than that of the soluble polymer, as illustrated in Figure 8. In addition, the oscillation amplitude may have been suppressed due to the force applied by the weight of the cantilever, although the effect was not quantitatively evaluated. The amplitude and frequency were controlled by the concentration of reactant,

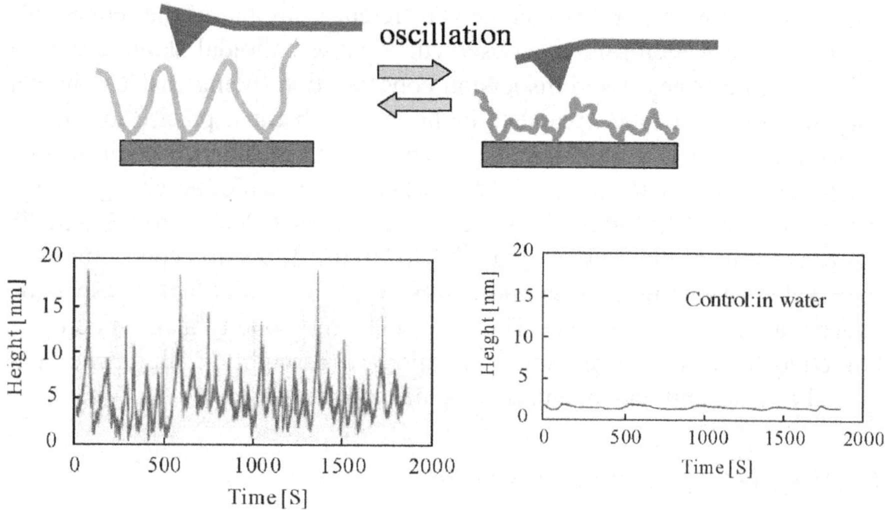

Figure 8. Self-oscillating behaviour of immobilized polymer in BZ substrate solution measured by AFM [39].

as observed in the solution. Here nano-scale molecular self-oscillation was observed for the first time. The oscillation polymer chain may be used as a component of a nano-clock or nano-machine.

6.2. FABRICATION OF MICROGEL BEADS MONOLAYER

As discussed in the previous section, we have been interested in the construction of nano-conveyers by functionalizing surfaces with self-oscillating gels or polymer beads. For this purpose, a fabrication method for organized monolayers of microgel beads was investigated [40]. A 2D close-packed array of thermosensitive microgel beads was prepared by double template polymerization. First, a 2D colloidal crystal of silica beads with 10 μm diameter was obtained by solvent evaporation. This monolayer of colloidal crystal can serve as the first template for preparation of macroporous polystyrene. The macroporous polystyrene trapping the crystalline order can be used as a negative template for fabricating a gel bead array. Functional surfaces using thermosensitive PNIPAAm gel beads were fabricated by the double template polymerization. It was observed that topography of the surface changed with temperature. The fabrication method demonstrated here was so versatile that any kind of gel beads could be obtained. This method may be a key technology to create new functional surface.

7. Attempts toward self-oscillation under biological conditions

7.1. SELF-OSCILLATION OF POLYMER CHAINS UNDER ACID-FREE CONDITIONS

So far, we had succeeded in developing a novel self-oscillating polymer (or gel) by utilizing the BZ reaction. However, the operating conditions for the self-oscillation are limited to conditions under which the BZ reaction occurs. For practical applications as functional bio- or biomimetic materials, it is necessary to design a self-oscillating polymer which acts under biological environments.

To cause self-oscillation of polymer systems under physiological conditions, BZ substrates other than organic ones, such as malonic acid and citric acid, must be built into the polymer system itself. Therefore, we took the next step, namely, to design novel self-oscillating polymer chains with incorporated pH-control sites, that is, novel polymer chains which exhibit rhythmic oscillations in aqueous solutions containing only the two BZ substrates, without using acid as an added agent. For this purpose, 2-acrylamido-2-methylpropanesulfonic acid (AMPS) was incorporated into the poly(NIPAAm-co-Ru(bpy)$_3$$^{2+}$) chain as the pH control site [41,42].

Figure 9 shows the self-oscillating transmittance change for the solutions of poly(NIPAAm-co-Ru(bpy)$_3$$^{2+}$-co-AMPS) (20 : 10 : 70 wt % in feed) at three constant temperatures (18°C, 21°C and 24°C). Under acid-free conditions and in the presence of only two BZ substrates (malonic acid and sodium bromate), we succeeded in causing soluble-insoluble self-oscillation of a polymer solution. Oscillating behaviors were remarkably influenced by the temperature, polymer concentration, and composition.

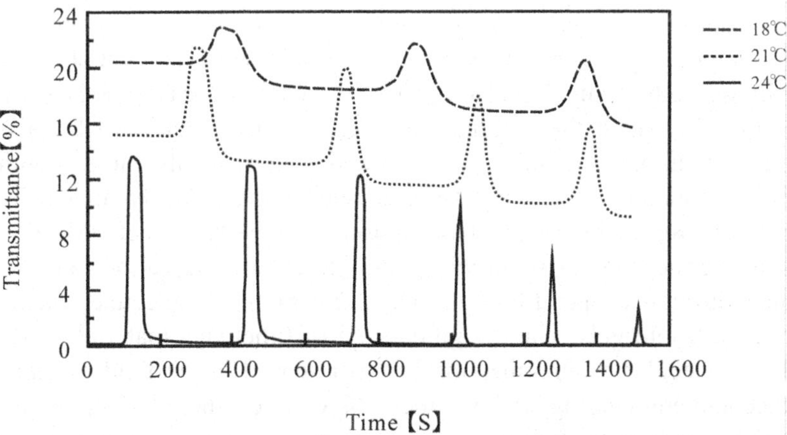

Figure 9. Oscillating profiles of optical transmittance for poly(NIPAAm-co-Ru(bpy)$_3$$^{2+}$-co-AMPS) solutions (polymer concentration=2.0 wt %) at several constant temperatures [41].

7.2. SELF-OSCILLATION UNDER OXIDANT-FREE CONDITIONS

As the next step, we attempted to introduce the oxidizing agent into the polymer. Methacrylamidopropyltrimethylammonium chloride (MAPTAC), with a positively charged group, was incorporated into the poly(NIPAAm-co-Ru(bpy)$_3$$^{2+}$) as a capture site for an anionic oxidizing agent (bromate ion) [43]. The bromate ion was introduced into the MAPTAC-containing polymer through ion-exchange. Under the conditions in which only two BZ substrates (malonic acid and sulfuric acid) were present, soluble-insoluble self-oscillation of the polymer was observed.

In the self-oscillating polymer solution system induced by the BZ reaction, self-oscillation was achieved without addition of oxidizing agent by utilizing the MAPTAC-containing polymer which included sodium bromate as a counter ion. The self-oscillating behavior was controllable by temperature. The polymer has two advantages because of the higher LCST; one is self-oscillation around body temperature, and the other is oscillation for a longer time without intermolecular aggregation among the polymer chains in the reduced state. A step toward practical use of self-oscillating polymers under biological conditions as novel smart materials has been established. Further, we have synthesized a quarternary copolymer which includes both pH-control and oxidant-supplying sites in the poly(NIPAAm-co-Ru(bpy)$_3$) chain at the same time. By using the polymer, self-oscillation under biological condition where only the organic acid (malonic acid) exists has been actually achieved (unpublished data).

8. Conclusion

As a material, gels have many unique characteristics which differ from other materials. Gels are not only soft materials which are wet and flexible like living body tissue, but also open materials which can exchange energy and substance with the external environment. Especially the latter characteristic is important, and it enables the design of intelligent gels with the combined functions of sensor, processor, and actuator. The dynamic behavior of gel is governed by cooperative motion of the polymer network. For this reason, the mechanism amplifying a minute external signal to macroscopic change through cooperation and synchronization of molecular interaction can be built into gels. The design concept of functional materials based on such a molecular synchronization has just started in the field of materials science and engineering, and polymer gels will become more important as a material which realizes the new concept. In that case, a living body serves as the best model, and it will be more important to design the materials from the standpoint of biomimetics and to clarify the molecular mechanism of the function.

As mentioned in this chapter, novel functional gels have been constructed by designing biomimetic gels with self-oscillating function. The self-oscillating gel may be useful in a number of important applications such as pulse generators or chemical pacemakers, self-walking (auto-mobile) actuators or micropumps with autonomous beating or peristaltic motion, devices for signal transmission utilizing propagation of chemical waves, oscillatory drug release synchronized with cell cycles or human biorhythms, etc. Further studies on the control of oscillating behavior as well as practical applications are expected.

References

1. P.J. Flory, *Principles of Polymer Chemistry*, Cornell University Press, Ithaca, NY (1953).
2. T. Tanaka, *Phys. Rev. Lett.*, **40**, 820 (1978); T. Tanaka, *Sci. Am.*, **244**, 124 (1981).
3. R. Yoshida, *Curr. Org. Chem.*, **9**, 1617 (2005).
4. R. Yoshida, K. Sakai, T. Okano and Y. Sakurai, *Adv. Drug Del. Rev.*, **11**, 85 (1993).
5. T. Okano (Ed.), *Biorelated Polymers and Gels – Controlled Release and Applications in Biomedical Engineering*, Academic Press, San Diego, CA (1998).
6. T. Miyata, Stimuli-responsive polymer and gels, in *Supramolecular Design for Biological Applications* (N. Yui, Ed.), CRC Press, Boca Raton, FL, pp. 191–225 (2002).
7. Y. Osada and A.R. Khokhlov (Eds.), *Polymer Gels and Networks*, Marcel Dekker, New York (2002).
8. N. Yui, R.J. Mrsny and K. Park (Eds.), *Reflexive Polymers and Hydrogels – Understanding and Designing Fast Responsive Polymeric Systems*, CRC Press, Boca Raton, FL (2004).
9. A.N. Zaikin and A.M. Zhabotinsky, *Nature*, **225**, 535 (1970).
10. R.J. Field and M. Burger (Eds.), *Oscillations and Traveling Waves in Chemical Systems*, Wiley, New York (1985).
11. I.R. Epstein and J.A. Pojman, *An Introduction to Nonlinear Chemical Dynamics: Oscillations, Waves, Patterns, and Chaos*, Oxford University Press, New York (1998).
12. R. Yoshida, T. Takahashi, T. Yamaguchi and H. Ichijo, *J. Am. Chem. Soc.*, **118**, 5134 (1996).
13. R. Yoshida, T. Takahashi, T. Yamaguchi and H. Ichijo, *Adv. Mater.*, **9**, 175 (1997).
14. R. Yoshida and T. Yamaguchi, in *Biorelated Polymers and Gels. Controlled Release and Applications in Biomedical Engineering* (T. Okano, Ed.), Academic Press, Boston, MA, Chap. 3 (1998).
15. R. Yoshida, M. Tanaka, S. Onodera, T. Yamaguchi and E. Kokufuda, *J. Phys. Chem. A*, **104**, 7549 (2000).
16. R. Yoshida, K. Takei and T. Yamaguchi, *Macromolecules*, **36**, 1759 (2003).
17. Y. Ito, M. Nogawa and R. Yoshida, *Langmuir*, **19**, 9577 (2003).
18. R. Yoshida, S. Onodera, T. Yamaguchi and E. Kokufuda, *J. Phys. Chem. A*, **103**, 8573 (1999).
19. R. Yoshida, G. Otoshi, T. Yamaguchi and E. Kokufuta, *J. Phys. Chem. A*, **105**, 3667 (2001).

20. K. Miyakawa, F. Sakamoto, R. Yoshida, T. Yamaguchi and E. Kokufuta, *Phys. Rev. E*, **62**, 793 (2000).
21. S. Tateyama, Y. Shibuta and R. Yoshida, *J. Phys. Chem. B*, **112**, 1777 (2008).
22. R. Yoshida, E. Kokufuta and T. Yamaguchi, *CHAOS*, **9**, 260 (1999).
23. R. Yoshida, T. Yamaguchi and E. Kokufuta, *J. Intell. Mater. Syst. Structures*, **10**, 451 (1999).
24. S. Sasaki, S. Koga, R. Yoshida and T. Yamaguchi, *Langmuir*, **19**, 5595 (2003).
25. R. Aoki, M. Enoki and R. Yoshida, *Key Eng. Mater.*, **321–323**, 1036 (2006).
26. Y. Takeoka, M. Watanabe and R. Yoshida, *J. Am. Chem. Soc.*, **125**, 13320 (2003).
27. R. Yoshida, K. Omata, K. Yamaura, M. Ebata, M. Tanaka and M. Takai, *Lab Chip*, **6**, 1384 (2006).
28. R. Yoshida, K. Omata, K. Yamaura, T. Sakai, Y. Hara, S. Maeda and S. Hashimoto, *J. Photopolym. Sci. Tech.*, **19**, 441 (2006).
29. O. Tabata, H. Hirasawa, S. Aoki, R. Yoshida and E. Kokufuta, *Sensors and Actuators A*, **95**, 234 (2002).
30. O. Tabata, H. Kojima, T. Kasatani, Y. Isono and R. Yoshida, *Proceedings of the International Conference on MEMS 2003*, pp. 12–15 (2003).
31. R. Yoshida, T. Sakai, O. Tabata and T. Yamaguchi, *Sci. Tech. Adv. Mater.*, **3**, 95 (2002).
32. S. Maeda, S. Hashimoto and R. Yoshida, *Proceedings of the IEEE International Conference on Robotics and Biomimetics (ROBIO 2004)*, p. 313 (2004).
33. S. Maeda, Y. Hara, R. Yoshida and S. Hashimoto, *Macromol. Rapid Commun.*, **29**, 401 (2008).
34. S. Maeda, Y. Hara, T. Sakai, R. Yoshida and S. Hashimoto, *Adv. Mater.*, **19**, 3480 (2007).
35. R. Yoshida, T. Sakai, S. Ito and T. Yamaguchi, *J. Am. Chem. Soc.*, **124**, 8095 (2002).
36. T. Sakai and R. Yoshida, *Langmuir*, **20**, 1036 (2004).
37. T. Sakai, Y. Hara and R. Yoshida, *Macromol. Rapid Commun.*, **26**, 1140 (2005).
38. D. Suzuki, T. Sakai and R. Yoshida, *Angew. Chem. Int. Ed.*, **47**, 917 (2008).
39. Y. Ito, Y. Hara, H. Uetsuka, H. Hasuda, H. Onishi, H. Arakawa, A. Ikai and R. Yoshida, *J. Phys. Chem. B*, **110**, 5170 (2006).
40. T. Sakai, Y. Takeoka, T. Saki and R. Yoshida, *Langmuir*, **23**, 8651 (2007).
41. Y. Hara and R. Yoshida, *J. Phys. Chem. B*, **109**, 9451 (2005)
42. Y. Hara and R. Yoshida, *Langmuir*, **21**, 9773 (2005).
43. Y. Hara, T. Sakai, S. Maeda, S. Hashimoto and R. Yoshida, *J. Phys. Chem. B*, **109**, 23316 (2005).

AN EXCURSION IN THEORETICAL NON LINEAR CHEMISTRY: FROM OSCILLATIONS TO TURING PATTERNS

P. Borckmans[*]
*Non Linear Physical Chemistry Unit, Service de Chimie
Physique et Biologie Théorique, Université Libre de Bruxelles,
CP 231 – Campus Plaine, 1050 Brussels, Belgium*

S. Métens
*Matière et Systèmes Complexes, UMR 7057 CNRS, Université
Paris 7 – Denis Diderot, 10, rue Alice Domon et Léonie
Duquet, 75205 Paris cedex 13, France*

Abstract. Some basic principles of theoretical nonlinear chemistry are introduced to explain an origin of temporal oscillations and spatial bistability, that have both been used in other chapters in conjunction with gels sensitive to such chemical behavior. Because the use of inert gels played such importance in their experimental discovery, elements of the theory of Turing stationary spatial periodic patterns are also discussed.

Keywords: chemical oscillations, reaction-diffusion patterns, (spatial) bistability, (pitchfork, transcritical, saddle-node) bifurcations, amplitude equations

1. Introduction

The field of non linear chemistry, as it has been dubbed, deals with the behaviors, once considered exotic, exhibited by reactions involving feedbacks, such as autocatalytic or inhibitory steps as defined by De Kepper et al. in Chapter 1.

The history of this field of research started out, early in the 1950s by two dichotomous milestones results.

---

[*] To whom correspondence should be addressed. e-mail: pborckm@ulb.ac.be.

P. Borckmans et al. (eds.), *Chemomechanical Instabilities in Responsive Materials,*
© Springer Science + Business Media B.V. 2009

The first, in 1951, is the *experimental* discovery by Belousov of long lived time periodic oscillations of the concentrations of the species from the cerium-catalyzed oxidation of citric acid by bromate. Because, at the time, it seemed to challenge thermodynamic wisdom, only in 1958 was he able to publish these results in an obscure Russian medical journal [1,2]. Discouraged, he definitely gave up this line of research.

The second, in 1952, was the first *theoretical* formulation of the necessary conditions for a spatial symmetry breaking leading, through the interplay of chemical reactions and mass transport (diffusion), to stationary periodic spatial concentration patterns [3]. Nevertheless, this result from the reputed British mathematician and computer pioneer, Alan Turing was not pursued much further due to his untimely death. This result, presented as a "chemical basis for morphology", was also criticized by biologists.

In a nutshell one could say that the development of the field until the early 1990s went from the corroboration and acceptance of Belousov' result to the experimental realization of Turing's idea.

It started with the theoretical demise of the opposing, false, thermo-dynamic arguments by Prigogine and his group. Starting already in 1955, this group went beyond thermodynamic argumentation to apply theoretical steps to describe such chemical oscillations and patterns [4,5]. Also beginning in 1961 came the continuation of Belousov's work by Zhabotinsky [6,7], who replaced citric by malonic acid that helped apprehending some mechanistic elements. The next landmark arose in 1972 when Field, Körös and Noyes (FKN) [8] described a detailed chemical mechanism of the Belousov-Zhabotinsky (BZ) reaction. This made the topic much more acceptable to "classical" chemists. Thereafter the control and design of new oscillators owes much respectively to the use of the continuously fed stirred tank reactor (CSTR), and the theoretical approach based on the so-called "cross-shaped diagram" [9] and realized experimentally by Epstein, Orban, De Kepper and colleagues [10]. Finally, near the end of 1989, the first Turing structure was obtained by De Kepper and collaborators [11] using the chlorite-iodide-malonic acid reaction in a continuously fed *hydrogel* reactor. Such a short summary naturally does not make justice to numerous other players in the game, whether also dealing with homogeneous liquid phase chemistry, or heterogeneous systems (especially catalysis or surface chemistry) or chemical engineers, but references may be found in textbooks and collective works [5,12–18]. Neither should we forget the precursors, the works of which came to the light when the field became more mature [7,19].

In the following we will stick to the specialization of liquid phase chemistry, although the methodology may be applied to a wider class of systems. The dynamical equations describing the systems are then

$$\frac{\partial \mathbf{c}}{\partial t} = \mathbf{f}(\mathbf{c}) + \boldsymbol{D}\ \nabla^2 \mathbf{c} - \mathbf{U}\nabla\mathbf{c} \qquad (1.1)$$

where $\mathbf{c}$ represents the vector of the concentrations of all species taken into account in the reaction kinetics $\mathbf{f}(\mathbf{c})$. $\boldsymbol{D}$ is the matrix of mass diffusion coefficients (in an ideal mixture picture as the experimental solutions are in general dilute). On the other hand $\mathbf{U}$ is the fluid velocity. Fluid flow may either be imposed, as in the case of a tubular reactor, or be a "parasitic" effect due to the necessary feeding of the chemical reactor to keep it functioning at finite distance from equilibrium. It may also be generated by the reactions themselves if reactants of different densities come into play. Whatever the origin, Eq. (1.1) must be complemented by the equation describing such flow; for our conditions, the Navier-Stokes equation. In the context of this school, where we wish to assess the impact of reactions on gels sensitive to their chemical environment, we will neglect the flow effects (third term in the right hand side of Eq. (1.1)) because of their natural quenching induced by the gels. In the case of the well mixed reactor, we will make the hypothesis (confirmed by experiments) that all spatial concentration gradients are wiped away. In that case the dynamics is driven only by the kinetic term $\mathbf{f}(\mathbf{c})$ and input-output balance conditions. All auxiliary conditions (boundary, initial) to solve Eq. (1.1) must also be given.

As discussed in Chapter 1, $\mathbf{f}(\mathbf{c})$ is a non linear function of the c's involved in the reaction. As our condensed history already leads us to guess, the determination of $\mathbf{f}(\mathbf{c})$ is a complex specialized task. Despite considerable kinetics works there are still dark corners left in the actual BZ reaction mechanism. Nevertheless one often proceeds by using empirical models derived from experiment, sometimes valid only in a restricted region in parameter space. From the theoretical point of view such models often include to many species to be able to carry out analytical calculations. One possible way forward is to take the differences in time scales into account to reduce the number of relevant chemical species (quasi-steady approximation) with due care. Furthermore if diffusion processes are present, it should be taken into account for this reduction in order to be sure not to violate mass conservation. Therefore theoreticians have developed "toy models" as a compromise between chemical realism and mathematical tractableness. Such are, for instance the Oregonator [20,21] (a contraction of the FKN model for the BZ reaction), the autocatalator [13] or the Brusselator [5,22]

$$A \to X$$
$$B + X \to Y$$
$$2X + Y \to 3X \qquad (1.2)$$
$$X \to P$$

where A, B are the initial reactants, P the final product and X, Y the intermediates. The initial reactants are considered in large excess (i.e. constant). This, characterizing a constant supply, is the so-called *pool chemical* approximation. Together with the fact that P is constantly eliminated, ensures that the system may be considered open to the external environment. The corresponding kinetics is then

$$\frac{dX}{dt} = A - (B+1)X + X^2Y$$
$$\frac{dY}{dt} = BX - X^2Y$$

(1.3)

where X and Y are respectively an activator and inhibitor. The corresponding reaction-diffusion system is then obtained by the addition of the transport terms as in Eq. (1.1).

One therefore sees that in the chemical world the first principal source of nonlinearities leading to instabilities arises from the local reactive dissipative processes. They may thus already be expressed in the absence of spatial degrees of freedom, as in a well-mixed reactor for instance, leading to a wealth of nonlinear behaviors. On the contrary in hydrodynamics and related fields inertia plays the key nonlinear role (at least in Newtonian fluids) and spatial variables are always important. Paraphrasing Nicolis [23]: "The intrinsic parameters k (the characteristic inverse time scale of the kinetics) and D in Eq. (1.1) have dimensions of, respectively, $[time]^{-1}$ and $[(length)^2/time]$. It follows that a reaction-diffusion system possesses intrinsic time (k^{-1}) and space $((D/k)^{1/2})$ scales contrary, again, to hydrodynamics where one has a whole spectrum of time and space scales for the selection of which the boundary conditions and the size – two extrinsic factors – play the key role. This places nonlinear kinetics at the forefront for understanding the origin of endogenous rhythmic and patterning phenomena".

2. Methodology

As shown in Eq. (1.3), the main characteristic of autocatalytic reactions is that they lead to *nonlinear equations*. The corresponding mathematical difficulty arises from the fact that the solutions of such equations may be multivalued. The mathematical tools to study the properties of nonlinear partial differential equations we are confronted with have grown with the subject and the most standard ones may now be found in numerous textbooks (e.g. [24,25] make for a good starting point, [26–28] are more mathematically oriented) while others remain hidden in research papers (e.g. [29]) or simply are *not available yet*. A detailed presentation obviously lies outside the scope of this text.

The aim is thus to find the solutions of a set of nonlinear kinetic equations or else nonlinear reaction-diffusion equations (if spatial factors are of the essence). The ideal situation may seem to try and solve these sets analytically. This may be done in some situations, sometimes at the price of a heavy mathematical investment. The mathematical field devoted to this approach goes under the name of *soliton* theory. However because of the possible multiplicity of solutions, one may need to work very hard to obtain a solution that is not relevant to the conditions of an experiment. The other way, most used in the field, is to start from a known solution. For instance in the Brusselator system, the stationary states $X_{ss} = A$, $Y_{ss} = B/A$ is such a branch of solutions that is followed when the values of the "control" parameters A or B are changed. Now, in order to be observed (had we considered a kinetic model pertaining to a real reaction) this branch of states needs to possess certain properties of stability, as a system in the true world is always affected by small perturbations due to imperfections, if not the natural occurring fluctuations.

The loss of stability of a state to *infinitesimal perturbations* signals a qualitative change in the dynamic behavior of the system, most often related to the modification of the number of branches of solutions. Any such change in the qualitative structure of solutions to differential equations is called a *bifurcation*. The notion of bifurcation is central to the qualitative theory of dynamic systems.

The principle of this study may be outlined in three steps.

- *Linear stability analysis and determination of the active modes*

Neglecting the convection contributions, Eq. (1.1) may be condensed as

$$\frac{\partial \mathbf{c}}{\partial t} = \mathbf{N}(\mathbf{c}, B, \nabla^2) + \text{b.c.} \tag{2.1}$$

where B stands for the chosen control (bifurcation) parameter we use. Let us also consider that we know a nonequilibrium uniform steady state $\mathbf{c_0}(B)$ of (2.1), solution of

$$\mathbf{N}(\mathbf{c_0}, B) = 0 \tag{2.2}$$

To test its linear stability, we study the response of the system to an infinitesimal perturbation to the reference state $\mathbf{c_0}(B)$

$$\mathbf{c}(\mathbf{r}, t) = \mathbf{c_0}(B) + \delta\mathbf{c}(\mathbf{r}, t) \tag{2.3}$$

We linearize Eq. (2.1), as well as the boundary conditions (b.c.), around the reference state $\mathbf{c_0}(B)$, by inserting (2.3) and keeping only the linear terms in $\delta\mathbf{c}$. This leads to

$$\frac{\partial \delta \mathbf{c}}{\partial t} = (\mathbf{L} + \mathbf{D}\ \nabla^2)\delta \mathbf{c} \qquad (2.4)$$

where the matrix elements L_{ij} of the linear (*Jacobian*) operator $\mathbf{L}$ are given by

$$L_{ij} = \frac{\partial N_i}{\partial c_j}\bigg|_{\mathbf{c}=\mathbf{c}_0} \qquad (2.5)$$

Then, if $\mathbf{L}$ and $\mathbf{D}$ are time-independent, Eq. (2.4) admits solutions of the form

$$\delta \mathbf{c}(\mathbf{r},t) = \mathbf{A}\Phi(\mathbf{r})e^{\omega t} \qquad (2.6)$$

where the amplitude $\mathbf{A}$ expresses the vectorial character of $\delta \mathbf{c}$ in concentrations space. As linear stability implies stability with respect to all possible infinitesimal perturbations, it is useful to choose $\Phi(\mathbf{r})$ from the complete set of normal modes of the Laplacian operator, for the given boundary conditions:

$$\nabla^2\Phi_n(\mathbf{r}) = -k_n^2\Phi_n(\mathbf{r}) \qquad (2.7)$$

Then the general solution of the linear system may then be written as

$$\delta \mathbf{c}(\mathbf{r},t) = \sum_n \mathbf{A}_n e^{\omega_n t}\Phi_n(\mathbf{r}) \qquad (2.8)$$

that when substituted in Eq. (2.4) leads to a set of homogeneous algebraic equations, the solvability condition of which is given by the *characteristic equation* ($\|..\|$ stands for the determinant)

$$\left\| \mathbf{L} - \mathbf{D}\ k^2 - \omega \right\| = 0 \qquad (2.9)$$

This constitutes a polynomial of order n in ω, the roots of which are either real or pairs of complex conjugate and are functions of the control parameter B. These roots are equivalently the eigenvalues (the spectrum) of the linear operator $\mathbf{L} - \mathbf{D}k^2$. From Eq. (2.9) it results that the stability of the reference state $\mathbf{c}_0(B)$ is determined by the sign of the real part, $Re\ \omega$, of the n roots. If $Re\ \omega_n < 0$ ($\forall$ n), the state $\mathbf{c}_0$ is said to be *asymptotically stable*, as all perturbations regress in time. When B is varied, as soon as one root becomes such that $Re\ \omega_n > 0$, that typical perturbation starts to increase exponentially with a time constant $\tau \propto (Re\ \omega_n)^{-1}$, and the state becomes *unstable* when $Re\ \omega_n(B_c) = 0$. This conditions defines the *bifurcation point* $B = B_c$. As the size of the perturbation increases, its becoming is to be determined taking into account the terms neglected in the linearizing process (see below). When diffusion is not involved, the number of roots corresponds to the number of chemical species in the concentration vector. However, unless very specific conditions are met, only a very small number of roots acquire a positive real part: most usually a single, or a pair if they are complex conjugates as we will see. To some extend, a classification has been reached for a small number (n ≤ 4). When diffusion is involved and space symmetry breaking occur, the real part of the roots may also go to zero at a bifurcation point but for a given wavenumber k_c (if the system is isotropic and the linear operator thus depends only on k^2. We will elaborate in Section 5.

Therefore two kinds of actors enter the game of bifurcation:

The *active modes* are those responsible for the loss of stability of the reference state by growing exponentially. They do so by drawing on the imposed driving force and thus end up competing non-linearly to create the new solutions, as soon as they have grown sufficiently to make the linear approximation obsolete. These modes are the eigenvectors $\mathbf{e_n}$ of the Jacobian matrix for which $Re\ \omega_n > 0$ when $B > B_c$. Because these eigen-modes involve all the δc_n, all the concentrations feel the bifurcation. Near B_c they evolve on the time scale $\tau \propto (Re\ \omega_n)^{-1}$.

On the other hand there exist so called *passive modes*, the linear frequencies of which are still damped above B_c. They however also come into the determination of the new states that are building up near B_c as they are continuously regenerated by the nonlinear interactions between members of the active set. Their dynamics results from the balance between this regeneration and their rapid linear decay. Their amplitudes may therefore be algebraically related to those of the active modes as a result of an adiabatic elimination process that is reminiscent of a quasi-stationary approximation. They are therefore often termed "slaved" modes as they feed on the stress source only through the active modes.

- *Amplitude equations for the active modes*

As mentioned above, there exist standard bifurcation techniques (e.g. [24]) that allow one to obtain evolution equations for the amplitudes of the active modes. We here only sketch the principles of this derivation referring the reader to typical works on the subject for the technical details. Near threshold, B_c, one proceeds by expanding the difference of the concentration fields $\mathbf{c(r},t)$ from their reference state $\mathbf{c_0}$ as an asymptotic series

$$\mathbf{c(r},t) - \mathbf{c_0} = \eta\,\mathbf{c_1} + \eta^2\,\mathbf{c_2} + \eta^3\,\mathbf{c_3} + \cdots \tag{2.10}$$

where the small expansion parameter η is related to the distance from threshold by

$$\mu \propto B - B_c = \eta\,B_1 + \eta^2\,B_2 + \eta^3\,B_3 + \cdots \tag{2.11}$$

and $\mathbf{c_1}$ is a linear combination of the active modes resulting from the linear stability analysis. The amplitudes A_i depend on the natural slow time scale of the critical mode that is proportional to $1/\mu$ ($T = \mu t$). The higher order corrections $\{\mathbf{c_n}\}$ (n > 1) are then determined by substituting the expansions Eqs. (2.10) and (2.11) in the nonlinear reaction-diffusion equations. Equating the successive powers of leads to a *set of linear inhomogeneous equations that are solved recursively*. The solvability conditions of these inhomogeneous equations (Fredholm alternative theorem) then lead to a set of ODE's: the *amplitude equations* for the active modes

$$\frac{dA_i(T)}{dT} = \mu A_i + G_i(\{A_j\}) \qquad (2.12)$$

where $G_i(\{A_j\})$ are nonlinear polynomials in the active amplitudes. As we shall see later, each given type of bifurcation is characterized by a particular structure of the polynomial and thus acquires a universal character. The information concerning a particular system is solely contained in their coefficients.

- *Determination of the stable bifurcated states – bifurcation diagram*

As the resulting amplitude equations, Eq. (2.12), have a lower dimension they are simpler to analyze than the original reaction-diffusion systems (Eq. (2.1)) remembering however that they are only valid in some neighborhood of the point where the reference state linearly looses its stability. Because of this relative simplicity they allow to scrutinize the key nonlinear effects that govern the structure of the bifurcation diagram. In the first place however they allow one to obtain the bifurcated solutions and to discuss their stability.

The investigation of the stability of the new states emerging at a bifurcation is again carried out through linear stability analysis, now however around the newly bifurcated state. The difficulty in the analysis arises from the fact that for each state one has to discuss the stability with respect to *all* possible types of perturbations: amplitude, modulus and orientation of the wavevector (for possible periodic states if space symmetry breaking occurs) and also resonant perturbations to other structures with different symmetries. Usually also, the first bifurcated state is only know at some approximation and the consistency of successive approximations has to be weighted carefully.

A more "static" definition relates to the parametric evolution of available, say, steady states. A branch of steady solutions is obtained by continuously changing a certain parameter of the system. All solutions belonging to the same branch retain the same qualitative character (it may be a branch of stationary homogeneous or of inhomogeneous solutions with eventually a certain symmetry). Branches terminate or intersect at bifurcation points.

Varying several parameters of the problem, one can obtain bifurcation surfaces in the parametric space that would separate parametric regions differing by the number and nature of states and/or their stability. Any change of the parameters of the system leads to the bifurcation when the corresponding trajectory in the parametric space intersects a bifurcation surface.

Identifying bifurcation surfaces and studying process dynamics in their proximity is a decisive step in the comprehensive investigation of the dynamics of a nonlinear system. Since dynamics is qualitatively the same throughout a domain in the parametric space bounded by bifurcation surfaces, one can obtain results applicable, in a qualitative sense, to the entire domain by studying dynamics close to a bifurcation, where this task is greatly simplified due to the possibility of exploiting the wide separation between relaxation times of marginal modes responsible for the loss of stability and other perturbations of the state. The most classical tools that have prevailed throughout these analyses are the reduction to normal form (amplitude) equations using perturbation techniques and/or symmetry arguments that allow to compute such bifurcation diagrams that record the possible states of the system, characterized by some variable(s), as a function of the convenient parameters. When possible such diagrams also help organize the results obtained by straightforward numerical integration. Put in another way, numerical results may help bridge the gap between various occurring bifurcations around which analytical methods are sometimes possible and in the vicinity of which numerics "suffers" from the critical slowing down of the marginal modes. Both information may then be used to interpret the experimental results.

3. Oscillations

In this section, we wish to discuss one of the main mechanisms for the appearance of sustained periodic oscillations of the concentrations in a well stirred chemical system (quenched diffusion).

To proceed we will make use of a well known model (the so-called $\lambda-\omega$, or Poincaré model) that has been used for instance to extensively study the behavior of chemical waves (then in the presence of diffusion). This model reads

$$\frac{dX}{dt} = AX + BY - \alpha X(X^2 + Y^2)$$

$$\frac{dY}{dy} = -BX + AY - \alpha Y(X^2 + Y^2)$$

$$(3.1)$$

Two caveats are in order. First it is fair to admit that this system cannot represent a chemical system in the true sense as the "concentrations" X and Y may become negative. However as it allows for exact solutions, it will enable the introduction, in not too formal a way, of some important concepts. This cannot be done, for instance with the FKN (even in its two variable approximation) or Brusselator models alluded to above. We will discuss some of the differences that appear in such systems. The other pertains to the coefficients A and B, that as in the Brusselator, can be though of as representing the concentrations of other species that evolve on a much longer time scale and that may be considered as constant on the time scale we consider. In this pool chemical approximation Eq. (3.1) can be considered as representing an open system, the parameters of which may be varied to control the distance from some reference state. Parameter α measures the strength of the non linear coupling that as in many other empirical kinetic schemes, whether arising from experiments or as theoreticians toy models, involve cubic terms in the concentrations. Here we will use A as control parameter.

By looking at Eq. (3.1) it is clear that $X = X_o = Y = Y_o = 0$ is a steady solution *whatever the values of the parameters*; can we find another one? Let us therefore test its stability. By linearizing around (X_o, Y_o) we obtain the characteristic equation from the determinant of the Jacobian matrix

$$\begin{Vmatrix} A - \omega & B \\ B & A - \omega \end{Vmatrix} = 0 = \omega^2 - 2A\omega + (A^2 + B^2) \tag{3.2}$$

Its roots are thus the complex conjugate pair $\omega_\pm = A \pm iB$. Thus $Re\,\omega_\pm$ remains negative, hence the state $X_o = Y_o = 0$ stays asymptotically stable, as long as $A < 0$ (for $\alpha < 0$, see below).

In order to discover what occurs when A becomes positive, i.e. when the state becomes unstable, let us introduce the following change of variables: $X = R \cos\theta$ and $Y = R \sin\theta$, then Eq. (2.1) can be rewritten as

$$\frac{dR}{dt} = R(A - \alpha R^2)$$

$$\frac{d\theta}{dt} = -B \tag{3.3}$$

where $R^2 = X^2 + Y^2 > 0$. The radial and angular dynamics are uncoupled (an advantage of the model) and so can be analyzed separately.

To proceed we first consider $\alpha > 0$. When $A > 0$, the radial equation hints at the fact that, besides the value $R = 0$ (that we know to exist but is unstable), another state with $R_{lc} = \sqrt{A/\alpha}$ may have some significance. In order to be more precise we may integrate the system Eq. (3.3) exactly.

The general solution, given an initial set of conditions (at $t = 0$, $R = R_i$ and $\theta = \theta_i$), is then (the most expedient way to solve the first equation is to write it for the variable R^2):

$$R^2(t) = \frac{AR_i^2}{\alpha R_i^2 + (A - \alpha R_i^2)exp - 2At} \qquad (A \neq 0) \qquad (3.4)$$

$$\theta = -Bt + \theta_i$$

First one may check that, when $A < 0$, the system "falls" spiraling to $R = 0$. When $A > 0$ all trajectories in the X,Y plane tend, for $t \to \infty$, to the *circle* of radius R_{lc}. If $R_i < R_{lc}$ the spiraling is outwards, and inwards if $R_i > R_{lc}$. The spiraling motion is the resultant of the increase (decrease) of R combined with the time dependence of the angle (phase) θ at constant angular velocity.

It may be shown easily that the new state $X(t) = R_{lc} \cos Bt$ and $Y(t) = -R_{lc} \sin Bt$ is stable and represents a situation where the concentrations vary periodically in time. The linear stability analysis for this new state can be done easily on the radial equation only as a perturbation of the angle merely results in a neutral phase shift.

The new state, that emerges at $A = 0$ and stably exists for $A > 0$, is an example of a *limit cycle* that coexists with the unstable state $R = 0$. Its radius increases with A. This may be represented by the bifurcation diagram (Figure 1a). When the real part of the least stable single pair of complex conjugate roots of the characteristic equation changes sign, we say the system undergoes a *Hopf bifurcation*. The bifurcation is *supercritical* as the limit cycle is born as the control parameter increases above its marginal value ($A = 0$).

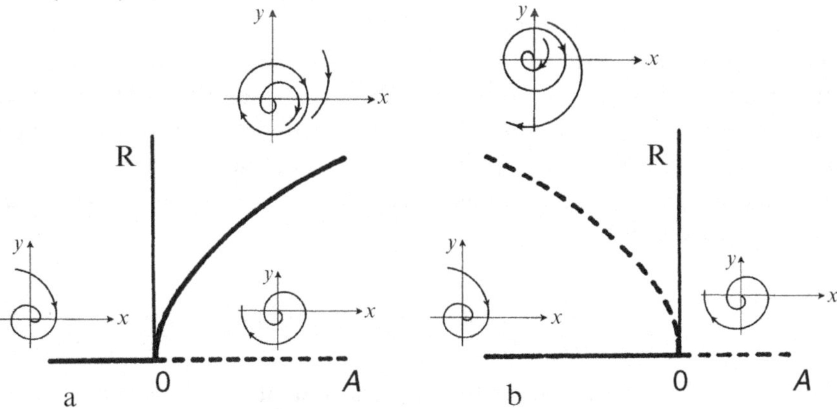

Figure 1. Super- and subcritical Hopf bifurcations for the λ-ω system. The insets show the trajectories of the system in concentration space and the existence of limit cycles (*circle*).

When $\alpha < 0$, a similar analysis may be carried out. The fate of the steady state $X_o = Y_o = 0$ remains the same. Stable for $A < 0$, undergoing a Hopf bifurcation at $A = 0$ and becoming unstable for $A > 0$. However the limit cycle R_{lc} now exists only for $A < 0$ (Figure 1b) and may furthermore be shown, as explained previously, to be *unstable*. Indeed all trajectories in the (X,Y) plane may be calculated from the exact solution (Eq. (3.4)). If $R_i < R_{lc}$ the spiraling is inwards to $X_o = Y_o = 0$ and if $R_i > R_{lc}$ the spiraling is outwards and the distance from the limit cycle growth unbounded. In such case, when the unstable limit cycle collapses into the steady state as the bifurcation parameter increases we have a *subcritical Hopf bifurcation*. These conclusions call for two remarks. First, albeit our calculation and bifurcation behavior are correct from the mathematical point of view, such result, by itself, can usually not represent the behavior of a true natural system. We knew from the start that the $\lambda-\omega$ model does not represent such case; but the same behavior could be found with a kinetic model and then signals some limitation in the kinetics used. Secondly, from the point of view of stability, it shows a limitation of the linear stability analysis. Although the state $X_o = Y_o = 0$ is linearly stable, it may become unstable for large enough perturbations. It is not asymptotically stable as roots with $Re \, \omega_n > 0$ now exist in the $A < 0$ region.

As a last remark, the dynamics of our system may also be written in terms of $Z = X + iY$ that interprets the X,Y plane as the complex plane. Eq. (3.1) then reads

$$\frac{dZ}{dt} = iBZ + (A - \alpha|Z|^2)Z \tag{3.5}$$

In more realistic kinetic models, one may use the techniques described in Section 2 to study the appearance of Hopf bifurcations if a steady reference state is known. For instance $X_o = B$, $Y_o = B/A$ is such steady state for the Brusselator model (Eq. (1.3)) that undergoes a supercritical Hopf bifurcation at $B_{lc} = (1 + A^2)$, if we use B as control parameter. In the vicinity of a supercritical Hopf bifurcation point it may also be shown that the dynamics obeys the *generic form* that is the well known as the complex Ginzburg-Landau equation [30]

$$\frac{dZ}{dT} = i\omega_c Z + \left[\mu - (g_r + ig_i)|Z|^2\right]Z \tag{3.6}$$

where ω_c is the critical phase (the square root of the determinant of the Jacobian matrix), T the slow time scale of the active modes, μ the reduced distance from the bifurcation point ($A - A_{lc}/A_{lc}$ in our example, where $A_{lc} = 0$) and the complex parameter g has to be calculated by the nonlinear analysis.

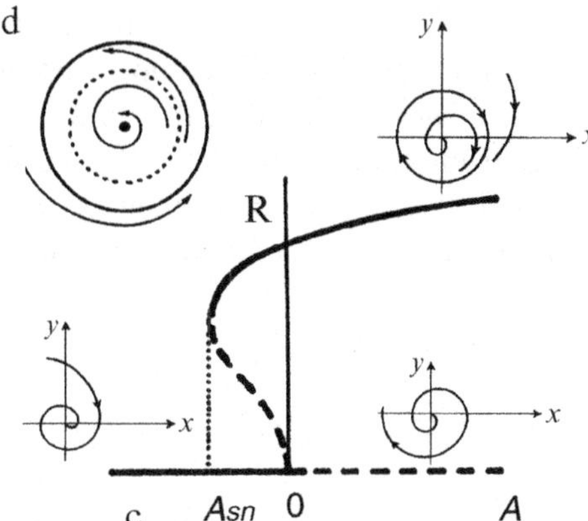

Figure 2. Subcritical Hopf bifurcation stabilized through the interaction with a saddle-node bifurcation of cycles.

The structure of Eq. (3.6) being generic, as already mentioned, and the identity of each system emerges from the particular values of ω_c, μ and g. The limit cycle, although still a closed curve in the space spanned by all concentrations is not a circle anymore. It is traveled in time with a speed that now depends on the amplitude as the equations for R and θ are no longer decoupled. Also the oscillations loose their harmonic character. These considerations may be extended if diffusion may come into play (unstirred reactor) and wave behavior may occur.

Finally a last remark concerning the subcritical Hopf bifurcation we encountered and where the stability of the system seemed at risk. What often takes place in realistic kinetic models is described by the bifurcation diagram (Figure 2), where another bifurcation (new in the present context) comes into play, at A_{sn}, to stabilize the system. In the subcritical region, when $A_{sn} < A < 0$, we then have besides the steady state $X_o = Y_o = 0$, two limit cycles of different amplitudes. That with the smallest radius is unstable (it corresponds to that one we have studied before), and that with the larger radius that is stable and assures the stability of the system (even in the supercritical region). For A_{sn} the radii two cycles become equal and they collapse in what is called a saddle-node type bifurcation that we will discuss in a simpler situation in the next section. All this may be analyzed within the framework of our model if we add quintic terms.

4. Bistability

To illustrate this phenomenon we will considered a continuously stirred tank reactor (CSTR) where the Iodate-Arsenous Acid (IAA) reaction occurs. An experimentally well established empirical kinetic scheme exists for this reaction.

We recall that the CSTR has played a crucial role in the development of the field of nonlinear chemistry when the competition between reaction and diffusion is not involved (well mixed reactor). It was the first reactor used to control the distance from equilibrium, and allowed the discovery of new families of chemical oscillators [31], and even became so reliable that it enabled the first, non electronic device, demonstration of a chaotic dynamic [16].

The dynamics of a CSTR is given by

$$\frac{d\mathbf{c_o}}{dt} = \mathbf{f}(\mathbf{c_o}) + k_o(\mathbf{c_{in}} - \mathbf{c_o}) \tag{4.1}$$

The first term is the usual kinetic contribution, while the second describes the effect of the inflow of the reagents and outflow of the reacting mixture. The inverse residence time of the species injected in the CSTR at concentration $\mathbf{c_{in}}$ is k_o. For isothermal conditions, $\mathbf{c_{in}}$ and k_o are the only experimental parameters tunable to control the distance from equilibrium. Nevertheless changing k_o is easier to carry out experimentally.

If the input flow is large, that is if the residence time (k_o^{-1}) is much shorter than the typical reaction times, the extents of the reactions are small and, in a stationary regime, the concentrations are close to the compositions in the inflows (*flow branch*, F_{CSTR}). In fact the flow branch is asymptotic to c_{in}. On the contrary if the residence time is much longer than the reaction time, the extent of the reaction is large and the composition in the reactor is nearer (see below) that of the thermodynamic equilibrium that one would obtain in a closed reactor with the same initial composition (*thermodynamic branch*, T_{CSTR}). In standard reactions, the branches of states F_{CSTR} and T_{CSTR} are smoothly connected at intermediate flow rates. However when auto-catalytic reactions are present, as in the IAA reaction we will discuss shortly, there exist sets of parameters for which the two states can coexist for a same set of flow rates, when $k_o^* < k_o < k_o^{**}$ (if the $\mathbf{c_{in}}$ are kept constant) (Figure 3). Similar behavior may be obtained if one of the c_{in} is varied while the others and k_o are maintained constant. Their stability domains overlap over a range of values of the control parameters, clearly defining two distinct branches F_{CSTR} and T_{CSTR} and the transition from one state to the other occurs with hysteresis when scanning the values of k_o. The reaction in the CSTR exhibits *bistability*. The transition from the monostable to the bistable situation usually proceeds through a critical point where the transition from F_{CSTR} to T_{CSTR}, although smooth, has a vertical tangent.

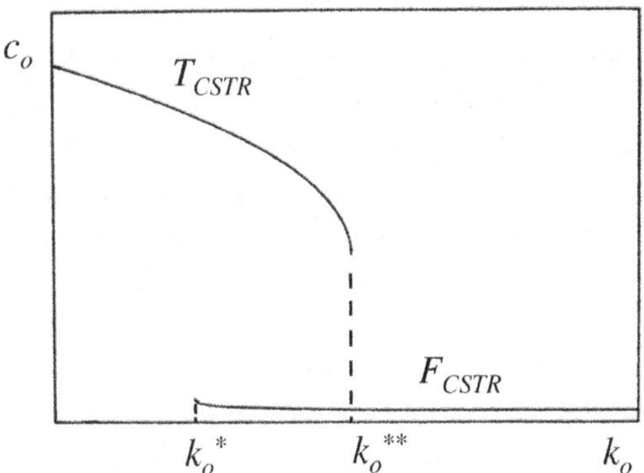

Figure 3. Bistable steady states in a CSTR.

A word of caution is perhaps needed. When dealing experimentally the $k_o = 0$ state indeed corresponds to the thermodynamic equilibrium state. However when performing theoretical work one mostly uses "toy" or else empirical models derived from the experimental kinetics studies valid in general for a limited set of conditions. As the latter may not contain all involved kinetic steps, the $k_o = 0$ limit should be carefully weighted and usually does not correspond to the true thermodynamic equilibrium state (for instance all reactions and their reverse should come into play as one nears the equilibrium conditions). Such work is however valid at a given distance from $k_o = 0$. Only comparison with experimental results allows a real assessment.

To study the dynamics of the IAA reaction we do not have to rely on a pool chemical approximation, as in the preceding section, to control the distance from equilibrium. To describe the behavior of the IAA reaction, Papsin et al. [32] have suggested a mechanism with seven elementary steps involving nine species. However here (and in the next section) we will only be interested in the situation where arsenous acid is in stoichiometric excess and when the pH is buffered. Under these conditions, the following empirical rate law has been established [33] that indicates that iodide (I^-) is an autocatalyst

$$f([IO_3^-],[I^-]) = \frac{d[I^-]}{dt} = (k_a + k_b[I^-])[I^-][IO_3^-] \tag{4.2}$$

where IO_3^- is the iodate and k_a and k_b are kinetic constants multiplied by $[H^+]^2$.

When the IAA reaction is operated in a *stirred batch* (closed reactor) it behaves as a "clock reaction" [16]. Starting from small amounts of iodide in the presence of iodate and arsenous acid, the reaction presents a well defined induction time, characterized by a low conversion rate, beyond which the reaction rate speeds up and the system switches rapidly to its thermodynamic equilibrium state. The solution suddenly emits a flash of color change – transparent to black, to transparent again. The flash occurs at the time of maximum rate of production of I_2 revealed by the presence of starch. The duration of this induction time (or time at which the flash occurs) depends on the initial concentration of iodide, hence the name of clock reaction.[1] For the same type of initial conditions, if the reaction is operated in an *unstirred batch*, it may give rise to a propagating wave that converts the reaction mixture from the initial state containing little or no iodide to the equilibrium state where iodide is the predominant species. The bulk of the chemical reaction occurs within a narrow reaction zone where iodine (colored by the presence of starch that creates a complex with the I_3^- species) is present. The wave typically propagates with a constant velocity and form [34]. Furthermore for some conditions this front may undergo a morphological instability [35]. All these phenomena may be semi-quantitatively described using the kinetics given by Eq. (4.2).

In the CSTR, the dynamics is thus described by [36]

$$\frac{du_o}{dt} = f(u_o,v_o) + k_o(u_{in} - u_o)$$

$$\frac{dv_o}{dt} = -f(u_o,v_o) + k_o(v_{in} - v_o)$$

$$(4.3)$$

where $u_o = [I^-]$ (autocatalyst), $v_o = [IO_3^-]$, $f(u_o,v_o) = (k_a + k_b u_o)u_o v_o$ (cf. Eq. (4.2)) and u_{in} and v_{in} the inflows concentrations.

Note that when the CSTR is working in "regime" the following conservation law $u_{in} + v_{in} = S_{in} = S_o = u_o + v_o$ applies and Eq. (4.3) can be reduced to a single equation. Indeed using as variables u_o and S_o, Eq. (4.3) may be rewritten as [37]

$$\frac{du_o}{dt} = (k_a + k_b u_o)(S_o - u_o)u_o + k_o(u_{in} - u_o)$$

$$(4.4)$$

$$S_o(t) = S_{in} + e^{-k_o t}(S_o(t=0) - S_{in})$$

---

[1] When the reaction is carried out without excess of arsenous acid, the iodine produced is not reduced by it and the switch in color is permanent. But the kinetic description requires four concentrations variables and is thus more complex to manage mathematically.

Therefore for times $t \gg k_o^{-1}$, $S_o \to S_{in}$ and we are left, to describe the dynamics with a single equation (and $v_o = S_{in} - u_o$)

$$\frac{du_o}{dt} = (k_a + k_b u_o)(S_{in} - u_o)u_o + k_o(u_{in} - u_o) \qquad (4.5)$$

This one variable model can exhibit bistability, as indeed $du_o/dt = 0$ is a polynomial of degree 3 (cubic), but never temporal oscillations since two dynamical variables with two time scales at least are necessary for this. The experimental results for u_o, varying u_{in} and k_o, shown in Figure 4 [38], validate the possibility of bistability for the system. The homogeneous steady states under the CSTR constraints obtained with the model for the same parameters as in the experiments are shown in Figure 5 [38]. The comparison between the experimental and theoretical curves (that bear clear analogy with a Van der Waals type diagram) is semi-quantitative.

From Figure 5, we see that as u_{in} is varied, we switch, at curve (d) from one to three steady solutions as may be expected for any cubic form. Thus in Figure 3 at k_o^* and k_o^{**} when the number of solutions change, bifurcations are involved. Indeed the bistability hysteresis loop occurs through two back to back, so-called, *saddle-node bifurcations* connected by their unstable states (the dotted line). Notice that this unstable state is not trivially observed experimentally but may nevertheless be traced by applying perturbations [32,39] to the steady states as shown in Figure 6. When u_{in} and k_o are fixed and the system has attained its steady state (say the lower one, i.e. the flow branch), we inject with a syringe a given quantity of iodide. If it is small the

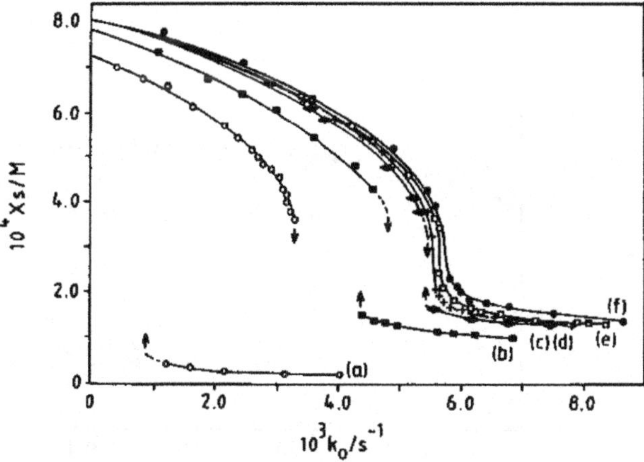

Figure 4. Experimental results for bistability of the IAA reaction in a CSTR. Depending on the input flow concentrations ((a),...(f)) and the reciprocal residence time of the reactor the system may exhibit bistability between two stable homogeneous steady states or monostability. Case (d) corresponds to a critical point. $k_a = 4.5 \ 10^3 \text{M}^{-3}\text{s}^{-1}$, $k_b = 4.5 \ 10^8 \text{M}^{-4}\text{s}^{-1}$, $[H^+] = 6.61 \ 10^{-3}\text{M}$. (Reproduced from [38], with permission.)

system the system will return to the steady state it just left. On the contrary, if its sufficiently large (meaning we overtake the elusive intermediate state), the system transits to the other steady state (on the thermodynamic branch) for the same value of k_o. By applying this procedure stepwise we may narrow down the position of the third state. As the system is never captured on this state it is a signature of its instability. Changing the value of k_o, the whole unstable branch may be traced with a good approximation. We could have proceeded the other way around, applying the perturbation on the thermodynamic branch injecting a scavenger of iodine to lower its concentration.

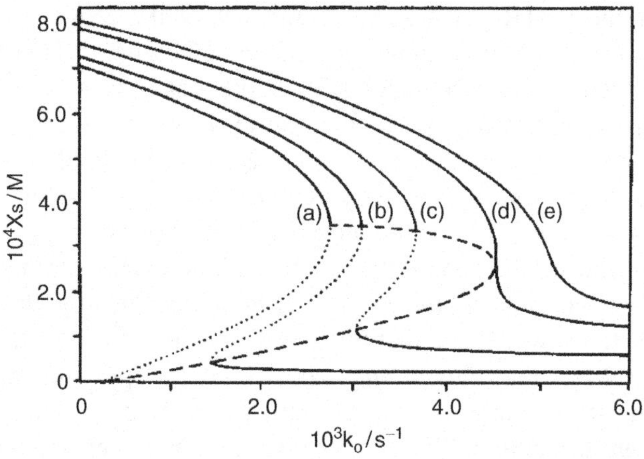

Figure 5. Theoretical bistability and monostability curves for the IAA reaction operated in a CSTR calculated from Eq. (4.5). Parameters are as in Figure 4. (Reproduced from [38], with permission.)

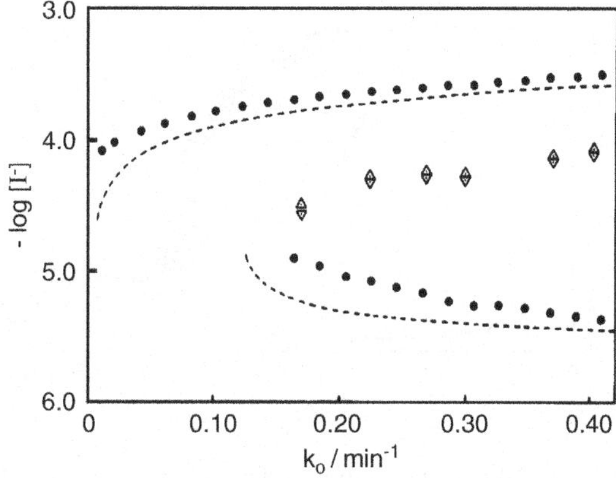

Figure 6. Iodide concentration as a function of ko. Subcritical (∇) and supercritical (Δ) perturbations are used to delineate the unstable steady state for a IAA system in the bistable region. (Reproduced from [32], with permission.)

Using similar techniques, a detailed study of the relaxation behavior of the bistable IAA reaction near its hysteresis limits [37,38], in particular the existence of plateau relaxation behavior [37,40], has been carried out.

Let us now come back to the *saddle-node* or *limit-point bifurcation*, two of which frame the hysteresis loop in the IAA system. One variable suffices to describe this system and it is therefore quite easy to obtain the (universal) amplitude equation for this bifurcation. Indeed in the neighborhood of k_o^* (Figure 3), the dynamics of the system may be written as [24]

$$\frac{dz}{dT} = \mu - gz^2 \tag{4.6}$$

where the right hand side is nothing more than the parabolic approximation of Eq. (4.5) in a coordinate system fixed at the bifurcation point. For the case we consider, $\mu \propto k_o^* - k_o$ and g is a negative constant function of the parameters of the system that may be obtained by elementary calculus. From Eq. (4.6), the two steady states may be obtained and their stability determined. One is stable and the other unstable as predictable. As this equation may readily be integrated, the complete dynamics in the neighborhood of the bifurcation may be obtained. This bifurcation corresponds to a single real root changing sign at marginality. When more concentration variables are involved, the reduction methods, alluded to in Section 2, may be applied to obtain Eq. (4.6). The saddle-node bifurcation we encountered in Section 3 is the interaction between limit-cycles and not steady states as here.

For more complex kinetics, the F_{CSTR} and T_{CSTR} states may interact with other states through a variety of bifurcations. Also the kinetics term of the IAA reaction under the conditions presented above may be written, with a trivial change of variable as $f(x) = -x^3 + \alpha x + \beta$. If one introduces a kinetic coupling with another species y such as

$$\frac{dx}{dt} = -x^3 + \alpha x + \beta - y$$
$$\frac{dy}{dt} = \gamma(x - \delta y) \tag{4.7}$$

For a suitable choice of parameters this system may among many other behaviors present either bistability of steady states or oscillations as shown in Chapter 1 (cross-shaped diagram) [9,29]. Near the center of the "cross", the behavior is particularly complex and it may be shown, that besides Hopf bifurcations, temporal oscillations of the concentrations may appear through other kinds of bifurcation that lie outside the scope of this short excursion. Model (4.7) is also a variant of the FitzHugh-Nagumo model [14], that finds its origin in electrophysiology, but has been extensively used in nonlinear chemistry to study pattern formation and wave behavior when diffusive processes are taken into account.

5. Spatial bistability

The experimental discovery of Turing patterns stands as a landmark in the field of nonlinear autocatalytic chemistry. This quest provided the driving force for the devising of novel reactors, as was mentioned earlier and in Chapter 1, that allow to control the asymptotic spatial concentration distribution far from equilibrium. Although the Turing patterns stand at the forefront, the clear and precise characterization of the much less glamorous phenomenon of spatial bistability is also of great importance in view of the scores of papers devoted to them in the engineers [41–43] and mathematicians community [44].

The aspect ratio is by definition the ratio between the size of the reactor part of interest and the characteristic size of concentration variations. For Turing structures in the disc OSFR that has large aspect ratio, the precise nature of the boundaries matter only at the level of the phase variable(s) of the patterns that are long ranged and allow for the description of defects in the structure. On contrary spatial bistability in the *annular* OSFR (Figure 7) (Chapter 1 and references therein) is the heir of boundary value problems specifically related to nonlinear reaction-diffusion problems [45–47].

The dynamics of an OSFR is given by the following set of equations, respectively for the CSTR and the gel

$$\frac{d\mathbf{c_o}}{dt} = \mathbf{f}(\mathbf{c_o}) + k_o(\mathbf{c_{in}} - \mathbf{c_o}) + \rho_V \frac{D}{L}\left[\frac{\partial \mathbf{c}}{\partial x}\right]_{x=0}$$

$$\frac{\partial \mathbf{c}}{\partial t} = \mathbf{f}(\mathbf{c}) + D\nabla^2 \mathbf{c}$$

$$(5.1)$$

where $\mathbf{c_{in}}$, $\mathbf{c_0}$ and $\mathbf{c}$ are the concentrations of the species respectively in the input flow of the CSTR, in the CSTR, and inside the gel; D is the corresponding matrix of diffusion coefficients, k_o the inverse of the residence time of the CSTR, L the depth of the gel measured from the feeding

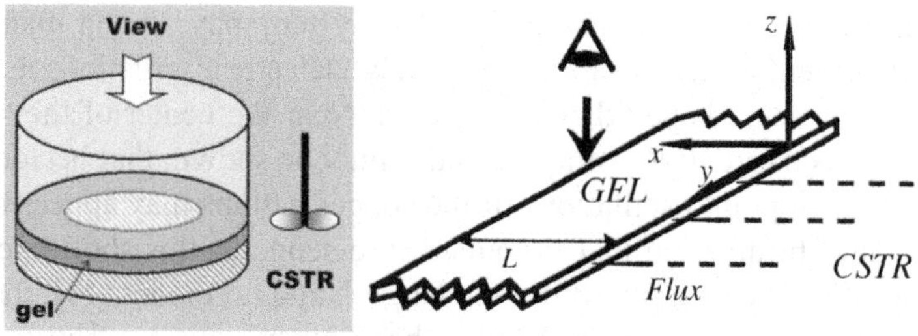

Figure 7. Sketch of an experimental annular OSFR (*left*). The straightened equivalent (zero curvature) used for theoretical calculations.

interface, ρ_V the ratio of the volume of the gel to that of the CSTR, and x is the direction normal to the CSTR/gel interface. The f's are the reaction rates, as before. On the right-hand side of Eq. (5.1), the second term represents the contribution of the input and output flows of the species in the CSTR. It contains all the expandable control parameters of the system (k_o, c_{in}). The CSTR therefore totally controls the feeding of the OSFR. The third term results from the diffusive flux of the species through the interface between the gel and the CSTR and represents the feedback of the gel contents on the CSTR dynamics. As usually $\rho_V \ll 1$, this last term can be neglected, as long as the normal concentration gradients are not too large. This is the case for the IAA model we will use again, so that the chemical state of the CSTR is, to a very good approximation, independent of the state of the gel. The concentrations in the CSTR act as Dirichlet boundary conditions for Eq. (5.1). At $x = 0$ ($\forall y, z$) (in contact with the CSTR) $c(x = 0, y, z) = c_0$. No-flux boundary conditions are applied at $x = L$ ($\forall y, z$) along the opposite wall, as well as on all the other impermeable boundaries.

The IAA reaction has never been operated in an OSFR. Because the agreement between the experiments and theory in the CSTR and in batch are almost quantitative, numerical integration of the system describing the OSFR dynamics given by Eq. (5.1), with $f(u,v)$ taken from Section 4, will be used to replace this missing phenomenology. Some information may also be gathered from other clock reactions discussed in Chapter 1.

For the case the IAA reaction we study here, the equations in the OSFR are then given by the combination of Eq. (4.3) and

$$\frac{\partial u}{\partial t} = f(u,v) + D\nabla^2 u$$

$$\frac{\partial v}{\partial t} = -f(u,v) + D\nabla^2 v$$

(5.2)

with the following boundary conditions:

- Dirichlet at the gel-CSTR interface ($x = 0$): $u(0, y, z) = u_o$; $v(0, y, z) = v_o$

where u_o and v_o are the concentrations in the CSTR and we have neglected the feedback of the gel on the CSTR dynamics $\rho_V \ll 1$.

- No flux boundary conditions at the other walls, particularly at $x = L$

$$\left(\frac{\partial u}{\partial x}\right)_{x=L} = \left(\frac{\partial v}{\partial x}\right)_{x=L} = 0$$

Before pursuing the analysis a comment is in order. Figures 4 and 5 show that when one allows a variation of the control parameters at hand, k_o and (u_{in}, v_{in}), the concentrations in the CSTR may take on a continuous spectrum of values. Naturally when it is in a bistability region there appears a gap in those values between the spinodal points. This is in contrast, if we remember the Van der Waals analogy, to the equilibrium phase transitions. There the fluctuations in the system lead to two coexisting phases whose compositions are determined by Maxwell's rule and who is driven either by nucleation or spinodal composition. Here the strong mixing overwhelms these fluctuations and the system is driven to one of the two HSS as in equilibrium systems with long range interactions. Therefore one should for each value of the triplet of parameters calculate the concentration in the CSTR. As in the end it does not influence the origin of spatial bistability we will, to go further, consider the concentrations in the CSTR as continuously variable bearing in mind that one can always find at least one triplet of parameters that can give rise to one such concentration.

Because the feed is uniform at the boundary of the gel, we may limit the discussion to one dimension along the x direction, as long as no transverse symmetry breaking arises. As all species involved in the IAA reaction are small ions, their diffusion coefficients may be taken equal in absence of large molecular weight complexing agents (see Chapter 1). Then, as iodide and iodate are the only stoichiometric significant iodine-containing species, the following conservation law applies when the OSFR is working in regime: $u + v = u_o + v_o = u_{in} + v_{in} = S_{in}$. System (5.2) then reduces to a one-variable model governed by the following evolution equation

$$\frac{\partial c}{\partial t} = c(1-c)(c+d) + D\frac{\partial^2 c}{\partial x^2}$$

$$c(x=0) = c_o \quad \text{and} \quad \left(\frac{\partial c}{\partial x}\right)_{x=L} = 0$$

(5.3)

We have defined the dimensionless variables $c = u/S_{in}$, $t = t'/\tau$, $x = x'/(D\tau)^{1/2}$ with $d = k_a/k_b S_{in}$ and $\tau = 1/k_b S_{in}^2$ and thereafter dropped the prime and $(0 < c < 1)$.

As simple as the model may be, we have various parameters at our disposal: the feeding concentration $0 < c_o < 1$ equal to the stationary concentration in the CSTR. Parameter d that may vary with temperature, as it is the ratio of kinetic constants, but also depends on the total quantity of iodine species present. Foremost in our analysis in relation to the actual experiments carried out with other clock reactions is the influence of the width L of the gel slab (or annulus in the experimental device).

We now present some characteristic results of the direct integration of the reaction-diffusion equation (5.3) that here, as we recall, will play the role of our "experimental data". For d we shall, at first, take the value at which the CSTR experiments, in Section 4, were performed ($d_{exp} \approx 0.0021$

$\ll 1$). As we switch to the theoretical discussion we will progressively relax the accessible values of parameter d, but also the concentrations, in order to understand the origin of spatial bistability.

5.1. NUMERICAL RESULTS

- *Steady concentration profiles*

Let us first consider the case of low values of c_o corresponding to situations where the CSTR is in its **flow state** F_{CSTR}.

When L is small, a theorem [48] and also the intuitive discussion on the behavior of a clock reaction in a spatial reactor (Chapter 1) explain that only one stationary state is possible. It is characterized by a very weak matter current at the feeding boundary, hence this profile is very *flat* as the concentration along it always remains close to c_o. Therefore we will name it the *F profile*. An example is drawn in Figure 8.

When L is very large, we know that the reaction will have time to evolve further as it reaches deeper in the gel through diffusion of the species, due to the clock mechanism. The incoming flux is now larger and, deep inside the gel, c reaches values of the same order as those on T_{CSTR}, the thermodynamic branch of the CSTR. Hence we call this type of profiled solution FT because it links concentrations at the F_{CSTR} level at the feeding to T_{CSTR} at the deepest part. An example is also given in Figure 8.

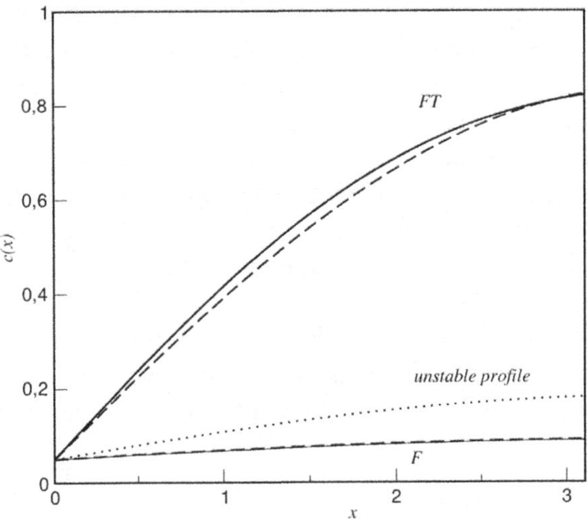

Figure 8. Numerical coexisting steady concentration profiles for the IAA system (plain and dotted lines) when the system is fed from the flow state of the CSTR ($c_o = 0.05$, $d = d^*$, $L = 3.5$). The *dashed lines* are the approximation obtained from the bifurcation calculations. (Reproduced from [47], with permission.)

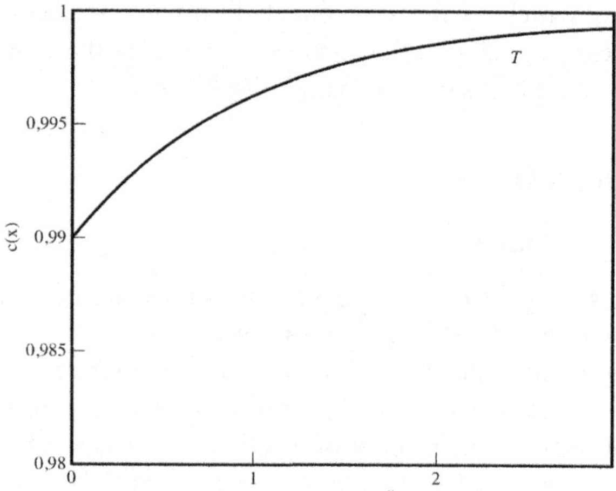

Figure 9. Numerical steady concentration profiles for the IAA system when the system is fed from the thermodynamic state of the CSTR. (Reproduced from [47], with permission.)

However, contrary to the experimentally observed and calculated profiles in the CDI and CT reactions [49,50], the *FT* profile grows smoothly and does not exhibit a steep switch at some intermediate position in the depth of the gel. This matter could be related to the existence of only one chemical characteristic time. In fact a similar smooth behavior is also found for the simplest reaction kinetics for the CT reaction (see Chapter 4).

If we now feed the gel with concentrations in the T_{CSTR} range, the reaction has already switched in the CSTR to values of concentration very close to those one would obtain in a batch near equilibrium (very little iodate left). The CSTR lives in a stationary state close to equilibrium. In the gel, the extent of the reaction can proceed a little further because of the diffusion time and hence be closer to equilibrium concentration. Therefore $c(x)$ also exhibit a very flat behavior shown in Figure 9. The difference in scale on the concentration axis should be noted. The larger L, the further $c(x)$ tends to the value one. This stationary profile is denoted as T type, but because it lies entirely close to equilibrium and is thus not expected to develop any instability, it is discarded from any further discussion. We have also checked that the three kinds of profiles, we have characterized, are always monotonously increasing functions of x, whatever the size of the system.

- *Bistability as a function of L*

For a range of values of L, the two profiles F and FT may coexist for the same feeding concentration c_o ,lying of F_{CSTR}. This is the *spatial bistability*, the origin of which lies at the core of this section.

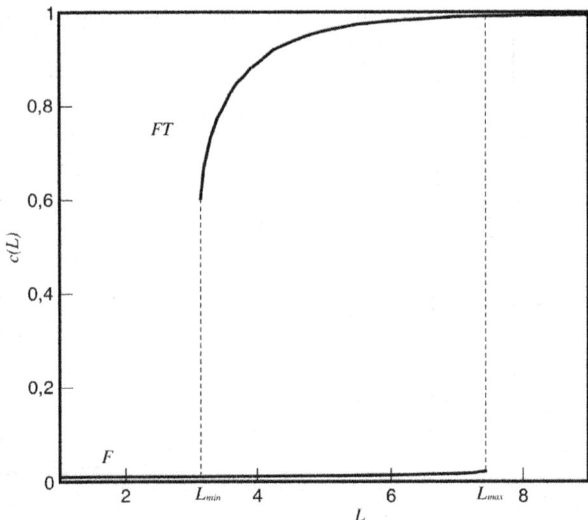

Figure 10. Bistability diagram exhibiting the two back-to-back saddle-node bifurcations for the IAA system. The concentration $c(x = L)$ at the impermeable boundary is represented as a function of L; $c_o = 0.05$, $d = d^*$. (Reproduced from [47], with permission.)

We may trace a bifurcation diagram by plotting for instance the value of the concentration at any given point inside the gel as our profile are monotonously varying. We choose to work with its largest value $c(x = L) = c(L)$, that at the impermeable back boundary. Figure 10 shows the hysteresis characterizing spatial bistability between the F and FT states. It seems to result from two back-to-back saddle-node bifurcations implying that they are connected by their unstable manifold, as for the steady states in the CSTR. Numerical analysis allows the tracing of the intermediate unstable steady profile (Figure 8). Spatial bistability therefore occurs for $L_{min} < L < L_{max}$.

- *Hysteresis as a function of c_o*

The previous subsection dealt with spatial bistability as a function of L, the depth of the gel. This is a geometrical parameter that relates to the construction of the OSFR, and it is not easily varied. It is therefore a critical parameter for the comparison between the works originating from different laboratories. However for a given material OSFR, thus a given L, the most obvious parameter to vary is c_o as it may itself be controlled by the feeding of the CSTR. Spatial bistability is also present.

- *Phase diagrams in (c_o, L) and (d, L) planes*

We have thus numerically confirmed the existence of spatial bistability for the IAA system in an OSFR. Complementary information can also be gathered from planar cuts in our three dimensional parameter space spanned by L, c_o and d.

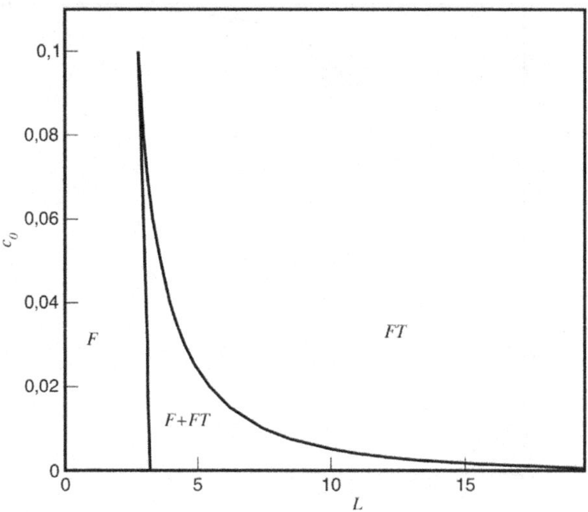

Figure 11. Spatial bistability between the *F* and *FT* states occurs inside the cusped region in the (c_o, L) parameters plane ($d = d^*$). Outside the cusp, only one of the profiles, either *F* (to the left) or *FT* (to the right), occurs. (Reproduced from [47], with permission.)

For instance Figure 11 summarizes all the information obtained from diagrams such as Figure 10 as we vary also c_o. As could have been expected, we recover a cusped region where *F* and *FT* coexist. The two curves delineating the cusp are the loci $L_{min}(c_o)$ and $L_{max}(c_o)$. We see that starting from a critical point that appears at a quite low value of c_o, the width of bistability region increases toward the lower values of c_o (it remains finite even at $c_o = 0$). We could also have show the (d, L) plane. There, for the first time, we stray from the line we have followed until now. This diagram shows that the value of d, although it is more difficult to control than c_o, plays as important a role as the two other parameters. This will come to light in the theoretical developments that now follow.

5.2. THEORETICAL APPROACH TO SPATIAL BISTABILITY

- *Exact solution*

When the diffusion coefficients are equal, the dynamics of the IAA system in an OSFR is governed by Eq. (5.3).

Because of the boundary conditions, there exists no homogeneous steady state in the gel (except $c = 0$ if $c_o = 0$ – not an interesting case in the chemical context – but that will prove important, as will be seen below). The steady concentration profiles that can develop in the depth of the gel are thus given by

$$\frac{\partial^2 c}{\partial x^2} + c(1-c)(c+d) = 0 + \text{b.c.} \tag{5.4}$$

This equation can be formally be solved by multiplying Eq. (5.4) by $(\partial c/\partial x)$ and integrating. We obtain

$$\frac{1}{2}\left(\frac{\partial c}{\partial x}\right)^2 + U(c) = E \text{ with the "potential" } U(c) = -\frac{c^4}{4} + \frac{(1-d)}{3}c^3 + \frac{d}{2}c^2,$$

where E is an integration constant determined by the boundary conditions: $E = U(c_L)$ where $c_L = c(L)$. In the phase plane $(c, (\partial c/\partial x))$, Eq. (5.4) defines an Hamiltonian flow (x is then considered as an equivalent time variable) where the fixed points $(0,0)$, $(1,0)$ and $(-d,0)$ (here lies a hint that taking $d = 0$ would introduce some degeneracy as two fixed points collapse) are res-pectively a center and two saddle points. From the last equation the so called "time" map may be written giving implicitly the solutions to our problem:

$$L = \int_{c_o}^{c_L} \frac{dc}{\sqrt{2(U(c_L) - U(c))}} \tag{5.5}$$

Since U has a quartic form, Eq. (5.5) can be solved [51] in terms of Jacobi elliptic functions. The corresponding boundary value problem can how-ever not be solved explictly. Nevertheless numerous works of "mathematical" character strive to study the properties of the corresponding time map. The time map can naturally also be evaluated numerically. However as for other conditions (different diffusion coefficients), we also wished to consider the two variables IAA case, we will now rather resort to the use of bifurcation techniques that lead to approximate but analytical results.

- *Bifurcation theory approach* [47]

We now explain the origin of spatial bistability using bifurcation theory. If we define $w = c - c_o$ and $\xi = x/L$ ($\xi \in [0,1]$), Eq. (5.3) reads

$$w_t = -[w^3 + (3c_o - 1 + d)w^2 + (3c_o^2 + 2c_o(d-1) + d)w$$

$$-c_o(1-c_o)(c_o + d)] + L^{-2}w_{\xi\xi} \tag{5.6}$$

with $w(0) = w_\xi(1) = 0$, and w_i and w_{ii} respectively correspond to the first and second partial derivatives with respect to i.

If $c_o = 0$ and $d = 1$, the system of Eq. (5.6) is invariant for the $w \rightarrow -w$, and also $c \rightarrow -c$ transformation. In this case $w = 0$ is the *only homogeneous steady state of the problem for any choice of L*. This is a situation reminding us of the starting point of the bifurcation calculation for the $\lambda-\omega$ system.

Let us then proceed with the linear stability analysis of this solution, to detect eventual bifurcations to new states, by investigating the fate of a small amplitude perturbation to it. We thus write $w = 0 + \delta w(\xi)\, e^{\Lambda t}$. We are then led to solve the following boundary value problem

$$L^2 \delta w + \delta w_{\xi\xi} = 0$$

$$\delta w(0) = \delta w_\xi(1) = 0$$

(5.7)

The corresponding solution is given by

$$\delta w = \sum_n A_n \sin(k_n \xi) \quad \text{with} \quad k_n = n\frac{\pi}{2L}$$

(5.8)

The spectrum of the eigenvalues of the linearized operator, $L^2 + \partial_{\xi\xi}$, is thus *discrete* and *real*, and given by

$$\Lambda_n = L^2 - n^2 \frac{\pi^2}{4}$$

(5.9)

The basic state $w = 0$ is marginal when $\Lambda_n = 0$, i.e. $L_n = n\pi/2$, and becomes unstable when L crosses the value of the smallest L_n. Such condition is naturally verified when $n = 1$, and thus the state $w = 0$ is asymptotically *stable* for $L < \pi/2$ and becomes *unstable* for $L > \pi/2$. Notice however, that while the system involves only a single concentration, the spectrum of the linear problem has an infinite spectrum of roots because of the presence of the diffusion operator.

The spectrum suggests that *nontrivial solutions* of the nonlinear problem *may bifurcate* from the point $(L, c_o, d) = (\pi/2, 0, 1)$.

In order to calculate the solutions which bifurcate from this point, we introduce a small parameter η and expand w, L, d, and c_o in power series of η:

$$w = 0 + \eta w_1(\xi,T) + \eta^2 w_2(\xi,T) + \eta^3 w_3(\xi,T) + ...,$$

$$L = \frac{\pi}{2}(1 + \eta L_1 + \eta^2 L_2 + ...),$$

$$d = 1 + \eta d_1 + \eta^2 d_2 + ...,$$

$$c_o = 0 + \eta c_{o1} + \eta^2 c_{o2} + \eta^3 c_{o3} + ...,$$

$$T = \eta^2 t + ...$$

(5.10)

Substituting the preceding expansion into Eq. (5.6) and equating to zero the coefficients of successive powers of η , we find, to the leading order

$$c(x) = c_o + A\sin(\pi x/2L) + O(\eta^2)$$

(5.11)

for the possible bifurcated profiles, where A is the solution of the *stationary amplitude equation* (obtained at the third order as a non trivial solvability condition)

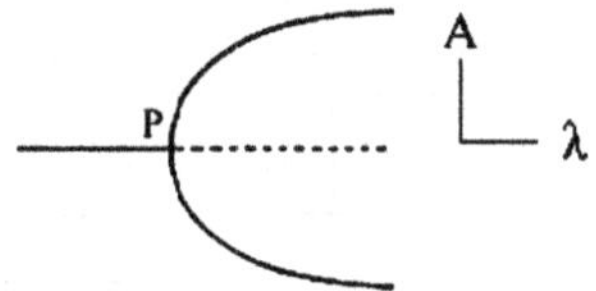

Figure 12. Schematic pitchfork bifurcation diagram for $c_o = 0$ and $d = 1$.

$$\frac{3}{4}A^3 - \lambda A - \frac{4}{\pi}c_o + \frac{8}{3\pi}(d-1)A^2 = 0 \qquad (5.12)$$

$$\text{with} \quad \lambda = 4L/\pi - 3(d-1)^2/4 + (d-3)$$

Equation (5.12) contains numerous behaviors as we vary c_o and d. The resulting bifurcation diagrams are conveniently given by representing the amplitude A as a function of λ (that, if $d = 1$, is essentially L). They are sketched as insets in Figure 14.

- Let us start from the case $c_o = 0$ and $d = 1$. Then from Eq. (5.11), besides the solution $c(x) = 0$, we also have

$$c(x) = \pm 4\sqrt{\frac{L - \pi/2}{3\pi}}\,\sin(\frac{\pi}{2L}x) + O(\eta^2) \qquad (5.13)$$

that bifurcates (see Figure 12) when $L > \pi/2$. From the amplitude equation we see that $c(x) = 0$ is still also a solution. The result is what is known (for obvious reasons) as a *pitchfork bifurcation*. As we have done in the case of the Poincaré model, when we studied the Hopf bifurcation, the next step would be to determine the (linear) stability of the new branches of solutions by using the slow time amplitude equation, but this will not be done here.

The universal amplitude equation for a pitchfork bifurcation is given by [24]

$$\frac{dz}{dT} = \mu z - g z^3 \qquad (5.14)$$

from which all properties in the neighborhood of the bifurcation may be calculated, as it is also integrable.

- Keeping first $c_o = 0$ and varying d, the pitchfork is transformed into a *transcritical bifurcation* (Figure 13). The saturation is then provided by a secondary *saddle-node bifurcation* (as we discussed for the subcritical Hopf bifurcation). The sign of $(d - 1)$ determines the direction of the $c \rightarrow -c$ symmetry breaking.

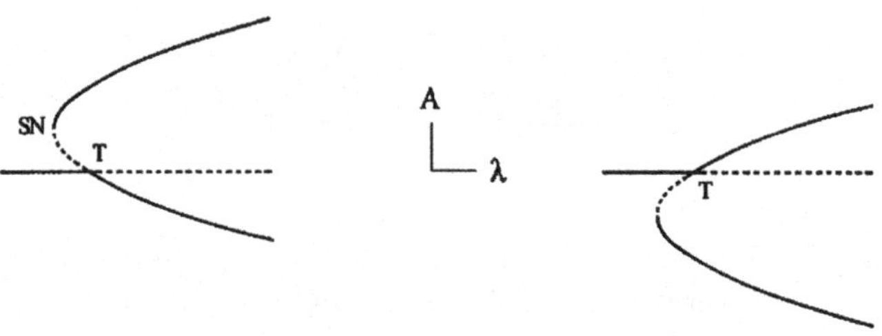

Figure 13. Schematic transcritical bifurcation diagram(T) for $c_o = 0$ and $d \neq 1$, saturated by a saddle-node bifurcation (SN). (*left*: $d < 1$ and *right*: $d > 1$).

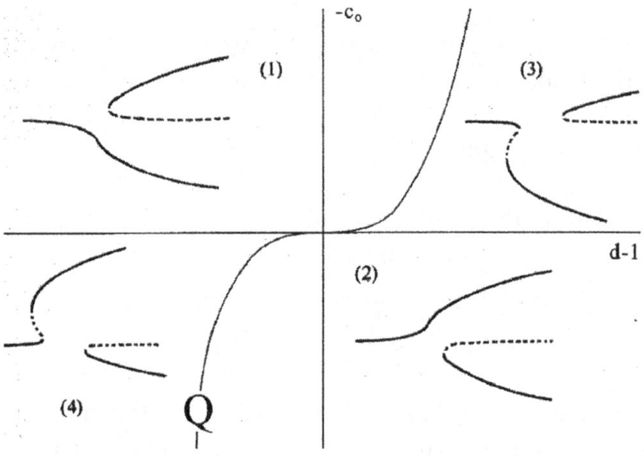

Figure 14. Schematic bifurcation diagrams for $c_o \neq 0$ and $d \neq 1$. See discussion in the text. (Adapted from [52], with permission.)

The universal amplitude equation for a transcritical bifurcation is given by [24]

$$\frac{dz}{dT} = \mu z - gz^2 \tag{5.15}$$

- When $c_o \neq 0$, the complete unfolding [52] appears as represented in Figure 14. As c_o plays the role of an imperfection parameter [24], we are left only with *saddle-node bifurcations* the only robust bifurcation-type in the presence of an imperfection.
- The plane is divided into four regions by the two curves $c_o = 0$ and $c_o = [(32/9\pi)^3(1 - d)^3]/27$. The pitchfork case corresponds to the intersection of the two curves. As announced the typical bifurcation diagrams (A as a function of λ) are drawn in the four regions in the $(-c_o, d - 1)$ plane.

Since $d_{exp} \ll 1$ and $c_o > 0$, the chemically valid region corresponds to the quadrant Q. As only branches corresponding to positive values of the concentrations are chemically relevant, the origin of spatial bistability lies in the crossing of the curve $c_o = [(32/9\pi)^3(1-d)^3]/27$ by varying c_o at the given value of d. This crossing corresponds to a critical point. The stability of the various branches may be determined from the slow time amplitude equation.

The physical mechanism of the spatial bistability can thus be apprehended through the study of the unfolding of an *organizing pitchfork bifurcation*. Mathematically, the simplest case of spatial bistability corresponds to the transcritical-saddle-node pair of bifurcation diagrams where one has the coexistence of the trivial $c(x) = 0$, F profile, with an FT profile for the same boundary condition, $c_o = 0$. This case is obviously experimentally unreachable, since its corresponds to no input concentration in the gel, but has proved useful to simplify the determination of the possible bifurcations to time-periodic regimes when the diffusion coefficients are such that $D_u > D_v$ (long range activation) [47].

6.   Turing patterns

In the preceding section, we have considered a first type of system where diffusion comes into play with reaction, as it allows the feeding of reactants to the full depth of the gel part of the OSFR. We have also noticed that, because of diffusion, the number of roots of the linear operator increases dramatically, even in a system containing a single chemical species, but remain well separated. We now show that this aspect may become much worse.

To conclude our brief trip, we now return to the problem of the once experimentally elusive Turing patterns [11,53] that were mentioned in the introduction, and the phenomenology of which is discussed in Chapter 1. The interest and large size of the subfield of Turing patterns is principally due to three main reasons: (a) the counterintuitive, short sided, idea that diffusion can help sustain a spatial organization and not disperse it, (b) their eventual role, still not proven, in biological morphogenesis, (c) the fact that the principle has become a paradigm as diffusive instabilities remain pertinent to the formation of structures in many other fields such as electron-hole plasmas in semiconductors, gas discharge devices, semiconductor structures (p-n junctions, p-i-n diodes), heterogeneous catalysis, electrochemistry, nonlinear optics, materials irradiated with energetic particles or light (see [45] for references). Therefore there exist numerous reviews in the literature [14,54–57] and we will be satisfied to stress a few important points in a line following the previous sections.

As always, we suppose we know some steady uniform reference state, and the first step is to proceed with its linear stability analysis. Now, because the diffusion operator is present, the characteristic equation, Eq. (2.9), provides a relation $h(B, k_n^2) = 0$, between the control parameter B and the eigenvalues of the Laplacian, k_n^2, that account for the spatial characteristics of the perturbations. Solving $h(B, k_n^2) = 0$ leads to the so-called marginal stability curve $B = B(k_n^2)$, that is expected to have at least one extremum for an instability to occur. Thus if $\Delta(k^2, B) = \|L - Dk^2\|$, then the conditions for a Turing instability to occur are

$$\Delta(k_c^2, B_c) = 0$$

$$\left. \frac{d\Delta(k_c^2, B_c)}{dk^2} \right|_{k_c} = 0 \tag{6.1}$$

These relations allow the simultaneous determination of the critical values of the bifurcation parameter B_c and the critical wavenumber k_c (not the wavevector!).

For a two species system,

$$k_c^2 = \sqrt{\frac{\Delta(0, B_c)}{D_a D_i}} \tag{6.2}$$

where $\Delta(0, B_c)$ is solely a function of the elements of the Jacobian matrix (thus of the kinetic parameters), while D_a and D_i are respectively the diffusion coefficients of the activator and inhibitory species. These developments imply that $D_i > D_a$ is a necessary condition (see Chapter 1). Relation (6.2) emphasizes the competition between reaction and diffusion and the intrinsic character of the wavelength.

To proceed a distinction is to be made between so called small and large aspect ratio systems. This is now defined as the ratio of the geometrical size of the system to that of the characteristic size of the phenomenon we study, here the intrinsic (Section 1) wavelength of the Turing pattern. For *small systems*, the spectrum of k_n is discrete. Because small systems have a boundary, where Dirichlet or Neuman conditions apply, the values of k_n will reflect the eventual geometrical symmetries of the systems, and thereby affect those of the pattern resulting from the bifurcation. For B sufficiently close to B_c , only one (or a small number of) mode(s) will become unstable. The corresponding $\Phi_n(\mathbf{r})$ will then determine the geometry of the spatial pattern in the nonlinear range also. For *extended systems*, the spectrum of the linear operator becomes continuous and new features are to be taken into account. Then periodic boundary conditions are used to simulate an unbounded system, and the $\Phi_n(\mathbf{r}) \propto \exp(i\mathbf{k}_m\mathbf{r})$ are plane waves, with $k_n \sim (2\pi/L)$.

This is the case for the experimentally observed Turing patterns, as their wavelength is of the order 10^{-2} cm. Figure 3e and f of [58], representing only a part of the experimental field, justify the feeling that we are dealing with large systems.

The supplementary features one has to take into account (with respect to the calculations in Section 4) arise from two degeneracies [25,57]. The first is related to the isotropic nature of the system. Indeed the active modes depend solely on k_c^2. Therefore an infinity of modes (plane waves) may become unstable (we may choose their modulation pointing in any direction of space). Further, as soon as the system crosses the bifurcation, because of the continuous character of the spectrum of the linear operator, a whole continuous band of modes, with their wavenumber framing k_c also become unstable (active). Because of the nonlinear terms all these modes will interact to determine the possible solutions. Once those are determined, their stability (and eventual metastability) have to be assessed: all this constitutes the lengthy *pattern selection problem*. We refer the reader to the specialized literature and merely discuss here some outcomes of this complex procedure.

Let us first discuss the standard results that arise in a 2D chemical system. From the isotropy argument, mentioned above, and because only one wavenumber comes into play, we expect to solutions to belong to the regular planforms [25,57,59]. Figure 15 represent the standard bifurcation diagram, when only those solutions that have some stability domain are represented (a pattern made of a tiling of squares also arises from the Turing bifurcation but is never stable for chemical systems – it is nevertheless known to appear for some hydrodynamic system). The simplest is that made of stripes that corresponds to a structure with a single active mode. We have represented the modulus of the amplitude in the figure. Indeed the latter is complex as it also contains a phase factor that fixes the position of the stripes with respect to some reference system. This phase is free. The bifurcation of stripes is supercritical. The other possible planform is an array of regular hexagons. Such an array may, geometrically, be characterized by two vectors in k-space making an $\pi/3$ angle. But here the dynamics of modes comes into play and a third mode is also activated by it, in such a way that the sum ϕ of the phases of the three modes is locked to 0 or π. The moduli of the three amplitudes are equal. Hence two kinds of hexagonal patterns arise (they look like inverse video pictures of one another). The bifurcations are transcritical at the lowest order. Now the sign of the quadratic terms in the amplitude equation comes in (not given here). It may be positive or negative. Figure 15 relates to the positive case. At the next order of approximation (cubic term in the amplitude equation), the $\phi = 0$ hexagons undergo a saddle-node bifurcation and become stable. The $\phi = \pi$ hexagons are always unstable. We observe other results on the diagram. When the

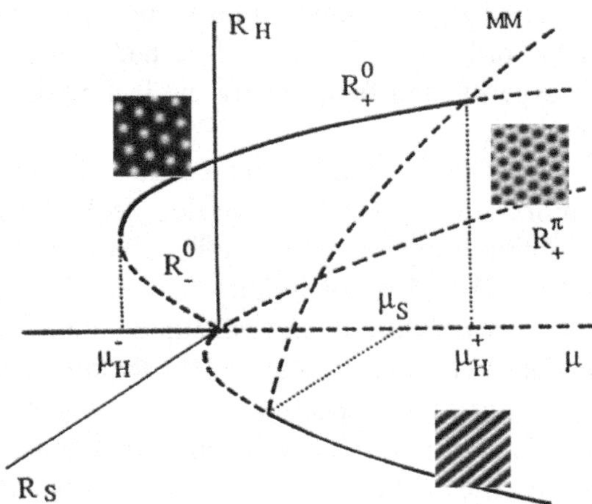

Figure 15. Standard bifurcation diagram for Turing patterns in the plane displaying the moduli R of the various structures (see text) versus the bifurcation parameter μ. *Full* and *dashed lines* respectively denote stable and unstable states.

stripes bifurcate they are unstable because of the presence of the $\phi = 0$ hexagons. However, as the bifurcation parameter μ increases, they stabilize and may coexist with the $\phi = 0$ hexagons, before these loose their stability themselves because of the stripes. This behavior is also related to the existence of a branch of mixed modes, that has two equal amplitudes while the third differs, and that is always unstable.

Figure 16 exhibits structures in a larger system, that show that the values of the phases are not rigidly fixed due of the existence of the band of modes framing those with k_c. The phases evolve on the longest time and length scales. This allows to explain the existence of the various types of defects in the structures.

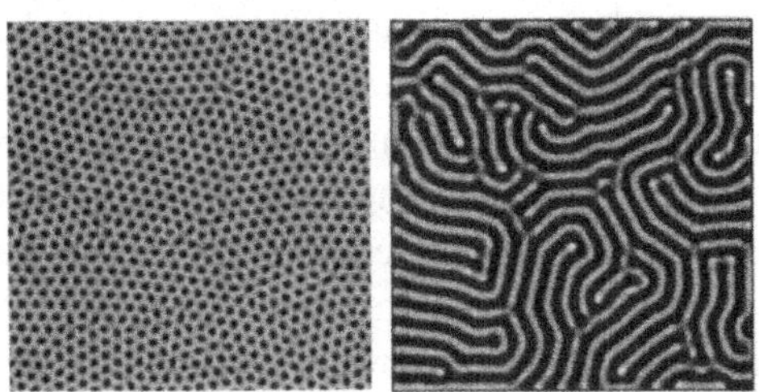

Figure 16. Hexagonal and striped Turing patterns in a large aspect ratio system exhibiting defects because of the lability of the phases of the active modes. Obtained from the numerical integration of the Brusselator. (Reproduced from [57], with permission.)

As may be guessed, in 3D, the situation is even more complex. To keep it at its simplest, the succession is (in real space) body centered cubic patterns (that also involve specific phase relations and hence appear subcritically), hexagonally packed cylinders and lamellae (the trivial extension to 3D of the 2D patterns) [60–62]. This, for example, closely resembles the spinodal structures of block copolymers. With the increasing computer power, numerical simulations soon produced a lamellae structure containing a twist grain boundary involving a Scherk surface [63]. The discovery of such pattern opened the door to the zoology of patterns relying on minimal surfaces that had been found in block copolymers. This program has now been carried out for chemical Turing patterns also [64–66].

Further, if the initial uniform steady state is part of a bistable system, new bifurcation diagrams and structures may occur because other homogeneous modes related to the existence of the initial bistability come into play [67]. Rhombic and superlattice-type patterns may then arise [68]. These concepts may also be useful in the context of nonlinear optics [69].

All this is fascinating, but if we take a step back to look at the complete picture, we soon remember that Turing patterns are observed in OSFRs where the steady states are NOT uniform because of the feeding of the gel. Therefore, the Turing pattern has to find a niche along the profiles where it can develop and survive. Very little work has been done along those lines, even numerically, to determine which symmetries are able to do so [46,59,70]. The observed experimental patterns often only present "layers" of stripes and/or hexagons distributed along some part of the profile (Chapter 1). So it seems the more complex structures are too sensitive to the ramps (and therezby deformations) imposed by the feeding. But a caveat applies as they are observed through a macrolens that integrates over the whole depth of the gel. Nevertheless hope remains, as a "spotty" pattern was observed among the earliest presented Turing patterns [54,71].

7. Conclusion

In this too short excursion, we have principally addressed the problem of the origin of chemical oscillations and spatial bistability as these two phenomena have already found their way into work involving gels sensitive to these chemical reactions. The short discussion on Turing patterns has been added because of the beauty of the concept and its historical importance in the field; ... and perhaps the dream of some day to be able to use them to spatially *sculpt* gels in a sustained way, as mechanical buckling does transiently during swelling or shrinking [72,73].

References

1. B.P. Belousov, *Sbornik Referatov po Radiatsionni Meditsine* (*in Russian*), p. 145 (1958).
2. B.P. Belousov, in *Oscillations and Travelling Waves in Chemical Systems* (R.J. Field, M. Burger, Eds., Wiley, New York, 1985) p. 605.
3. A. Turing, *Phil. Trans. R. Soc.* **B237**, 37 (1952).
4. I. Prigogine and R. Balescu, *Bull. Acad. R. Belg.* **41**, 917 (1955); **42**, 256 (1956).
5. G. Nicolis and I. Prigogine, *Self-Organization in Non Equilibrium Systems* (Wiley, New York, 1977).
6. A.M. Zhabotinsky, *Biofizika* **9**, 306 (1964).
7. A.M. Zhabotinsky, *Chaos* **1**, 380 (1991).
8. R.J. Field, E. Körös and R. Noyes, *J. Am. Chem. Soc.* **94**, 8649 (1972).
9. J. Boissonade and P. De Kepper, *J. Phys. Chem.* **84**, 501 (1980).
10. M. Orban, P. De Kepper, I. Epstein and K. Kustin, *Nature* **292**, 816 (1981).
11. V. Castets, E. Dulos, J. Boissonade and P. De Kepper, *Phys. Rev. Lett.* **64**, 2953 (1990).
12. R.J. Field and M. Burger, Eds., *Oscillations and Travelling Waves in Chemical Systems* (Wiley, New York, 1985).
13. P. Gray and S.K. Scott, *Chemical Oscillations and Instabilities* (Clarendon Press, Oxford, 1990).
14. J.D. Murray, *Mathematical Biology*, 3rd Ed. (2 vol.) (Springer, Berlin, 2002).
15. R. Kapral and K. Showalter, Eds., *Chemical Waves and Patterns* (Kluwer, Dordrecht, 1995).
16. I.R. Epstein and J.A. Pojman, *An Introduction to Nonlinear Chemical Dynamics* (Oxford University Press, New York, 1998).
17. G. Nicolis and F. Baras, Eds., *Chemical Instabilities: Applications in Chemistry, Engineering, Geology and Materials Sciences* (Reidel, Dordrecht, 1984).
18. P. Gray, G. Nicolis, F. Baras, P. Borckmans and S.K. Scott, Eds., *Spatial Inhomogeneities and Transient Behaviour in Chemical Kinetics* (Manchester University Press, Manchester, 1989).
19. A. Pacault and J.-J. Perraud, *Rythmes et Formes en Chimie* (Presses Universitaires de France, Paris, 1997) – Que sais-je? # 3235.
20. R.J. Field and R. Noyes, *J. Chem. Phys.* **60**, 1877 (1974).
21. J.J. Tyson, *J. Chem. Phys.* **68**, 905 (1977).
22. I. Prigogine and R. Lefever, *J. Chem. Phys.* **48**, 1695 (1968).
23. G. Nicolis, *Faraday Discuss.* **120**, 1 (2002).
24. Nicolis, G. *Introduction to Nonlinear Science* (Cambridge University Press, Cambridge, 1995).
25. L.M. Pismen, *Patterns and Interfaces in Dissipative Systems* (Springer, Berlin, 2006)
26. J. Guckenheimer and P. Holmes, *Nonlinear Oscillations, Dynamical Systems, and Bifurcations of Vector Fields* (Springer, Berlin, 1983).
27. G. Iooss and D.D. Joseph, *Elementary Stability and Bifurcation Theory*, 2nd Ed. (Springer, Berlin, 1990).
28. S. Wiggins, *Introduction to Applied Nonlinear Dynamical Systems and Chaos*, 2nd Ed. (Springer, Berlin, 2000).
29. J. Guckenheimer, *Physica D* **20**, 1 (1986).
30. I.S. Aranson and L. Kramer, *Rev. Mod. Phys.* **74**, 99 (2002).
31. I. Epstein, K. Kustin, P. De Kepper and M. Orban, *Sci. Am.* **248**(3), 112 (1983).
32. G.A. Papsin, A. Hanna and K. Showalter, *J. Phys. Chem.* **85**, 2576 (1981).
33. H.A. Liebhafsky and G.M. Roe, *Int. J. Chem. Kinet.* **11**, 693 (1979).

34. H.A. Hanna, A. Saul and K. Showalter, *J. Am. Chem. Soc.* **104**, 3838 (1982).

35. D. Horvath and K. Showalter, *J. Chem. Phys.* **102**, 8 (1995).

36. N. Ganapathisubramanian and K. Showalter, *J. Phys. Chem.* **87**, 1098 (1983).

37. G. Dewel, P. Borckmans and D. Walgraef, *J. Phys. Chem.* **88**, 5442 (1984)

38. N. Ganapathisubramanian and K. Showalter, *J. Chem. Phys.* **84**, 5427 (1986).

39. P. De Kepper, I.R. Epstein and K. Kustin, *J. Am. Chem. Soc.* **103**, 6121 (1981).

40. N. Ganapathisubramanian, J.S. Reckley and K. Showalter, *J. Phys. Chem.* **91**, 938 (1989).

41. P.B. Weisz and J.S. Hicks, *Chem. Eng. Sci.* **17**, 265 (1962).

42. R. Aris, *Chem. Eng. Sci.* **24**, 149 (1969).

43. D. Luss in ref. 18, p. 103, and references therein.

44. J. Shi, *Topics in Nonlinear Elliptic Equations* (Ph. D. Thesis, Brigham Young University, UT, U.S.A. 2003).

45. P. Borckmans, G. Dewel, A. De Wit, E. Dulos, J. Boissonade, F. Gauffre and P. De Kepper, *Int. J. Bif. Chaos.* **12**, 2307 (2002).

46. K. Benyaich, *Bistability, temporal oscillations and Turing patterns in a spatial reactor* (Ph. D. Thesis, Université Libre de Bruxelles, Belgium, 2005).

47. K. Benyaich, T. Erneux, S. Métens, S. Villain and P. Borckmans, *Chaos* **16**, 037109 (2006).

48. R. Iglish, *Math. Ann.* **108**, 161 (1933).

49. P. Blanchedeau, J. Boissonade and P. De Kepper, *Physica D* **147**, 283 (2000).

50. J. Boissonade, E. Dulos, F. Gauffre, M.N. Kuperman and P. De Kepper, *Faraday Discuss.* **120**, 253 (2001).

51. H.T. Davis, *Introduction to Nonlinear Differential and Integral Equations* (U.S. Atomic Energy Commision-GPO, 1960/also Dover 1962).

52. M. Golubitsky and D.G. Shaeffer, *Singularites and Groups in Bifurcations Theory* – Vol. 1 (Springer, Berlin, 1985).

53. Q. Ouyang and H.L. Swinney, *Nature* **352**, 610 (1991).

54. J. Boissonade, E. Dulos and P. De Kepper, in *Chemical Waves and Patterns* (R. Kapral and K. Showalter, Eds., Kluwer, Dordrecht, 1995), p. 221.

55. Q. Ouyang and H.L. Swinney, in *Chemical Waves and Patterns* (R. Kapral and K. Showalter, Eds., Kluwer, Dordrecht, 1995), p. 269.

56. I. Lengyel and I.R. Epstein, in *Chemical Waves and Patterns* (R. Kapral and K. Showalter, Eds., Kluwer, Dordrecht, 1995), p. 297.

57. P. Borckmans, G. Dewel, A. De Wit and D. Walgraef, in *Chemical Waves and Patterns* (R. Kapral and K. Showalter, Eds., Kluwer, Dordrecht, 1995), p. 323.

58. Q. Ouyang and H.L. Swinney, *Chaos* **1**, 411 (1991).

59. P. Borckmans, A. De Wit and G. Dewel, *Physica A* **188**, 137 (1992).

60. D. Walgraef, G. Dewel and P. Borckmans, Phys. Rev. A **21**, 397 (1980).

61. A. De Wit, G. Dewel, P. Borckmans and D. Walgraef, *Physica D* **61**, 289 (1992).

62. T. Callahan and E. Knobloch, *Nonlinearity* **10**, 1179 (1997).

63. A. De Wit, P. Borckmans and G. Dewel, *Proc. Nat. Acad. Sci. (USA)* **94**, 12765 (1997).

64. T. Leppanen, M. Karttunen, R.A. Barrio and K. Kaski, *Brazilian J. Phys.* **34**, 368 (2004).

65. H. Shoji, K. Yamada and T. Ohta, *Phys. Rev. E* **72**, 065202(R) (2005).

66. H. Shoji, K. Yamada, D. Ueyama and T. Ohta, *Phys. Rev. E* **75**, 046212 (2007).

67. S. Métens, G. Dewel, P. Borckmans and R. Engelhardt, *Europhys. Lett.* **37**, 109 (1997).

68. M. Bachir, S. Métens, P. Borckmans and G. Dewel, *Europhys. Lett.* **54**, 612 (2001).

69. L.A. Lugiato, *Chaos, Solitons and Fractals* **4**, 1251 (1994).

70. S. Setayeshgar and M.C. Cross, *Phys. Rev. E.* **58**, 4485 (1998); **59**, 4258 (1999).

71. J.-J. Perraud, K. Agladze, E. Dulos and P. De Kepper, *Physica A* **188**, 1 (1992).
72. T. Tanaka, S-T. Sun, Y. Hirokawa, S. Katayama, J. Kucera, Y. Hirose and T. Amyia, *Nature* **325**, 796 (1987).
73. E.S. Matsuo and T. Tanaka, *Nature* **358**, 482 (1992).

CHEMOMECHANICS: OSCILLATORY DYNAMICS IN CHEMORESPONSIVE GELS

J. Boissonade (boisson@crpp-bordeaux.cnrs.fr),
P. De Kepper (dekepper@crpp-bordeaux.cnrs.fr)
*Université de Bordeaux and CNRS, Centre de recherche Paul Pascal, 115,av.
Schweitzer, F-33600 Pessac*

Abstract. We review the different strategies to produce mechanical oscillations by coupling a gel which swells/deswells as a function of the chemical composition of its solvent ('chemo-responsive gel') with an autocatalytic reaction kept far from equilibrium. Afterwards we focus on the case of oscillations obtained by coupling the gel with a reaction that exhibits spatial bistability. The principles are illustrated with a simple swelling and reaction-diffusion model and experimental data obtained with the chlorite-tetrathionate reaction an a polyelectrolyte gel.

Keywords: gel, reaction-diffusion, swelling, chemical oscillations, autocalalytic, bistability, excitability, chemomechanics, polyelectrolyte, actuator

1.   State of the art

1.1.   INTRODUCTION

To study experimentally non-equilibrium chemical patterns induced by reac-tion-diffusion processes, it is necessary to simultaneously keep the system far from equilibrium by a permanent feed of fresh reactants, to eliminate convection, and to control diffusion. The standard way to achieve this goal is to use inert hydrogels as support media. The gel is immersed in a per-manently refreshed reservoir of well mixed reactants – like a continuous stirred tank reactor (CSTR) – where it diffusely exchanges matter at the gel/reservoir interfaces. Various examples are reported in [1]. Recently, sev-eral authors have replaced the inert gel by an active gel that can swell/shrink by absorption/expulsion of the reactive solution as a function of the chemical composition in the network. Their purpose is to obtain mechanical motions, such as oscillations or waves induced and controlled by the chemistry. Such systems can be used as autonomous actuators controlled by their chemical environment rather than by external fields or to deliver periodically drugs in a specific physiological environment. One can distinguish the case where the gel is coupled with an oscillating reaction from the case where the reaction cannot oscillate in the absence of the active gel. A brief review of the first case is given in this introduction. More developments can be found in this volume in [2]. The rest of our contribution to this book is mainly devoted to the second case where the oscillatory character results from the sole coupling of a non oscillating reaction with the mechanical properties of the gel.

1.2.　DYNAMICS CONTROLLED BY A CHEMICAL OSCILLATORY ENVIRONMENT

This is the most straightforward way to couple an active gel to an oscillating reaction. The piece of gel is immersed in a solution where a standard oscillating reaction takes place. pH oscillators coupled to polyelectolyte gels are especially appropriate. In order that significant variations of the amount of absorbed solution occur during an oscillation, the diffusion time of the species and of the solvent into the core of the gel must be short enough in regard to the period (typically a few minutes for the classical chemical oscillators). This imposes that at least one dimension of the gel be very small (typically less than 1 mm).

The first experiments were performed by Yoshida et al. with the (H^+, SO_3^{2-}, H_2O_2, $Fe(CN)_6^{4-}$) pH oscillator and a thin ribbon of poly(N-isopropyl-acrylamide-co-acrylic acid-co-butyl acrylic acid) (NIPAA-co-AA-coBMA) [3]. Experiments on small gels of poly(methylmethacrylic acid) (PMMA) in another pH oscillator, namely, the (H^+, BrO_3^-, SO_3^{2-}, $Fe(CN)_6^{4-}$) reaction, were achieved by Crook et al. [4] and completed by direct force measurements [5]. Similar results were obtained by the same group with triblock copolymers (PMMA/ PMAA/PMMA) – where PMAA is poly(methylacrylic acid) – with the same chemical oscillator. Recently, this group also designed a bilamellar actuator based on the opposite swelling properties of two triblock copolymers, respectively a polyacid and a polybase [6,7].

1.3.　DYNAMICS CONTROLLED BY AN IMBEDDED CHEMICAL OSCILLATOR

The Belousov-Zhabotinskii (BZ) reaction is the most popular chemical oscillator and is extensively described in monographs and textbooks [8–10]. In the original version of this reaction, a metal ion that plays the role of a catalyst is necessary to produce oscillations. Yoshida and Takahashi have covalently bonded a Ru(II) bipyridine complex $Ru(bpy)_3^{2+}$, to a polymer network of NIPAA [11]. This makes oscillations possible only within the gel network. The changes of the degree of oxidation of the Ru induce a swelling/shrinking process associated to the changes in the hydrophilic character of the polymer chains. When the piece of gel is small enough, it oscillates as a whole [12], but, when the piece is larger, traveling expansion/contraction waves are associated to chemical waves [13, 14]. On these basis, Yoshida and coworkers have proposed various applications to sensors and actuators (see [2] in this volume). Several authors have developed models and numerical simulations for this system. Yashin and Balazs made simulations in a system constrained to one dimension [15] on theoretical basis similar to those described in section 3. Villain et al. [18] applied equations derived from a generalisation of the Sekimoto theoretical approach of transport in gels [16,17] to the case of a spherical gel and achieved comparison with the experimental results of Yoshida. More recently Yashin and Balasz have proposed a lattice spring model to perform the computations in two dimensions [20, 21].

1.4. DYNAMICAL BEHAVIOR BY COUPLING WITH A NON OSCILLATORY REACTION

The oscillations described above were always induced by an oscillatory reaction. The instability originates in the sole chemical kinetics. Oscillations would occur even in a non active gel. Thus, the mechanical oscillations of the gel are *slaved* to the chemistry. However, even if there are now a growing number of such reactions, they still belong to a limited number of families. Only a few autocatalytic reaction can lead to oscillatory dynamics, since the chemical feedbacks necessary for the primary instability to occur are rather complex. However, all autocatalytic reactions can amplify fluctuations and have a potential to create instabilities. For instance, almost all reactions that present autocatalysis or substrate inhibition can exhibit temporal bistability in a CSTR or spatial bistability in a gel provided that the characteristic times, respectively the residence time and a feeding time of the gel core are appropriate. The latter can be related to the ratio of a the characteristic diffusion length to the typical distance to the core. One may anticipate that application of non chemical feedbacks to this larger class of systems can be at the origin of the instabilities responsible for the oscillations. Two different feedback mechanisms related to the swelling of the gel have been successfully proposed. Siegel and coworkers use a glucose driven chemomechanical oscillator made of a system of two compartments connected through a membrane made of a pH responsive gel. The first compartment is directly connected to an input/ouput flow and acts as the feed for the second compartment with which it exchanges species through the membrane. The pH changes in the second compartment make the gel to swell/shrink, inducing large permeability changes of the membrane [22,23]. This exerts a feedback on the feed of the second compartment, leading to an oscillatory process as initially predicted in a model [24] (see also [25] in this volume).

Another approach, to which this contributed paper is devoted, relies on the fact that, because the core of the gel is fed by diffusion from the boundaries, the spatial distribution of concentrations within this gel depends on its size and its geometry. Therefore, volume changes caused by the composition of the gel contents induce changes in this distribution which in turn create a feedback on the volume (and consequently on the size and/or shape) of this gel. This feedback can generate an oscillatory instability, even if the reaction mechanism does not allow for an oscillatory behavior in any container of fixed geometry.

In these two cases, it is necessary that at least one subprocess exhibits bistability and hysteresis so that the stationary state is located on the unstable branch, forcing the system to permanently switch between the two other branches. In Siegel's system, the hysteresis is associated to a first order volume phase transition of the gel. In the second case, we show that it can be provided by a spatially bistable chemical reaction. In the following, we first report a series of experiments by the Bordeaux group, which demonstrates that an original dynamical oscillatory behavior can be obtained by the association of a pH sensitive gel with a non oscillatory chemical system, namely, the chlorite-tetrathionate reaction. Then we show theoretically and numerically on a model how such instabilities can occur. The equations and modeling developed in this section could also be applied to the case of a gel forced by an oscillating reaction. In regard to our former publications [26,27], this approach is more rigorous by including additional features but delivers qualitatively identical results.

2.   Chemomecanical instabilities: An experimental example

2.1.   THE EXPERIMENTAL SYSTEM

We shall briefly summarize the principles of the experiments performed by the Bordeaux group. For more technical details and precise input data, the reader is invited to turn to the original publications [28–31].

2.1.1.   *Experimental set-up*
A sketch of the apparatus is given in Figure 1. The core of the reactor is made of a cylinder of poly(NIPAA-co-AA) hydrogel fixed at one end. Typical dimensions are R_c ~1 mm (radius) and l ~2 cm (axis). This well-known polyelectolyte gel swells in alkaline media and shrinks in acid media. The gel is immersed in a CSTR which is continuously fed by piston pumps. The dynamics of the gel are monitored by video cameras connected to a frame grabber for subsequent image analysis. Shrunk parts of the gel are turbid and are

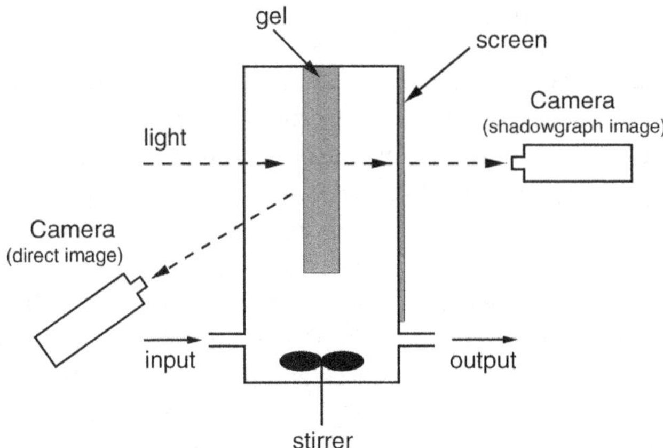

Figure 1.   Scheme of the experimental set-up.

revealed by their milky color. The limits of the transparent regions are best detected by shadowgraph techniques. Observations of the chemical states are made with an appropriate pH color indicator.

2.1.2.   *The reaction*

The reaction used in the experiments is the chorite-tetrathionate reaction (CT) that has been extensively studied for a number of remarkable dynamical properties, including temporal and spatial bistability, long range activation phenomena [31–35], and cellular front instabilities [36, 37]. The reaction is autocatalytic with H^+. During the reaction, a basic state is converted into an acid state in accordance with the following balance equation:

$$7\,ClO_2^- + 2\,S_4O_6^{2-} + 6\,H_2O \rightarrow 7\,Cl^- + 8\,SO_4^{2-} + 12\,H^+ \tag{1}$$

In a CSTR, this reaction exhibits bistability but no oscillations. When the CSTR is kept in the basic state by a sufficiently fast renewal of input reactants, the contents of a nonresponsive gel, like agarose, can either remain in a state with a low extent of reaction (i.e. basic) or switch to an acid state, except for a boundary layer that insures continuity with the concentrations in the CSTR. Over a significant domain of parameters, these two different distributions of concentrations in the gel can be both stable (spatial bistability) for a same set of parameter values. We come back later on the capabilities that this behavior can bring about. With the responsive gel used in the experiments, if the whole contents remain in a basic or neutral state, the gel is swollen and remains transparent, but when the core of the gel is in the acid state, it shrinks and becomes turbid.

2.2. OSCILLATORY BEHAVIOR

The parameters (gel composition, temperature, CSTR residence time, concentrations in the input flow, and the radius of the cylinder) are chosen in order that the gel be highly pH responsive and that the dynamical chemical state belong to the spatial bistability domain. Various dynamical phenomena, including different oscillatory regimes, unexpected in a nonresponsive gel with similar conditions, could be observed [28–31]. In Figure 2, we show one period of a simple mechanical oscillation of the gel cylinder. In this series of snapshots, the motions of the gel are revealed by the turbid core of the gel. The length of the gel increases and decreases by about 15% during one period ($\simeq 80\,\mathrm{min}$). Simultaneously, the diameter of the cylinder – that can be followed by shadowscopy – was found to change by a similar amount [29]. Due to gel inhomogeneities and to the constraints exerted by tip gluing, these diameter oscillations are not strictly in phase along the gel. Other more complex oscillations were also observed and are reported in [30].

This experimental result clearly demonstrates that a non oscillatory reaction coupled with the volume changes of a chemoresponsive gel can induce new instabilities leading eventually to mechanical oscillations. In the next section, we show that a simple model of a swelling gel, coupled to a kinetic toy model of a non oscillatory but spatially bistable reaction, can actually give rise to simultaneous periodic oscillations of the concentrations and the gel size. We do not claim that this provides an exact explanation for the reported experimental observations, which are affected by a number of additional problems, but that a non trivial behavior can be expected as a consequence of the proposed coupling. In Section 4, we report on experimental observation of chemomechanical excitability, a related dynamical phenomenon, for which we give a consistent theoretical interpretation.

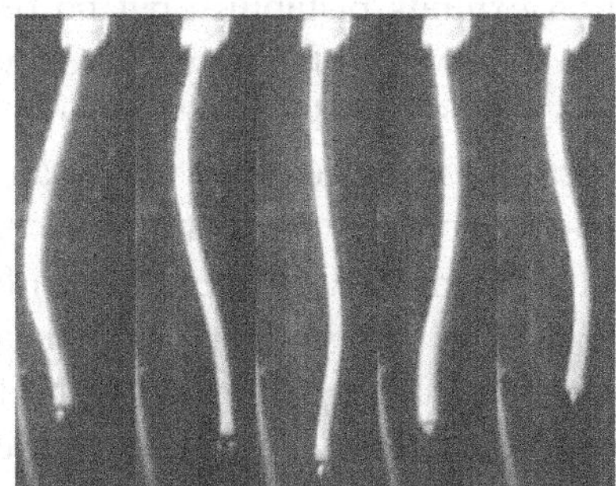

Figure 2. Volume oscillations of the cylinder of gel over one period of oscillation: Length variations are revealed by the turbid core (milky color) along the axis. (From [29], with permission.)

3. Chemomechanical oscillations with a non oscillatory reaction: a first theoretical approach

3.1. THE PRINCIPLES

To make computations easier, the following developments are applied to a spherical gel but can obviously be extended to other simple geometries. The proposed mechanism relies on spatial bistability. Let us first briefly remind the origin of this phenomenon. For a more extensive understanding in the context of chemistry and/or different experimental examples see [1, 31, 32, 38–43]. Let us consider a reaction which presents an autocalalytic step so that the reaction rate increases tremendously when the concentration c of the autocatalytic species, the "activator", becomes large. Then, at low values of c, the extent of reaction is low, and at high values of c, the extent of reaction is large. For brevity, we call these two states respectively "unreacted" and "reacted". We assume that the sphere is immersed in a CSTR operating at short residence time in order to keep it in the unreacted state by a sufficient renewal of the reactants. The volume of the CSTR is assumed to be large enough for its composition not to be significantly influenced by the gel contents. Within the gel, the capability to sustain the unreacted state depends on the competition between the reaction rate and the rate at which the fresh reactants are transported to the core by diffusion from the CSTR/gel interface. Thus, the extent of reaction within the gel depends on its size. For a small size, here the radius of the sphere, the gel contents remain almost unreacted, whereas for a large one, these contents are in a reacted state, except for a boundary layer at the surface in contact with the bath (we still call this spatial distribution a "reacted" state). One can show that, in many cases, there is a domain of spatial bistability, i.e. a range of intermediate sizes where both spatial concentration profiles are stable. The selected state depends on the initial amount of autocatalytic species able to switch on and to propagate the reaction to the core of the gel. For a sphere of radius R_s, this domain corresponds to a range of radii $R_{\text{inf}} < R_s < R_{\text{sup}}$. Thus, if one could continuously change radius R_s back and forth, crossing the values R_{inf} and R_{sup} one could travel along an hysteresis cycle switching alternatively the reaction on and off and commuting the gel contents between the "reacted" and the "unreacted" state.

In a chemoresponsive gel, these changes of the radius will be precisely insured by the volume changes induced by the chemistry, exerting in turn a feedback on the gel contents composition. One assumes that the gel tends to swell in the unreacted state and to shrink in the reacted state. The initial size of the gel is chosen in such a way that, when immersed in the unreacted state of the CSTR, the radius of the sphere would be $R_s > R_{\text{sup}}$. Then, the gel switches to the reacted state which causes the gel to shrink. If the radius decreases until $R_s < R_{inf}$, the gel switches back to the unreacted state and swells again until $R_s > R_{sup}$. Thus, the process repeats periodically. Although the reaction is only bistable, the system enters into an oscillatory regime, both mechanically (i.e. volume oscillations) and chemically (i.e. concentrations in the core of the gel). We now switch to the presentation of our numerical model based on a kinetic toy model for the reaction and a Flory-Huggins approach to swelling [44].

3.2. THE KINETIC MODEL

The reaction-diffusion equations are

$$\frac{\partial u}{\partial t} = -u^2 v^2 + D_u \nabla^2 u$$

$$\frac{\partial v}{\partial t} = \frac{12}{7} u^2 v^2 + D_v \nabla^2 v$$

$$(2)$$

where v and u are respectively the concentrations of an autocatalytic species and of a substrate. This model was initially used as a reduced kinetic model for the CT reaction (from which the nontrivial stoichiometric coefficient originates). However, in regard to a number of oversimplifications and inadequacies, it is advisable to consider it as a simple toy model. When $D_u = D_v$, one of the species can be eliminated so that the system is reduced to one variable which excludes any oscillatory phenomenon at a fixed radius but still permits spatial bistability. In the following, we shall assume $D_u = D_v = 1$. The values at the CSTR/gel boundary $r = R_s$, where r is the distance to the center of the sphere are fixed to $u_0 = 1$, $v_0 = 0.05$ and the equations (2) are solved numerically. The system exhibits spatial bistability for $4.48 < R_s < 5.42$. The two concentration profiles $v(r)$ of the autocatalytic species that are stable for $R_s = 5$ are represented in Figure 3. Curve 1, where v remains almost

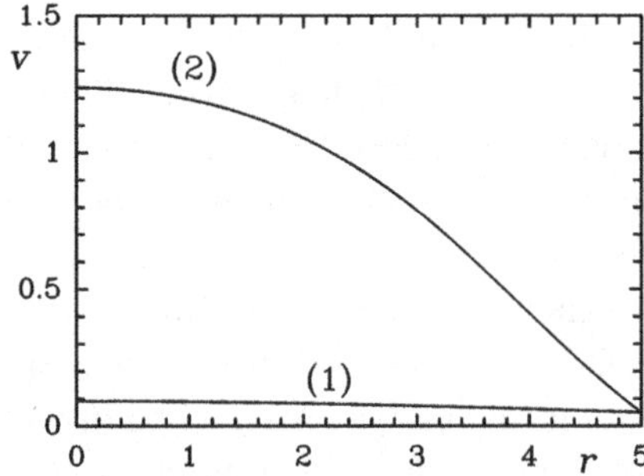

Figure 3.   Spatial bistability at $R = 5$. (1) "unreacted" state; (2) "reacted" state.

constant, is the "unreacted" state branch. Curve 2, where a large amount of v is produced in the core of the sphere, is the "reacted" state branch. For the CT reaction they would respectively correspond to an all basic state and to an acid state in the core of the core.

3.3. MODELING OF SWELLING DYNAMICS

Since the pioneering work of Tanaka and Fillmore [45], there have been a number of proposals to describe the dynamics of swelling [19, 21, 46–53]. Whatever their complexity, they all rely on unavoidable drastic simplifications and approximations so that the results are rather qualitative. We follow a simplified two fluid approach close to that developed in [50]. A slightly more general derivation can be found in [52].

We limit ourselves to neutral polymers. In the chemomechanical experiments, the reactants are very diluted (typically less than 10^{-2} mol/l). Thus, the volume of the reactants is negligible in regard to the volume of the other components, namely, the polymer with volume fraction ϕ, and the solvent (water) with volume fraction $1 - \phi$. The gel is also rich in water in order to keep the reaction rates as close as possible to those in pure water, so that, even in the shrunk state, ϕ keeps rather low values (say $\phi \simeq 0.1$). We also exclude a first order volume transition which would lead to a densely shrunk polymer. We first describe the swelling dynamics regardless of the solute concentrations. The chemical effects will be introduced afterwards through the dependance of the so-called Flory parameter with these concentrations. The dynamics at a given point within the gel express the permeation of the solvent moving at velocity $\vec{v}_S$ through the polymeric matrix moving at velocity $\vec{v}_P$ in the laboratory referential.

One also assumes that the gel is incompressible and that there is no transverse instability so that the solvent and polymer motions are radial. In these conditions, the mass balance in volumetric fluxes is given by:

$$\phi\vec{v}_P + (1 - \phi)\vec{v}_S = 0 \tag{3}$$

which allows to express $\vec{v}_S$ as a function of $\vec{v}_P$.

The local forces by surface unit that act on the polymer matrix are given by:

$$dF_i = \sigma_{ij} \cdot n_j dS \tag{4}$$

where the $\bar{\bar{\sigma}}_{ij}$ are the components of the stress tensor $\bar{\bar{\sigma}}$ due to both the polymer/solvent mixing and to the elastic forces. Expressed in terms of the density of volumic forces, (4) can be written:

$$\mathbf{f} = \nabla \cdot \bar{\bar{\sigma}} \tag{5}$$

Since the swelling process operates at very low Reynolds number, these driving forces are permanently balanced by the dynamic forces due to the friction of the solvent on the polymeric matrix (Stokes forces). These forces are proportional to the relative velocity of these two constituents, so that one can write:

$$\text{Friction forces} \;=\; \text{Swelling forces}$$

$$\zeta(\phi)(\vec{v}_P - \vec{v}_S) \;=\; \nabla \cdot \bar{\bar{\sigma}} \tag{6}$$

where $\zeta(\phi)$ is a friction coefficient which obviously increases with ϕ. On the basis of considerations on the correlation length in semidilute polymers, a number of authors assume a dependence $\zeta(\phi) \propto \phi^{3/2}$. This expression is certainly not valid at large ϕ's and alternatives or corrections have been proposed [48, 50]. It is also questionable at small ϕ's. An alternative expression, which might be a good approximation at small ϕ is given by the Ogston model [54], as proposed in [48]:

$$\zeta(\phi) = \frac{RT}{V_S} \frac{1}{D_0} \, \phi \, e^{\eta \sqrt{\phi}} \tag{7}$$

where η is of the order of the ratio of the radius of the polymer chains to the size of the solvent molecule, V_S the molar volume of the solvent and D_0 the autodiffusion coefficient of this solvent. This expression with $\eta = 5$ has been used in our computations.

From (3) and (6), one gets:

$$\vec{v}_P = \frac{(1 - \phi)}{\zeta(\phi)} \nabla \cdot \bar{\bar{\sigma}} \tag{8}$$

Taking into account the conservation law for the polymer

$$\frac{\partial \phi}{\partial t} + \nabla \cdot (\phi \vec{v}_P) = 0 \tag{9}$$

Equation (8) can be written:

$$\frac{\partial \phi}{\partial t} = -\nabla \cdot \left(\frac{\phi(1 - \phi)}{\zeta(\phi)} \nabla \cdot \bar{\bar{\sigma}} \right) \tag{10}$$

Equations (8) or (10) can be alternatively used to compute the motions of the polymer matrix. The first form is appropriate for computations in Eulerian mode, as done in this work, or in [27,48]. The second one is more appropriate for computations in Lagrangian mode as in [50].

The stress tensor will be expressed in the frame of the Flory-Huggins theory and the elementary Kuhn theory of elastomers [44,55,56]. We shall define a reference state of volume fraction ϕ_0 in which the network is assumed to be "relaxed". In this reference state, the coordinates of a point i attached to the polymer network are expressed as $\vec{R}^{(i)}$. When the gel swells or shrinks the point moves to a new position of coordinate $\vec{r}^{(i)}$.

The stress tensor can be decomposed into an isotropic part and an anisotropic part according to

$$\bar{\bar{\sigma}}_{ij} = \underbrace{-\delta_{ij}\Pi}_{\text{isotropic}} + \underbrace{\bar{\bar{\sigma}}_{ij}^{(\text{noniso})}}_{\text{nonisotropic}} \tag{11}$$

One assumes that the stress, which can be derived from the free energy of the system, is made of two additive contributions.

The first one results from the mixing of the monomers and the solvent and does not depend on the network structure. This contribution is obviously isotropic and, according to the Flory-Huggins theory, is given by:

$$\Pi_{\text{mix}} = -\frac{RT}{V_S}[\underbrace{\phi + \log(1-\phi)}_{\text{entropic}} + \underbrace{\chi\phi^2}_{\text{energetic}}] \tag{12}$$

The entropic terms come from the distribution of the constituents (monomer and the solvent molecules) in space, whereas the energetic term comes from the mutual interaction of these constituents and is characterized by a unique coefficient, the Flory parameter χ. Since the mixing contribution is isotropic, Π_{mix} is equivalent to a pressure and is commonly referred to as the "osmotic pressure".

The second contribution to $\bar{\bar{\sigma}}$ stems from the elastic forces that result from the extension of chains, when the system is submitted to deformations from the reference state. This contribution can be divided into an isotropic part and an anisotropic part. To understand the origin of these two terms, we shall consider the case of a unidirectional deformation. In a unidirectional deformation, the swelling is constrained along one coordinate axis by external forces, walls or geometrical constraints. In this transformation, the length of an infinitesimal piece of gel changes from dL to dl, changing the volume fraction from ϕ_0 to ϕ. But the volume of a gel or an elastomer is invariant during a shear deformation. Thus the swelling process must be understood as a virtual isotropic dilatation ($dL \to dL'$) and ($\phi_0 \to \phi$), followed by a virtual deformation $dL' \to dl$ at constant volume and constant density (Figure 4). In the first step, one has

$$\frac{dL'}{dL} = \left(\frac{\phi_0}{\phi}\right)^{\frac{1}{3}} \tag{13}$$

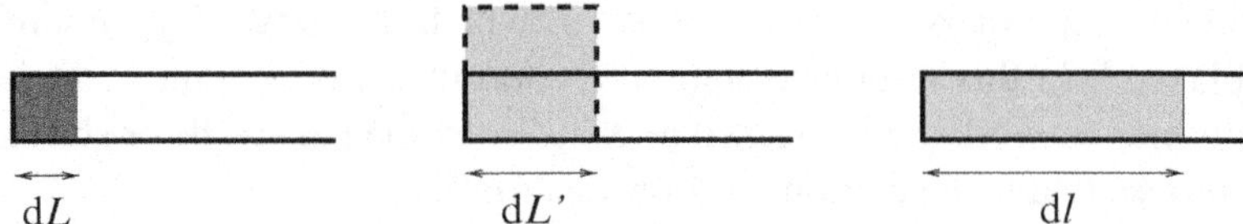

Figure 4.   Decomposition of unidirectional swelling into two virtual steps.

The extension factor λ during the deformation is

$$\lambda = \frac{dl}{dL'} = \frac{dl}{dL}\frac{dL}{dL'} \tag{14}$$

In a system constrained to remain unidimensional, one has $\phi_0\,dL = \phi\,dl$. Thus, from (13) and (14), one gets:

$$\lambda = \left(\frac{\phi_0}{\phi}\right)^{\frac{2}{3}} \tag{15}$$

In a spherical gel, in the absence of transverse instabilities, the elongation is purely radial. Equations (13) and (14) are still valid if one replaces l, L' and L by r, R' and R respectively. From these equations and the relation $\phi_0 R^2 dR = \phi r^2 dr$, one gets the following expression for λ:

$$\lambda = \left(\frac{\phi_0}{\phi}\right)^{\frac{2}{3}} \left(\frac{R}{r}\right)^2 \tag{16}$$

There are a number of theories to derive the elastic free energy and the stress tensor for a network [57]. All these theories rely on various strong hypothesis and have lead to large controversies. We have chosen to content ourselves with the results provided by a phenomenologic approach assuming that the free energy by unit volume and the elastic modulus $G(\phi)$ both scale like $(\phi/\psi)^n$ [50, 58]. To recover Flory's expressions [59], we use $n = 1/3$ and $\psi = \phi_0$ where the reference state is supposed to be in a relaxed state. For a unidirectional deformation, the following expression for the elastic part of the stress tensor can be derived [50]:

$$\bar{\bar{\sigma}}_{rr}^{(elas)} = K_{net}\left(\frac{\phi}{\phi_0}\right)^{\frac{1}{3}}\left[1 + C_\lambda\left(\lambda^2 - \frac{1}{\lambda}\right)\right] \tag{17}$$

where K_{net} is a constant which depends on the network properties and C_λ is a constant of order unity proportional to the shear modulus. The terms that contain λ represent the nonisotropic contribution $\bar{\bar{\sigma}}_{rr}^{(noniso)}$ to the elastic stress. In spherical coordinates:

$$(\nabla \cdot \bar{\bar{\sigma}})_{rr} = \frac{\partial \sigma_{rr}}{\partial r} + \frac{2\sigma_{rr} - \sigma_{\theta\theta} - \sigma_{\phi\phi}}{r} \tag{18}$$

For the isotropic contribution, the three components of the stress are equal so that $\bar{\bar{\sigma}}_{rr}^{(iso)} = \bar{\bar{\sigma}}_{\theta\theta}^{(iso)} = \bar{\bar{\sigma}}_{\phi\phi}^{(iso)}$. In (18) the corresponding last term vanishes. For the nonisotropic contribution, only the component in the extension direction is nonzero so that:

$$(\nabla \cdot \bar{\bar{\sigma}})_{rr} = \frac{\partial \sigma_{rr}}{\partial r} + 2\frac{\sigma_{rr}^{(noniso)}}{r} \tag{19}$$

Since $(\nabla \cdot \bar{\bar{\sigma}})_{rr}$ can be obtained from (12), (16), (17), (19), and $\zeta(\phi)$ can be obtained from (7), the swelling dynamics can now be computed in the absence of chemistry from (8) or (10).

In polyelectrolyte gels, the effects of chemistry is mainly associated to the distribution of charges within the gel. These are entropic effects due to the local excess of charges due to the dissociation of the polyelectrolyte and the energetic effects due to the changes in electrostatic interaction between ions attached to the chains (see Section 5). In a neutral gel, as assumed here, the effects of chemistry must be contained in the parameters that describe the energetic interactions. Since we account for these interactions by the sole Flory parameter χ, we assume that χ depends on the chemical composition. When multiple energetic parameters are considereded [15, 18], several choices are possible but they lead to similar results. In our model, we assume that χ is an increasing function of the concentration v of the autocatalytic species. Since we also have in mind a behavior similar to the polyelectolyte case in the presence of charges, we also assume that $\chi(v)$ saturates when v is large enough. A simple representation is given by a classical sigmoidal function:

$$\chi(v) = \frac{(\chi_{min} + \chi_{max})}{2} + \frac{(\chi_{max} - \chi_{min})}{2}\tanh(s(v - v^*)) \tag{20}$$

where the parameters χ_{min}, χ_{max}, v^* and s are chosen in such a way that χ changes from 0.30 (for $v = 0$) to 0.53 (for $v \gg v^*$) (Figure 5). It results from

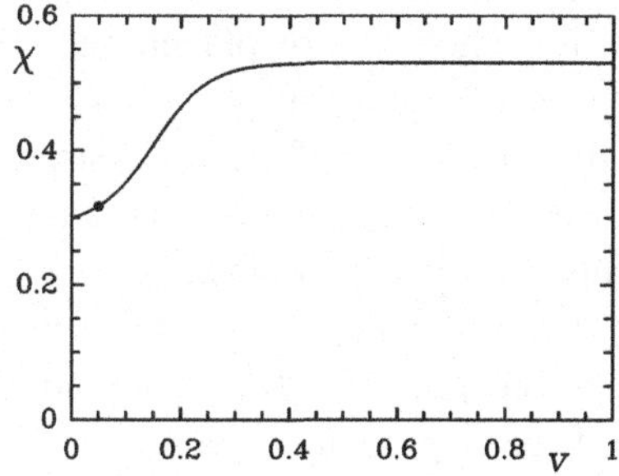

Figure 5.   Function $\chi(v)$. $\chi_{min} = 0.2885$, $\chi_{max} = 0.53$, $v^* = 0.15$ and $s = 10$. The point corresponds to $\chi(v_0)$, the value at the boundary.

the expansion in powers of ϕ of the Flory expressions that swelling becomes large only when χ crosses a value close to 1/2 [44], a condition which is met here.

The diffusion coefficient of the solutes depends on the volume fraction ϕ and should take the form:

$$D = D_0 \frac{\epsilon}{\tau} \qquad (21)$$

where D_0 is the diffusion coefficient in the pure solvent, ϵ is the permeability of the gel, which reflects the proportion of available space, and τ is the tortuosity which reflects the lengthening of the random walk to turn around the obstacles. For $\phi \ll 1$, one can take $\epsilon = 1 - \phi$ and $\tau = 1 + \alpha\phi$ which gives an effective diffusion coefficient:

$$D = D_0 \frac{1 - \phi}{1 + \alpha\phi} \qquad (22)$$

For a random network of spheres, an exact result, due to Maxwell, gives $\alpha = 1/2$ [60]. For a random system of fibers, which is more appropriate to a polymeric network, one can extrapolate the results of numerical simulations [61] and obtain $\alpha = 1$. Since in our simulations $\phi \ll 1$, we have neglected these effects to avoid unnecessary complexity. The diffusion of species is much faster than the collective diffusion of the polymer matrix at velocity $\vec{v}_P$ that governs the swelling process. Moreover, it results from (3) that, for $\phi \ll 1$, $\vec{v}_S \ll \vec{v}_P$ so that the convective transport by the solvent flow can also be neglected in regard to the diffusive transport.

Due to this large difference between the swelling and the solute diffusion times, the dynamical equations can actually be integrated in Eulerian mode by time splitting. A grid of moving points is attached to the polymer and allows for the computation of the density ϕ at each step. In agreement with a local equilibrium principle, the boundary value of ϕ is fixed at each step to a value ϕ_{eq} solution of the equation $\bar{\bar{\sigma}}_{rr}(\phi_{eq}, \chi(v_0), \lambda) = 0$ where λ is the value of the elongation at $r = R_s$ (boundary point). This value changes from step to step. For each step δt, the reaction-diffusion equations (2) are integrated on a fixed grid of size R_s. Then, the displacements of the grid points over δt are obtained from (8) and new ϕ and ϕ_{eq} values are computed from the distribution of points on the new grid. Then, the distribution of concentrations is interpolated on this new grid and the process is repeated. The stability of the algorithm can be checked by step control.

Although significant improvements have been brought in regard to our former published simulations [26, 27], these new computations provide qualitative similar results. These improvements are

-   The nonisotropic effects have been accounted for. However, they do not bring significant changes. The reason is that, in a sphere, starting from an equilibrium isotropic state, a step of χ causes a swelling/shrinking process that eventually leads to another isotropic state. Nonisotropic effects only appear transiently, changing the relaxation time but not the final state [50]. This is in contrast with systems constrained to one dimension in [15] where the nonisotropic stress is always present.

- The relaxed state of volume fraction ϕ_0 is a shrunk state, not a swollen state. This is more in agreement with the common proposal that this is the state at which the gel is prepared and that swelling normally occurs afterwards.

3.4.   THE PULSATING GEL

A sphere of gel of volume fraction $\phi_0 = 0.1$ is assumed to be at equilibrium for $\chi = 0.53$ (shrunk state). This corresponds to a value $K_{net} \simeq 6.05 \times 10^{-5}$. The initial radius is fixed to $R_0 = 3$. At equilibrium, the volume fraction in pure solvent ($\chi = 0.30$) would be $\phi = 0.121$, i.e. $R/R_0 = 2.02$ (swollen state). The sphere is immersed in a bath at $u_0 = 1$, $v_0 = 0.05$. In Figure 6, is represented the evolution of the radius R_s of the sphere and the concentration v_c at the center. When the system swells and the radius reaches $R_{max} = 0.54$, the unreacted state (low v_c) cannot any longer be sustained and the gel contents switch to the reacted state (high v_c). Then almost the whole gel takes χ values close to 0.53, the sphere shrinks and the radius R_s decreases down to $R_{min} = 0.45$ where the system switches back to the unreacted state and the sphere begins to swell again, thus entering in a regime of periodic oscillations. The chemical bistability hysteresis are essential. If there were no hysteresis, the system would find an intermediate value where the changes in χ and R_s would be balanced.

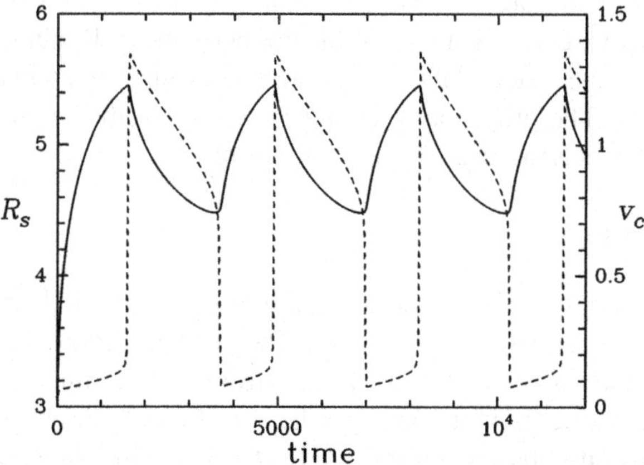

Figure 6.   Sphere radius $R_s(t)$ *(scale on left, full lines)* and concentration $v_c(t)$ at center *(scale on right, dotted lines)*.

We have demonstrated on this simple model, that a bistable but non oscillatory reaction can generate a chemomechanical instability in a responsive gel. However, the CT reaction does not meet all the required criteria. Due to an excessively long induction time, the bistability domain cannot be crossed since R_{max} becomes excessively large in realistic experimental conditions. Thus, the phenomena presented in Section 2.2 cannot be interpreted in the frame of this simple theory. Nevertheless, another observed behavior, namely, chemomechanical excitability can be understood in simple terms and is described in the next section.

4. Chemomechanical excitability: Experimental data and qualitative theory

4.1. THE EXPERIMENT

The experimental set-up is the same as in Section 2.2. The input concentrations are chosen in such a way that the input parameters correspond to a point within the bistability domain. Although both states are separately stable in this domain, it was previously found that the "reacted" state dominates the "unreacted" state, except in a very restricted domain located close to the bistability limit, so that the reacted state would grow at the expense of the "unreacted" state when they are brought into contact [35]. As a result, if a small part of the gel is in the acid state, this state should invade the whole gel. The cylinder of gel, initially in the basic state, is rapidly extracted from the reactor and the free tip is briefly immersed in an acid solution before reintroduction into the bath. In regard to the different time scales, this is equivalent to perform an acid perturbation on the tip without changing the nonequilibrium conditions. As a result of the perturbation, the tip of the gel switches in the acid state which begins to propagate but, in contrast to the case of non responsive gels, this transition is not permanent. Behind the acid front the system switches back to the basic state that eventually recovers within the whole cylinder. The propagating acid zone goes together with a contraction wave that appears as a traveling neck in Figure 7.

4.2. INTERPRETATION

The different steps (a–d) that describe the mechanism of chemomechanical excitability as a sequence of events are sketched in Figure 8. In the upper part of the figure, a dot represents the position and the qualitative nature of a state (black for acid, white point for basic) of a cylindrical piece of gel in the nonequilibrium phase diagram ($[OH]_0^-, R$). The abscissa $[OH]_0^-$ is the feed concentration in the CSTR and the ordinate R is the radius of the cylinder.

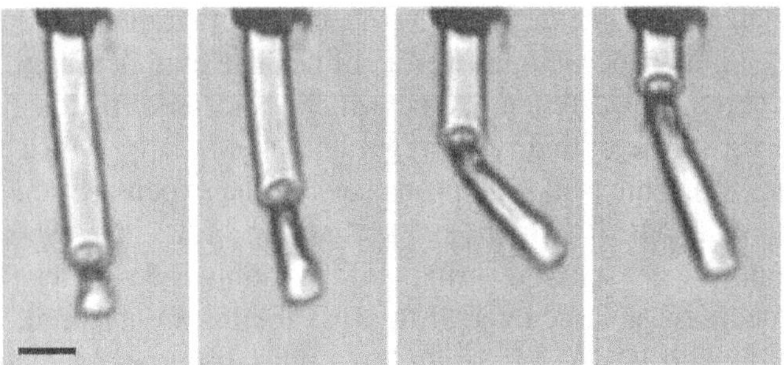

Figure 7. Chemomechanical excitability: propagation of a contraction wave. (shadowgraph view). The scale bar is 4 mm long. (From [30], with permission.)

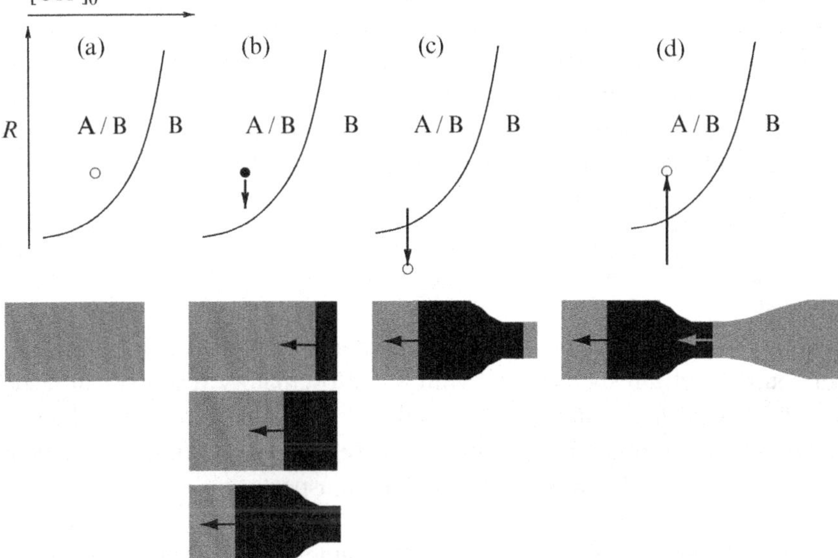

Figure 8. Chemomechanical excitability: formation and propagation of a contraction pulse.

The gel is bistable (domain AB) above the continuous line and monostable and basic ('unreacted') below the line (domain B). This schematic diagram is in agreement with the experimental and previously computed diagrams [32, 35]. In the different vignettes (a–d), the dot is associated to the state at the tip of the cylinder at the successive stages of the experiment. In the lower part of the figure, the drawings represent schematic side views of the extremity of the cylinder at the corresponding steps of evolution. Acid parts are drawn in black, basic parts are drawn in grey. In the reacted state, the thin basic boundary layer that connects the acid core to the basic environment is not represented.

During the experiment, $[OH]_0^-$ is fixed, so that, in the diagram, the dynamical evolution occurs on a vertical line. The cylinder is initially in the basic state (Figure 8a). When a supercritical acid perturbation is applied – here at the tip of the cylinder – the gel contents locally switch to the acid state (Figure 8b). This acid state propagates at the expense of the basic state since, in domain AB, it is more stable than the basic state. Shortly behind the acid front, the gel starts to shrink and the radius R decreases (Figure 8b). When the decrease is large enough to cross the bistability limit, the system enters into domain B. The acid state loses its stability and switches back to a basic state (Figure 8c). Behind the basic front, the system starts to swell, reenters the bistability domain and returns to the initial swollen basic state and initial radius (Figure 8d). This basic state is protected from the back propagation of the acid state by the neck where the latter is unstable so that the system completely recovers the inital unperturbed state. This differs from a nonresponsive gel where state A would invade the whole cylinder. In this qualitative explanation, we have not used a particular model for the mechanism of swelling. This is another clear example where the changes of geometry induced by the changes of composition in the responsive gel deeply modify the overall dynamics.

5. Conclusion

Although we have shown experimental evidences and theoretical predictions that coupling of a responsive gel with a non oscillatory reaction can lead to new dynamical instabilities, we are still far from full consistency between experimental observations and these theoretical predictions. Convergence should be achieved by improvements on both aspects.

First, to fit with predictions of Section 3, it is necessary to use a reaction that exhibits spatial bistability with both a large enough chemical change to cause large swelling/shrinking and a sufficiently narrow domain of bistability to allow for the characteristic size (the radius for a cylinder) to move across the width domain and follow the full hysteresis cycle. In this respect, the bromate-sulfite reaction, the spatial bistability of which has recently been studied experimentally and successfully modelled in numerical simulations, is especially promising [42].

Apart from the BZ reaction that presents very specific properties in all domains, the best candidates for generating chemomechanical instalities are pH oscillators and pH bistable reactions. Actually, polyelectrolyte hydrogels that are sensitive to pH are probably the most studied responsive gels. Although some energetic corrections of electrostatic nature can be considered, especially for highly charged gels, the main source of swelling is of entropic nature so that the above theory has to be extended and the distribution of charges has to be taken into account. The charged groups of the polyelectrolyte dissociate in agreement with the equilibrium equation

$$AH \leftrightarrows A^- + H^+ \tag{23}$$

where A^- is attached to the fixed chains whereas H^+ is a mobile ion. The mobile ions exert an additional osmotic pressure. At strong dilutions, this pressure is analog to a perfect gas and takes the form $\Pi_{ion} = RT \sum_i c_i$, where the c_i's are the concentrations of the mobile ions within the gel so that, the stress tensor $\bar{\bar{\sigma}}_0$ in the absence of charge has to be modified into

$$\bar{\bar{\sigma}} = \bar{\bar{\sigma}}_0(\phi) - \Pi_{ion}\bar{\bar{I}} \tag{24}$$

and the equilibrium condition into

$$\Pi_0 + \Pi_{ion} = RT \sum_i c_i' \tag{25}$$

where Π_0 is the isotropic part of the stress tensor and the c_i''s are the concentrations in the CSTR. According to the Boltzman law, the relation between the concentrations of the ions at the boundary are given by the relation

$$\frac{c_i}{c_i'} = \exp\left(-\frac{z_i e U}{k_B T}\right) = K^{z_i} \tag{26}$$

where U is the potential difference between the bath and the gel contents, z_i the number of electronic charges e of species i, and K is the so-called Donnan ratio. Following Rička and Tanaka [62], the values of K and ϕ_{eq} at equilibrium are solution of the nonlinear system of equations

$$\sum_i z_i K^{z_i} c_i' + z_a \frac{c_0 \phi_{eq}}{1 + K c_h'/K_a} = 0 \tag{27}$$

$$-\Pi_0(\phi_{eq}) - RT \sum_i c_i'(K^{z_i} - 1) = 0 \tag{28}$$

where K_a is the dissociation constant of equilibrium (23). Equations (27) and (28) hold respectively for the electroneutrality and the mechanical equilibrium. These equations can be easily generalized to the case of the local equilibrium in presence of a distorsion λ to compute ϕ_{eq} at each step. Then, since the boundary conditions are known at each step, the motion equations can be solved, provided that the coupled reaction-diffusion equation are simultaneously solved. The reaction-diffusion equations must be extended to take into account the motion of charges [63]. As a first approximation, one can assume that A^- is transported at the polymer velocity $\vec{v}_P$ and that all the other solutes are transported at the solvent velocity $\vec{v}_S$. Modeling of chemo-mechanical oscillators on these basis in a lagrangian formulation has been partially achieved in a one dimensional system and applied to the realistic reaction model of the spatially bistable Bromate-Sulfite reaction [64]. Similar theoretical approaches of dynamics of swelling in ionic solutions can be found in [49, 51].

In this contribution, we have reported experimental evidences of chemo-mechanical motions that do not directly result from the sole chemical dynamics but that are induced by the coupling of the reaction-diffusion processes with the changes of geometry caused by swelling. We have demonstrated, by simulation of swelling of a neutral gel in presence of a bistable reaction, the valididity of this concept and layed the foundations of a generalisation to ionic media. A more appropriate theoretical frame in view to our reported and planned experiments is presently developed.

Acknowledgements

This work has been supported by the CNRS and the Agence Nationale de la Recherche.

References

1. P. De Kepper, J. Boissonade, and I. Szalai, this volume
2. R. Yoshida, this volume
3. R. Yoshida, T. Yamaguchi, and H. Ichijo , *Mat. Sci. Eng. C* **4**, 107 (1996)
4. C.J. Crook, A. Smith, R.A.L. Jones and J. Ryan, *Phys. Chem. Chem. Phys.* **4**, 1367 (2002)
5. A.J. Ryan et al., *Faraday Discuss.* **128**, 55 (2005)
6. J.R. Howse, P.D. Topham, C.J. Crook, A.J. Gleeson, W. Bras, R.A.L. Jones, and A. J. Ryan, *Nano Lett.* **6**, 73 (2006)
7. P.D. Topham, J.R. Howse, C.J. Crook , A.J. Gleeson, W. Bras, S.P. Armes, R.A.L. Jones and A. J. Ryan, *Macromolecules* **40**, 4393 (2007)
8. R.J. Field and M. Burger, *Oscillations and Traveling Waves in Chemical Systems* (Wiley, New York, 1985)
9. R. Kapral and K. Showalter, *Chemical Patterns and Waves* (Klüwer, Amsterdam, 1995)
10. I.R. Epstein and J.A. Pojman, *An Introduction to Nonlinear Chemical Dynamics* (Oxford University Press, New York, Oxford, 1998)
11. R. Yoshida and T. Takahashi, *J. Am. Chem. Soc.* **118**, 5134 (1996)
12. R. Yoshida, M. Tanaka, S. Onodera, T. Yamaguchi, and E. Kokufuta, *J. Phys. Chem.* **A104**, 7549 (2000)

13. R. Yoshida, E. Kokufuta, and T. Yamaguchi, *Chaos* **9**, 260 (1999)

14. K. Miyakawa, F. Sakamoto, R. Yoshida , E. Kokufuta, and T. Yamaguchi, *Phys. Rev. E* **62**, 793 (2000)

15. V.V. Yashin and A.C. Balasz, *Macromolecules* **39**, 2024 (2006)

16. K. Sekimoto, *J. Phys. II* (*France*) **1**, 19 (1991)

17. K. Sekimoto, *J. Phys. II* (*France*) **2**, 1755 (1992)

18. S. Villain, *Comportement mécanique de gels soumis à des réactions autocatalytiques* (in French). PhD Thesis , Univ. Paris VII (2007)

19. S. Métens, S. Villain, and P. Borckmans, this volume

20. V. V. Yashin and A. C. Balasz, *Science* **314**, 798 (2006)

21. V. V. Yashin and A. C. Balasz, *J. Chem. Phys.* **126**, 124707 (2007)

22. A.P. Dhanarajan, G.P. Misra, and R.A. Siegel, *J. Phys. Chem.* **A106**, 8835 (2002)

23. G.P. Misra and R.A. Siegel, *J. Controll. Release* **81**, 1 (2002)

24. X. Zou and R.A. Siegel, *J. Chem. Phys.* **110**, 2267 (1999)

25. R.A. Siegel, this volume

26. J. Boissonade, *Phys. Rev. Lett.* **90**, 188302 (2003)

27. J. Boissonade, *Chaos* **15**, 023703 (2005)

28. F. Gauffre, V. Labrot, J. Boissonade, and P. De Kepper, "Spontaneous deformations in polymer gels driven by chemomechanical instabilities" in *Nonlinear Dynamics in Polymeric Systems*, ed. J.A. Pojman and Q. Tran-Cong-Miyata (ACS Symposium Series 869, ACS, Washington, DC, 2003), p. 80

29. V. Labrot, *Structures chimio-mécaniques entretenues: couplage entre une réaction à autocatalyse acide et un gel de polyélectrolyte* (in French). PhD Thesis, Univ. Bordeaux I (2004)

30. V. Labrot, P. De Kepper, J. Boissonade, I. Szalai, and F. Gauffre, *J. Phys. Chem.* **B109**, 21476 (2005)

31. J. Boissonade, P. De Kepper, F. Gauffre and I. Szalai, *Chaos* **16**, 037110 (2006)

32. M. Fuentes, M.N. Kuperman, J. Boissonade, E. Dulos, F. Gauffre and P. De Kepper, *Phys. Rev. E* **66**, 056205 (2002)

33. F. Gauffre, V. Labrot, J. Boissonade, P. De Kepper, and E. Dulos, *J. Phys. Chem.* **A107**, 4452 (2003)

34. D. Strier and J. Boissonade, *Phys. Rev. E* **70**, 016210 (2004)

35. I. Szalai, F. Gauffre, V. Labrot, J. Boissonade, and P. De Kepper, *J. Phys. Chem.* **A109**, 7849 (2005)

36. A. Toth, I. Lagzi, and D. Horvath, *J. Phys. Chem.* **100**, 14837 (1996)

37. D. Horvath and A. Toth, *J. Chem. Phys.* **108**, 1447 (1997)

38. P. Blanchedeau and J. Boissonade, *Phys. Rev. Lett.*, **81**, 5007 (1998)

39. P. Blanchedeau, J. Boissonade, and P. De Kepper, *Physica D* **147**, 283 (2000)

40. K. Benyaich, T. Erneux, S. Métens, S. Villain, and P. Borckmans, *Chaos* **16**, 037100 (2006)

41. I. Szalai and P. De Kepper, *Phys. Chem. Chem. Phys.* **8**, 1105 (2006)

42. Z. Virányi, I. Szalai, J. Boissonade, and P. De Kepper, *J. Phys. Chem.* **111A**, 8090 (2007)

43. P. Borckmans and S. Métens, this volume

44. M. Doi, *Introduction to Polymer Physics* (Clarendon Press, Oxford University Press, Oxford, 1996)

45. T. Tanaka and D.J. Fillmore, *J. Chem. Phys.* **70**, 1214 (1979)

46. A. Onuki, *Adv. Polym. Sci.* **109**, 63 (1993)

47. T. Tomari and M. Doi, *Macromolecules* **28**, 8334 (1995)

48. M.A.T. Bisschops, K.Ch.A.M. Luyben and L.A.M. van der Wielen, *Ind. Eng. Chem. Res.* **37**, 3312 (1998)

49. E.C. Achilleos, K.N. Christodoulou and I.G. Kevrekidis, *Comp. Theor. Polym. Sci.* **11**, 63 (2001)

50. B. Barrière and L. Leibler, *J. Polym. Sc.* **41**, 166 (2003)

51. J. Dolbow, E. Fried and H. Ji, *J. Mech. Phys. Solids* **52**, 51 (2004)

52. T. Yamaue and M. Doi, *Phys. Rev. E* **69**, 041402 (2004)

53. T. Yamaue, T. Tanigushi and M. Doi, *Mol. Phys.* **102**, 167 (2004)

54. A.G. Ogston, B.N. Preston and J.D. Wells, *Proc. R. Soc. Lond. A* **333**, 297 (1973)

55. A.Y. Grossberg and A.R. Khokhlov, *Statistical Physics of Macromolecules* (AIP Press, New York, 1994)

56. G. Strobl, *The Physics of Polymers* (Springer, Berlin, Heidelberg, New York, 1997)

57. B. Erman and J.E. Mark, *Structure and Properties of Rubberlike Networks* (Oxford University Press, Oxford 1997)

58. J. Bastide and S.J. Candau, "Structure of gels investigated by static scattering techniques" in *Physical Properties of Polymeric Gels*, ed. J.P. Cohen Addad (Wiley, Chichester, 1996), p. 143

59. P.J. Flory, *Principles of Polymer Chemisty* (Cornell University Press, Ithaca, NY, 1953)

60. E.M. Cussler, *Diffusion: Mass Transfer in Fluid Systems* (Cambridge University Press, Cambridge, 1997)

61. M.M. Tomadakis and S.V. Sotirchos, *AIChE J.* **39** 397 (1993)

62. J. Rička and T. Tanaka, *Macromolecules* **17**, 2916 (1984)

63. J. Newman and K.E. Thomas-Alyea, *Electrochemical Systems* (Wiley, New York, 2004)

64. J. Boissonade, *Eur. Phys. J. E* **28**, 337 (2009)

STRUCTURAL APPROACHES ON THE TOUGHNESS IN DOUBLE NETWORK HYDROGELS

Taiki Tominaga, Yoshihito Osada[*]
RIKEN, Advanced Science Institute, 2-1 Hirosawa, Wako, Saitama 351-0198 Japan

Jian Ping Gong
Graduate School Science, Hokkaido University, North 10 West 8, Sapporo 060-0810 Japan

Abstract. Most hydrogels are mechanically too weak to be used as any load bearing devices. We have overcome this problem by synthesizing hydrogels with a double network (DN) structure. Despite the presence of 90% water in their composition, these tough gels exhibit a fracture stress of 170 kg/cm^2, similar to that of cartilage. The relation between their mechanical strength and structure for a wide range of conditions should be analyzed to apprehend the origin of the toughness of the DN-gels. We recently reported some experimental results obtained by dynamic light scattering and small angle neutron scattering. Some new experimental results obtained by neutron scattering in both deformed and undeformed conditions provided for a new understanding of the origin of toughness. We review the studies on the structure of DN-gels towards understanding of the toughness origin. Studies on DN-gels for biomedical applications are also described.

Keywords: hydrogel, double network, dynamic light scattering, neutron scattering, wearing, biodegradation, biocompatibility

1. Introduction

Gel scientists have paid attention to synthetic hydrogels, which are made of swollen, cross-linked polymer networks and containing more than 90 vol % water. If hydrogels are functionalized with free chains on their surfaces [1], these gels exhibit low surface friction and thus serve as attractive candidates

---

[*] To whom correspondence should be addressed. e-mail: osadayoshi@riken.jp

as, for example, artificial replacements for damaged cartilage. We have reached a very low coefficient of friction of synthetic gels ($\mu \sim 10^{-3}$ to 10^{-4}), which cannot be obtained from the friction between two solid materials [2–5]. Unfortunately, however, most hydrogels made from either natural or synthetic sources suffer from lack of mechanical strength.

Recently, we have reported a novel method to overcome this problem by inducing a double network (DN) structure for various combinations of hydrophilic polymers [6]. These DN hydrogels made of poly (2-acrylamido-2-methylpropane sulfonicacid) (PAMPS) and polyacrylamide (PAAm)., containing about 90% water, exhibit fracture strength as high as a few to several tens of megapascals [6] and show high wear resistance due to their extremely low coefficient of friction.

The DN-gels are comprised of two independently cross-linked networks, and an optimal combination is found when the first network is a relatively rigid polyelectrolyte and the second one is a flexible neutral polymer. We also found that the cross-linking density of the two networks and the molar ratio of the two polymers are two crucial parameters in improving the resistance against stress. Thus, a gel with highest strength was obtained when the first network is highly cross-linked and the second is only slightly cross-linked or even without cross-linking [7–10]. The DN-gels acquire very strong mechanical properties only when the molar ratio of the second network to the first network lies in a range of several to a few decades. This is in strong contrast with the conventional interpenetrating polymer network (IPN) or semi-IPN hydrogels, which usually are equimolar in composition and therefore do not exhibit substantial improvement in mechanical strength.

In this review, recent studies of the structures of DN-gels by scattering methods and the resulting interpretation models are described. An example of biological application is also introduced.

2.   Structure of DN-gels by dynamic light scattering

The relation between the mechanical strength and the underlying structure in a wide range of conditions should thus be obtained for approaching the origin of the toughness of the DN-gels.

As the first highly cross-linked network has a high Young's modulus but is, in general, quite brittle on its own, we have assumed that the dramatically enhanced mechanical strength is due to an effective relaxation of stress through the loosely cross-linked second network which dissipates the fracture energy and prevents crack development. In this chapter, we concentrate our attention on the molecular dynamics of the second network by changing its cross-linking density while fixing the molar ratio between the two networks using the dynamic light scattering (DLS) technique. As a result of DLS, it was shown that the presence of a slow mode besides the "gel mode" (fast mode) correlates with the enhancement of the strength of DN gels at the low cross-linking density of the second network. The origin of the slow mode and its role in the dramatic increase of the strength is discussed [11,12].

Figure 1 shows the effects of cross-linking density (y_2) on the intensity-time correlation function (CF), $g^{(2)}(\tau)$, and the characteristic decay time distribution function (DF), $G(\Gamma)$ of the second network, i.e., PAAm single network (SN) gel. Around $y_2 = 0.2$ mol %, the behavior of DF changes. For $0 \leq y_2 \leq 0.2$ mol %, there are two modes, i.e., fast and slow modes, but there exists only one mode for $y_2 \geq 0.2$ mol %. The slow mode of DF in PAAm SN gel decreases and disappears by increasing y_2. Considering the time range of Γ^{-1} for the respective modes, it is reasonable to deduce that the fast mode corresponds to the so-called gel mode and the slow mode to the translational mode of the PAAm polymer. The variation of the slow mode with the increase in y_2 is thought to be related to the sol-gel transition of the PAAm gel, as studied by Shibayama et al. for poly(N-isopropylacrylamide) gels and solutions [13,14].

Figure 2 shows the effect of cross-linking density of second network (y_2) on CF and DF in the PAMPS/PAAm DN gels. As the first network PAMPS behaves as a background material in the time range of the DLS measure-ments, the DF of PAMPS/PAAm DN gels represents that of the second net-work PAAm. Similar to the DF behavior of PAAm SN gels, a slow relaxation has been observed besides the "gel mode" for $0 \leq y_2 \leq 0.25$ mol %, but the slow relaxation disappears for $y_2 \geq 0.5$ mol %. Compared with that of PAAm SN gels, the characteristic decay times of gel mode for PAMPS/PAAm DN gels ($\Gamma^{-1} \sim 10^{-2}$ ms) are much shorter than that for PAAm SN gels ($\Gamma^{-1} \sim 10^{-1}$ ms), indicating the trapping effect of highly cross-linked PAMPS (first network) to PAAm polymers in DN gels.

On the other hand, the slow mode in PAMPS/PAAm DN gels has a much wider relaxation distribution and a much larger $G(\Gamma)$ value in com-parison with that of PAAm SN gels. The area ratio of the slow mode to the "gel mode" has been calculated for DN-gels, and its relationship with the cross-linking density of PAAm (second network) is shown in Figure 3. The result shows that the cross-linking density dependence of the area ratio is very similar to that of the mechanical strength. This suggests that the increased mechanical strength of DN-gels correlates, in DN-gels, to the slow modes of $\Gamma^{-1} = 10^{-1}$ to 10^{3} ms.

The remarkable enhancement of the DN-gels cannot be explained in terms of existing theories proposed for different mechanisms enhancing the fracture energy G of the soft polymeric systems, such as Lake-Thomas' theory [15] and de Gennes' theory [16]. For example, if the fracture energy is estimated by the Lake-Thomas theory, G is around 10 J/m^2, two orders of magnitude smaller than the experimental value of tough DN-gels [2].

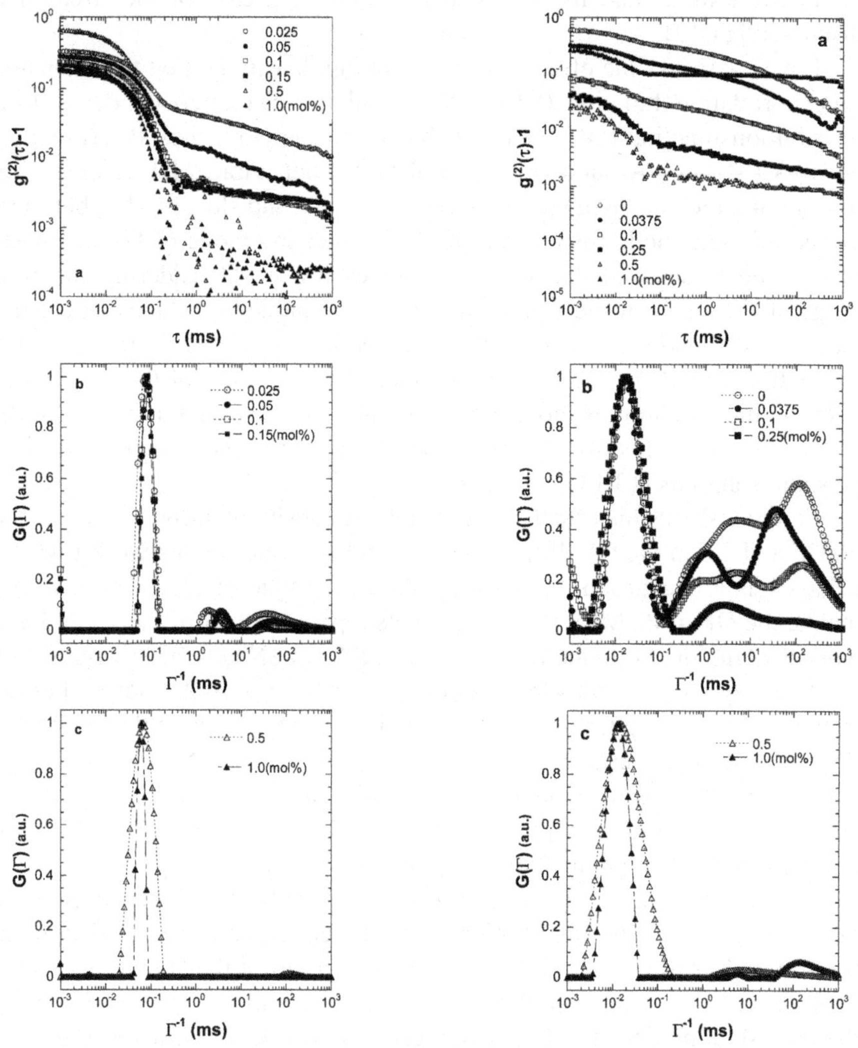

Figure 1. (left column) Effects of the cross-linking density (y_2) on (a) the CF and (b, c) the characteristic decay time distribution function (DF) in the PAAm SN gels for (b) $0 \leq y_2 \leq 0.2$ mol % and (c) $y_2 \geq 0.2$ mol %. The cross-linking densities (y_2, mol %) of PAAm SN gels are shown in the figures. (Reproduced from [11], with permission.)

Figure 2. (right column) Second network cross-linking density (y_2) dependency of (a) the CF and (b, c) the DF in the PAMPS-4-1/PAAm-2-y_2 DN-gels for (b) $0 \leq y_2 \leq 0.25$ mol % and (c) $y_2 \geq 0.5$ mol %. The second cross-linking densities (y_2, mol %) of PAMPS-4-1/PAAm-2-y_2 DN-gels are shown in the figures. (Reproduced from [11], with permission.)

Accordingly, we proposed a structural model for PAMPS/PAAm DN-gels showing a high mechanical strength. As shown in Figure 4, PAPMS networks (first component) are rigid and inhomogeneous, and large "voids" exist due to the specific radical polymerization mechanism [13,14,17].

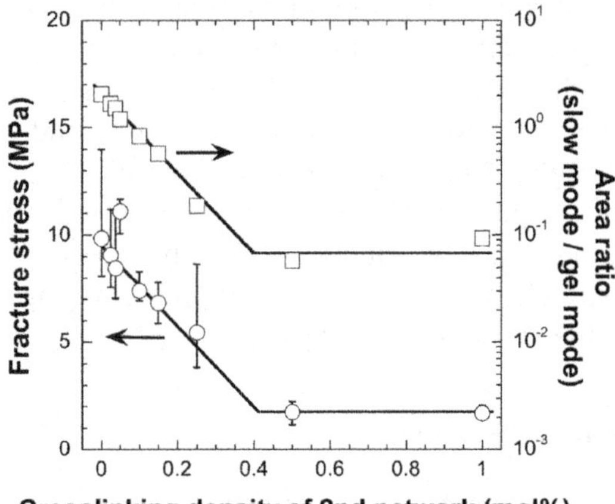

Figure 3. Effects of the cross-linking density of the second network (y_2) on fracture stress and on the area ratios of the slow mode to the "gel mode" obtained from Figure 2b, c. Sample: PAMPS-4-1/PAAm-2- y_2 DN-gels (swelling degree, Q_{swell} = 10). (Reproduced from [11], with permission.)

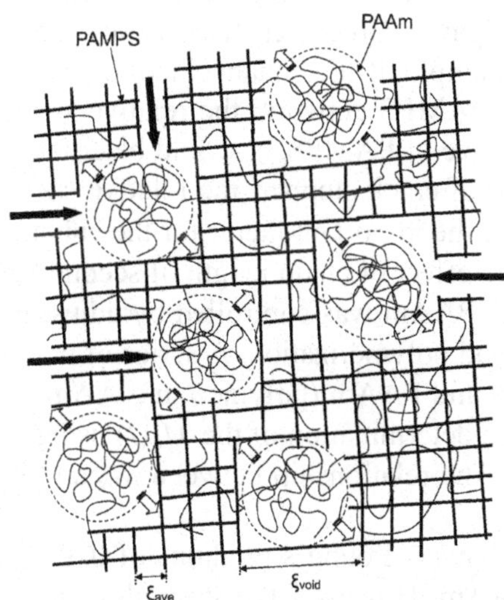

Figure 4. Structural model and mechanism to prevent crack development in PAMPS/PAAm DN-gel. For simplicity, this schematic representation especially shows the case of DN-gel in which the first PAMPS network is a rigid polyelectrolyte, and the second PAAm one is a flexible neutral polymer (cluster) without cross-linking, showing the highest fracture strength and strain. The void mesh size, ξ_{void}, is much larger that the average mesh size of the PAMPS network, ξ_{ave} ($\xi_{void} \gg \xi_{ave}$). (Reproduced from [11], with permission.)

When PAAm is polymerized in the PAMPS network, some PAAm interpenetrates this network. The remainder fills in the large "voids" of the PAMPS gel while partially entangling with the PAMPS network. The linear or loosely cross-linked PAAm in the "voids" effectively absorbs the crack energy either by viscous dissipation or by large deformation of the PAAm chains, preventing the crack growth to a macroscopic level. In other words, one possibility is that the increased mechanical strength of DN-gels results from the effective relaxation of locally applied stress and dissipation of the crack energy through diffusive fluctuation of the PAAm polymer (cluster). The other is that the part of PAAm polymer entangled within the PAMPS network can be an "anchor", and PAAm polymers with high molecular weight may be stretched very largely and ruptured during the fracture process, consuming the crack energy [15].

3. Mechanical measurements

3.1. MOLECULAR WEIGHT DEPENDENCE

We consider from Figure 4 that a key to elucidate the toughness of the DN gels is its dependence on the molecular weight M_w of the PAAm chains. Indeed M_w not only affects entanglement between PAAm and PAMPS but also the way PAAm chains are entangled with each other. These entanglements generate dissipative processes that makes the DN-gel resistant to crack growth. For example, if the enhancement of the G is due to the chain sliding of PAAm in PAMPS network, the relative size of the PAAm chain to the average mesh size of PAMPS, ξ, would be crucial; if, on the contrary, the voids play a role of crack stopper, the relative size of the PAAm chain to the voids would be the important factor for the enhancement [18].

To investigate the effect of chain length of second PAAm on the mechanical strength, M_w of PAAm was controlled by adjusting the amount of the initiator I in the second polymerization. A series of two-component polymer solutions of PAMPS and PAAm (polymerizing AAm in PAMPS solutions) were prepared for the determination of the $M_w(I)$ using gel permeation chromatography (GPC). The relationship between M_w and its fracture energy G was investigated.

All the samples have a Young's modulus of 0.3 MPa, regardless of the change in M_w of PAAm. However, the strength of the DN-gels is strongly dependent on the M_w of PAAm. A notable increase in the strength of DN-gel is observed when M_w is around a value of 10^6. On the other hand, the strength of the samples does not change when M_w is below the critical value. As mentioned above, the error on the fracture stress σ becomes larger for $M_w > 10^6$, i.e., for tough gels.

For the gels with $M_w > 10^6$, the tearing test was performed just after the second polymerization, i.e., without the swelling treatment, since the loss of PAAm by diffusion is negligible. We find that G increases sharply for M_w in the range $10^6 < M_w < 3 \; 10^6$, and above it, G saturates to some value.

According to the void model, the "void" size (ξ_{void}) is considered to be much larger than the PAAm radius (d) as well as the mesh size of the first network PAMPS (ξ), at the extreme as large as several micrometers, although ξ is around several nanometers. Therefore, the PAAm in the voids may play an important role.

Supposing that the voids are relevant for the strength of the DN-gels and that the PAAm which fill the voids behaves as a solution, we can discuss a physical meaning for the critical molecular weight $M_w = 10^6$ on the basis of the theory of polymer solution. The rheological properties of a polymer solution strongly depends on the concentration of the polymer; and the concentration can be categorized into the following three regions: (1) dilute solution region, where random coils of the polymer do not come into contact with each other; (2) semidilute region, where polymer chains overlap with each other; (3) concentrated region, where polymer chains are entangled with each other. The product of the polymer concentration c (g/cm^3) and the intrinsic viscosity $[\eta]$ (cm^3/g) gives the nondimensional parameter determining in which regime the polymer solution lies. In general, the condition $c[\eta] = 1$ is a criterion for the boundary between the diluted and semidiluted regions. For PAAm aqueous solution, it has been reported that $c[\eta] \sim 10$ would correspond to the boundary between the semidiluted and concentrated regions [19].

To characterize the region of the PAAm inside the void of the DN-gels that is effective for enhancing the mechanical strength of the gel, we calculated the equivalent concentration of PAAm in the DN-gels and plot Figure 5 in terms of $c[\eta]$. In estimation of c, the loss of PAAm due to the diffusion is considered [we ignore the change of M_w: if we take account of it, the left three data (filled circles) in Figure 5 shift right; however they should stay in the foot of the "hill"; hence, the following discussion is not modified]. The curves turned out to show dramatic transition in mechanical behavior around $c[\eta] \sim 20$, which coincides well with the critical entanglement point of PAAm solution. Therefore, the drastic increase in the mechanical strength of DN gels occurs when PAAm chains are entangled with each other. This result also helps us to understand why the ratio of AAm to AMPS of the DN-gels is another important criterion that influences the mechanical property. We found that the fracture energy saturates at a $c[\eta]$ value of about 100; a further increase in $c[\eta]$ does not enhance the strength.

We consider that the provided "void" and "entanglement" probably play two major roles at the crack tip; the first is making curvature larger to pre-

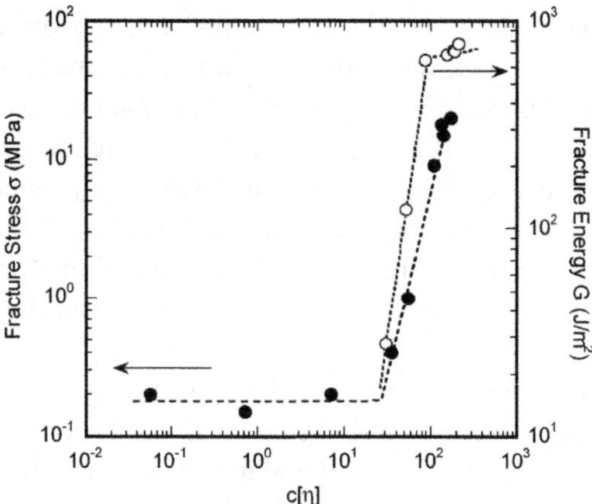

Figure 5. Strength of the DN-gels against $c[\eta]$ calculated from the data in Table 1 of [18]. Here, c is the concentration of the second PAAm in the DN-gels taking into account the PAAm lost by diffusion, $[\eta]$ is the intrinsic viscosity of PAAm (cm^3/g) calculated from the average molecular weight of PAAm (M_w) using $[\eta] = KM_w^{\alpha}$. (Reproduced from [18], with permission.)

vent stress concentration, and the second is energy dispersion or resistance of fracture energy. Thus, both effects reduce stress around the leading edge of crack propagation and prevent the crack from growing to a macroscopic level. That is, the void acts as a "crack stop". Generally, fractures in polymers are accompanied by various molecular processes, such as chain uncoiling, scission, and chain pullout at the interface of the two surfaces created by the crack propagation. We suppose the same is true inside the void, when physically entangled PAAm plays the major part of the energy-consuming process. This process becomes effective when the entanglement condition of PAAm is reached. This toughening mechanism gives us the tough hydrogel with high swelling degree.

3.2. NECKING PHENOMENA

By modification of the first network structure: either by reducing the cross-linker concentration or by adopting γ-ray radiation as cross-linking method, the necking of DN-gels can be obtained. Necking phenomena were found during elongation of the gels, i.e., constricted zones that appear in the sample and grow up with further elongation [20]. After the necking, the gels become very flexible: they sustain an elastic elongation as large as several tens of the original length. Figure 6 shows a typical loading curve of necking DN-gel at an elongation velocity of 500 mm/min. The vertical axis shows the nominal elongation stress, i.e., the elongation force divided by initial cross section of the sample, and the horizontal axis is the relative extension, i.e., the change of the sample length divided by the initial length, which is identical to the usual strain as long as the deformation of the sample is uniform. The letters a–e represent the correspondence between the pictures and the data points. At the early stage of the elongation, the stress monotonically increases with the extension; at this stage, the sample was uniformly elongated as shown in pictures a and b. Above a critical elongation c, the necked regions grow up, eating the un-necked region located in the middle part of the sample, while the elongation stress hardly increases (c–d) After the un-necked region has disappeared, the sample is uniformly stretched again (e), which corresponds to the re-increase of the stress in the loading curve. The necking propagation usually starts for an extension factor of 2–3 and finishes at a factor of 7–8. The gel remarkably softens after the necking. The Young's modulus of the softened gel is 0.015 MPa, which is smaller than 0.1 MPa, that of the prenecked gel.

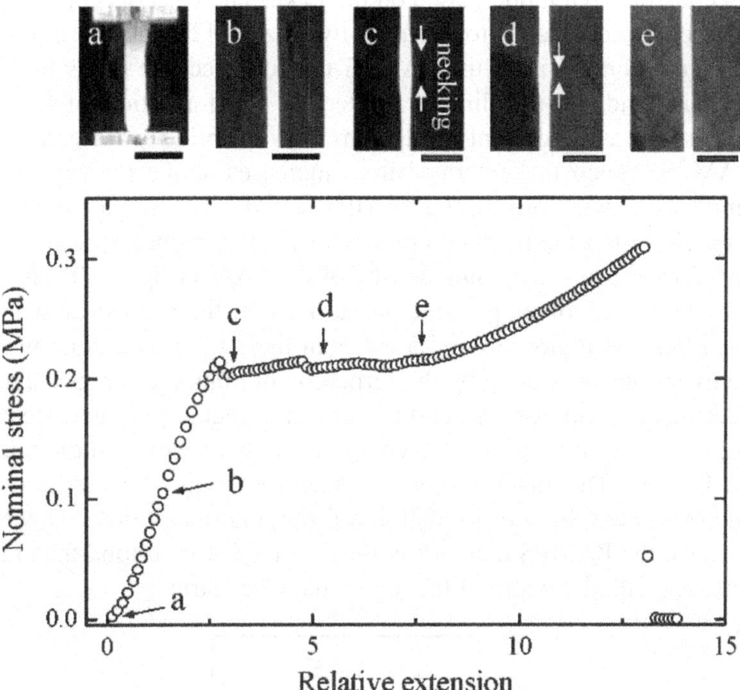

Figure 6. Loading curve of PAMPS-1-2/PAAm-2-0.02 DN-gel under uniaxial elongation at an elongation velocity of 500 mm/min and pictures demonstrating how the necking process makes progress. The inserted letters represent the correspondence between the pictures and the arrowed data points. Scale bars show 10 mm, and the width of the undeformed gel in picture a corresponds to the thickness of the sample (~4 mm). In pictures c and d, the upper and lower parts (necked regions) of the gel are slightly narrowed compared with the middle part (un-necked region). The necked regions grow up with the extension of the sample (the opposite arrows in the pictures). (Reproduced from [20], with permission.)

The finding of the necking phenomena has significance in discussions of the high toughness of the original (former) DN-gels that do not exhibit the necking *on the macroscopic scale*. It is suggested from some experimental observations and theoretical studies that the inhomogeneity of the first network plays an important role for the high toughness properties. However, the detailed mechanism has been an open question until now. The necking phenomena provides a thought direction for explaining the phenomenon: if the necking deformation occurs in a mesoscale region around the crack tip of the DN-gels, the stress concentration is remarkably reduced, resulting in the large toughness. The necking phenomenon is important for the fragmentation of the fracture process of the DN-gels. They can be regarded as a damage accumulation of the first network and suggests a new hypothesis for the anomalous toughness of the original DN-gels.

From the above experimental results, two models were produced. Brown constructed a model based on the Lake–Thomas concepts [15], crack propagation is considered to occur in two stages [21]. In the initial stage, failure occurs only within the PAMPS network because of its high cross-link density and the swelling generated by the formation of the PAAm networks. It is assumed that the PAAm network bridges the cracks within the PAMPS. When the external stress increases above the critical stress, then multiple cracks form in the PAMPS network, leaving a damaged zone of material around the primary crack with a much reduced elastic modulus controlled by the low cross-link density of the PAAm (Figure 7). The second stage of failure involves propagating a crack in the second network. It is assumed that the region around a crack in the PAAm is a zone where the first network has broken up by the formation of numerous cracks. This zone is rather like a yield zone around a crack in a ductile polymer. Because of the big reduction in modulus, the energy available is equal to the toughness of the PAAm. The macroscopic toughness is mainly a measure of the energy dissipated in the formation of the multiple cracks involved in breaking up the PAMPS network in the damaged strip around the crack and so increases with the width of the highly damaged strip.

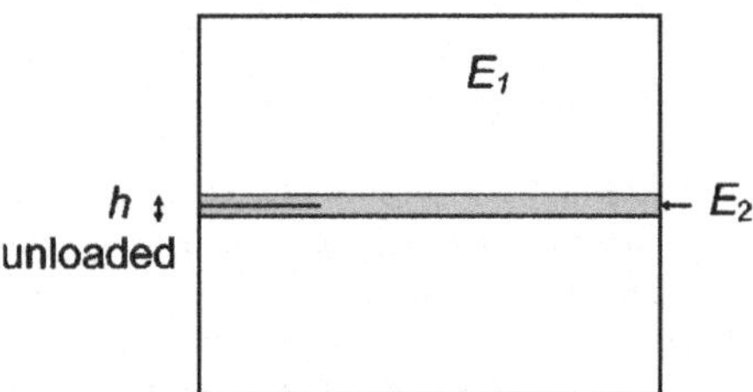

Figure 7. Geometry of the damaged zone around the crack. (Reproduced from [21], with permission.)

A local damage model assumes that the material locally softens around the crack tip due to the damage of the first network and then the crack extends within the softened zone [22]. The assumptions are as follows: (i) The yielding occurs at a critical condition characterized by a threshold stress, which corresponds to the critical tensile stress for the necking in the gels undergoing macroscopic necking; (ii) There is a sharp boundary between the damaged and undamaged zones, and the size of the damaged zone can be characterized by only one spatial scale h (h is the size in the reference (undeformed) state); (iii) The damaged zone behaves as a very soft and purely elastic material with intrinsic fracture energy (Figure 8). First order estimation indicates that energy dissipation by the softening greatly exceeds the "bare fracture energy" of the softened material, and then the effective fracture energy can reach the order of 100 J/m^2. This is consistent with the experimental values of $G \sim 400$ J/m^2.

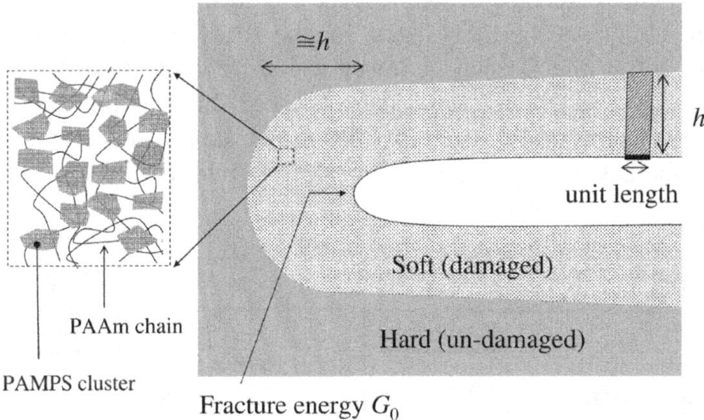

Figure 8. The structure of crack assumed in the proposed model. The DN-gel around the crack tip gets very soft due to the damage of the first PAMPS network. In the softened (damaged) zone, PAMPS clusters play a role of cross-linker of PAAm chains (*the left illustration*). (Reproduced from [22], with permission.)

4. Individual structure of DN-gels by small angle neutron scattering

To ascertain the PAMPS structure within DN-gels, small angle neutron scattering (SANS) measurements were performed at the 30 m NG3 and NG7 beamlines, and a Bonse-Hart type diffractometer for ultra-small-angle neutron scattering, beamline BT5, at the NIST Center for Neutron Research, USA. The measured reciprocal wavevector space, q corresponds to $10^{-3} \leq q$ (Å^{-1}) ≤ 0.15, and $5 \times 10^{-5} < q$ (Å^{-1}) $< 10^{-3}$ respectively. Structural information is thus probed over five orders of magnitude in $q = 2\pi/\underline{\xi}$, where $\underline{\xi}$ is the real-space correlation length, ranging from nanometers to micrometers [23,24]. Neutron scattering can determine individual network structures within DN-gels if prepared by deutrated agents.

4.1. STATIC INDIVIDUAL STRUCTURE WITHIN DN-GELS

In Figure 9a, we show the SANS data from 2 M PAAm linear chains in water and in DN-gel [25]. This figure shows that the scattering results of PAAm in water and within the DN-gel, hence in the presence of PAMPS, are quite different; particularly in the low q region there is a four orders of magnitude drop in the scattering intensity. In pure water, PAAm chains in the semi-dilute and concentrated regimes form aggregated clusters resulting in higher intensities at lower q. Since the PAAm concentration was kept constant, the decreased scattering intensity in the DN-gel shows that PAAm chains are better dispersed in the presence of PAMPS than in pure water.

Figure 9b shows the SANS data for the PAMPS network structure in water and within the DN-gel. A lower scattering intensity in low q region for PAMPS in the DN-gel was again observed. In addition, the scattering peak located near 0.04 Å^{-1} from PAMPS is also suppressed with addition of PAAm in the DN-gel. The scattering peak from pure PAMPS gel represents

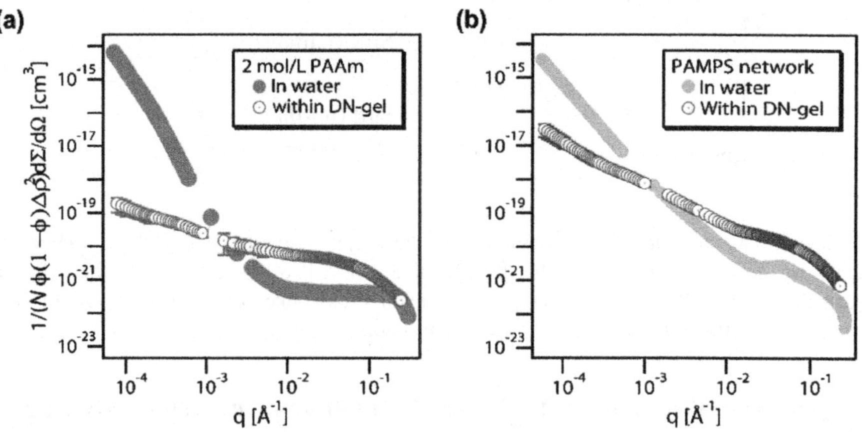

Figure 9. SANS from (a) PAAm linear chains and (b) PAMPS network within the DN-gels prepared at 2 M PAAm concentration and in pure water. The neutron scattering intensity in absolute units is normalized with respect to the contrast factor and volume fraction of scatterers. Uncertainty in measured intensity is smaller than the size of markers used. (Reproduced from [25], with permission.)

inter-chain correlations due to the repulsive interactions between the sulfonic acid groups; the suppression in the polyelectrolyte peak indicates the association between PAMPS and PAAm. Macromolecular complexation between oppositely charged weak polyelectrolytes in aqueous media via hydrogen bonding between proton-rich (e.g., carboxylic acid; –COOH) and electron-rich (e.g., amide; –$CONH_2$) groups is well established [26]. Similar association between the strongly charged sulfonic acid group with weakly charged amide group (–SO_3H and –$CONH_2$) in DN-gels might be therefore favorable.

The further fitting analysis of SANS results from PAMPS/PAAm DN-gels and their solution mixture analogs has shown that enthalpic association as Flory-Huggins parameters, between components 1 (polyelectrolyte) and 2 (neutral polymer), is predominant in DN-gels, $\chi_{12} \ll \chi_{1s} < \chi_{2s}$ ($0.05 \ll 0.32 < 0.45$) [27]. The interactions with the solvent (water) involves index s. Figure 10 shows the effect of $\chi_{12} \equiv \chi_{PE\text{-}NP}$ on the modeled scattering data from 1:7 (PAMPS:d_3PAAm) solution blend. The best fit values for $\chi_{PE\text{-}NP}$ are clearly within -0.1 and 0.1, a range that is well below the polymer-solvent interaction parameters for PAMPS and PAAm with water. It is also to our surprise, how well our ternary model explains the experimental data in both the single component PAMPS solutions and the solution blends of PAMPS and PAAm, especially at very low polyelectrolyte concentrations. The presence of the neutral polymer seems to improve the fit between the SANS data and the theory by comparing the results [27].

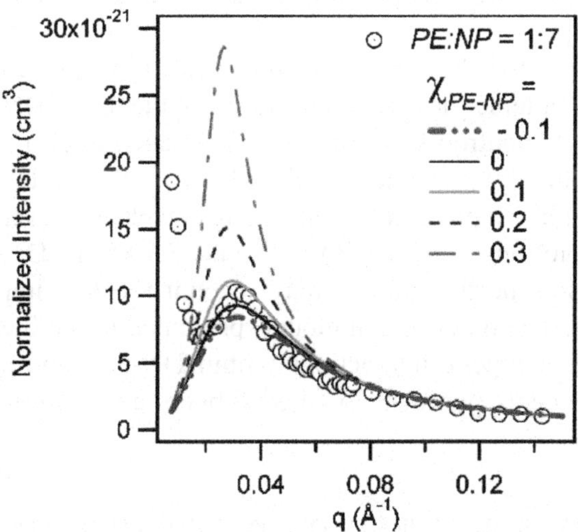

Figure 10. Influence of interaction parameter, $\chi_{PE\text{-}NP}$, on the polyelectrolyte structure factor S_{PE} in solution blends. Markers are experimental data presented for the PAMPS: d_3PAAm solution blend of 1:7 volume ratio and lines are estimates obtained by using the corresponding best fit parameters for different $\chi_{PE\text{-}NP}$ values. (Reproduced from [27], with permission.)

The association between PAMPS and PAAm may be driven probably by weak electrostatic interactions between sulfonic acid and carbonyl groups. Such weak interactions are perhaps necessary to minimize the formation of irreversible complexes that are obtained typically from mixtures of oppositely charged polymers [28]. Although this study is specific to the PAMPS/PAAm system, the results are consistent with the behavior of other charged polymer/neutral polymer blends in which PAAm is a constituent. For example, polymerizing acrylamide in the presence of anionic components, such as sodium dodecyl sulfate or poly(sodium p-phenylene sulfonate) results in a relatively homogenized network morphology as compared to that prepared in pure water [29,30]. Similarly, Durmaz and Okay reported that the swelling degree of copolymerized poly(AMPS-co-AAm) hydrogels unexpectedly attains a plateau region in the composition range 10 mol % PAMPS [31–33]. The observed results were attributed to the existence of counterions that do not contribute to Donnan osmotic pressure. The association between PAAm and the anionic polyelectrolyte in water may provide a physical explanation for the existence of "osmotically hidden" counterions [34]. Most of the polymer pairs used in reference 6 for the preparation of DN-gels may similarly be associated via hydrogen bonding between the carboxylic acid or sulfonic acid group of the polyelectrolyte and the amide group on the "neutral" polymer. The association between the pair of poly(acrylic acid)/poly(ethylene glycol) used recently by Frank and co-workers [35] in the preparation of artificial cornea was also well-established previously by Tanaka and co-workers [36].

We establish, for the first time, the weak (and possibly reversible) thermodynamic interactions between the constituents of DN-gels in water using small-angle neutron scattering measurements. Such interactions may account for the highly "entangled state" of PAMPS/PAAm DN-gels reported recently [12]. Unlike topological chain entanglements, however, enthalpically favorable interactions between PAMPS and PAAm in DN-gels offer an energy dissipation mechanism that may account for their improved mechanical properties. The deformation models presented in the literature thus far do not consider energetic interactions within DN-gels and instead rely on mechanistic concepts pertinent to filled rubbers and semi-interpenetrating networks.

4.2. INDIVIDUAL STRUCTURE UNDER DEFORMATION WITHIN DN-GELS

Next, SANS measurements of the structure of each component and the total network of the DN-gel under a pure shear deformation were conducted. A sample cell with quartz windows (Figure 11a) was designed to deform the DN-gel and to keep it in the deformed state during SANS measurements

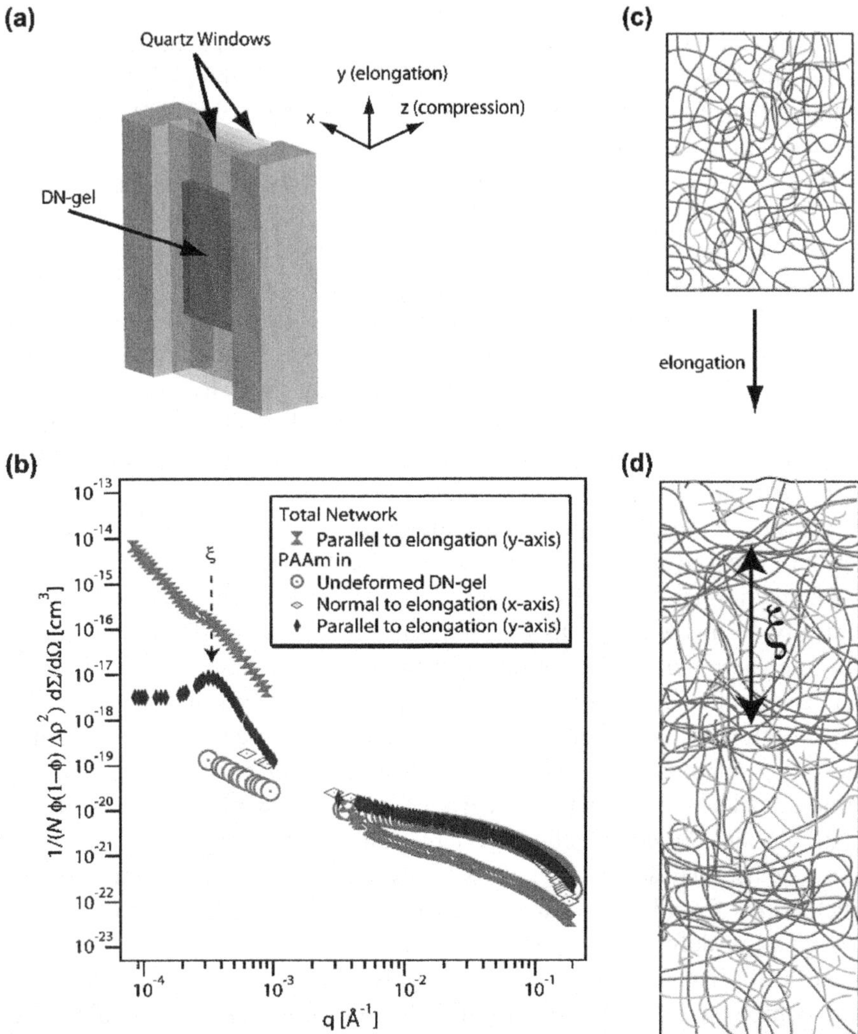

Figure 11. (a) Schematic of the sample cell used for *in situ* SANS measurements of DN-gels under pure shear deformation. (b) SANS from DN-gels measured in undeformed state and in pure shear deformation mode. Scattering data parallel and normal to the tensile deformation plane at ultra-low *q* are collected using a Bonse-Hart diffractometer. Uncertainty in measured neutron scattering intensity is smaller than the size of the markers used. Schematics represent the (c) static structure of DN-gel and (d) the change it undergoes when subjected to a 50% extension using the sample cell shown in (a). PAMPS and PAAm are represented in the schematics by line colors blue and red, respectively. The correlation length, ξ, corresponding to the peak position in SANS from deformed DN-gel is from the concentration fluctuations in the component structure of PAAm at ≈1.5 μm. (Reproduced from [25], with permission.)

[25,37]. The specimen was elongated along the y-axis by compressing the quartz windows along the z-axis while maintaining constant the width (x-axis). This design minimizes the potential for crack or micro-void formation, features that would also scatter neutrons, because there is no tensile stress exerted on the specimen.

For these measurements, the imposed strain was 50% in extension along y-axis. Figure 11b shows the SANS data taken *in-situ* on the deformed DN-gel at two contrast conditions that reveal the PAAm structure and the total network structure. Neutron scattering in the small-angle region ($10^{-3}\,\text{Å}^{-1} < q < 1\,\text{Å}^{-1}$) did not exhibit any anisotropic "butterfly" patterns that are commonly found in deformed polymer networks. In the ultra-small-angle regime ($5\;10^{-5}\,\text{Å}^{-1} < q < 10^{-3}\,\text{Å}^{-1}$), the scattering intensity is anisotropic as seen from the difference in the scattering curves between the data averaged parallel and perpendicular to the elongation direction. The most striking feature in the data is a pronounced scattering maximum at a q value corresponding to ~1.5 µm periodicity along the elongation axis only when it can be observed under deformation. This feature is grossly different from the conventional "butterfly" patterns where the scattering intensity monotonically decays with q. This feature is also present in the deformed structure of samples prepared with contrast conditions to reveal the PAMPS structure. Scattering intensity from the total network is also anisotropic in the low angle region but exhibited only a barely discernible scattering maximum in the direction parallel to the elongation axis (Figure 11b). The lack of a pronounced scattering maximum from the total network under deformation indicates that the compositional fluctuations of the components are mostly out-of-phase. These results clearly show the occurrence of compositional fluctuations from both linear PAAm chains and the PAMPS network when DN-gels are deformed (Figure 11d). Although the scattering approach has been performed on other tough hydrogels, such as slid ring (SR) gel and nanocomposite (NC) gel [38–40], this hidden macro-structure could not be observed except for DN-gels.

Revisiting the fracture toughness data of other DN-gels included in reference 2, it is interesting to note that the increase in fracture toughness of DN-gels prepared from two polymers with a favorable intermolecular association (PAMPS/PAAm, PAA/PAAm, PAMPS/TFEA, etc.) is always larger than those without favorable interactions (PAMPS/PAMPS, PAA/PAA or PAAm/PAAm). This observation suggests that the attractive enthalpic interactions between two components may play a role in the toughening of DN-gels. In the schematics, the deformation process depicted in Figure 11c–d calls for a large scale dissociation between PAMPS and PAAm, and the amount of energy dissipation during deformation can be related to the strength of inter-species association.

These SANS results are reproducible; however many unexpected results were obtained. For example, cross-linked PAMPS network structure becomes homogeneous within DN-gels, and the DN-gels under deformation have an out-of-phase structure. In addition, SANS results did not show void structure on PAMPS network within DN-gels. These experimental results are not consistent to the "void model". Work is ongoing to clarify these unexpected behaviors.

Based on SANS results, Wu proposed a stress transfer model. A mechanism for stress transfer from the PAAm chains, bridging the existing cracks, to PAMPS network next to the crack is considered [37]. The model is built upon the mechanism behind single fiber fragmentation (SFF) measurement used in fiber-reinforced composites to evaluate the interfacial strength between fiber and matrix [41,42]. In fiber-reinforced composites, the fiber typically has a higher modulus and a lower fracture strain relative to that of the matrix. The test sample in SFF measurements consists of one fiber with a sufficient aspect ratio embedded in a matrix and the sample cross section is greater than that of the fiber by several orders of magnitude. The tensile load is applied on the matrix and is transmitted to the fiber via fiber/matrix interface characterized by the magnitude of interfacial shear strength. When sufficient tensile force is applied, multiple cracks occur along the fiber till the length between cracks reaches a critical length below which load cannot be adequately transmitted from the matrix to the fiber via the interface. This fiber fragmentation length is related to the mechanical properties of the constituents.

For the case of DN-gels, the external load is exerted on the PAAm chains within a crack and is transmitted to the brittle PAMPS network next to the crack to induce additional cracks to occur (Figure 12). In this instance, PAAm chains interacting with PAMPS play the role of the matrix and the PAMPS network acts as the fiber. The model provides an explanation for the observed periodic compositional fluctuations (Figure 11b) in the micrometer range induced by large strain deformation.

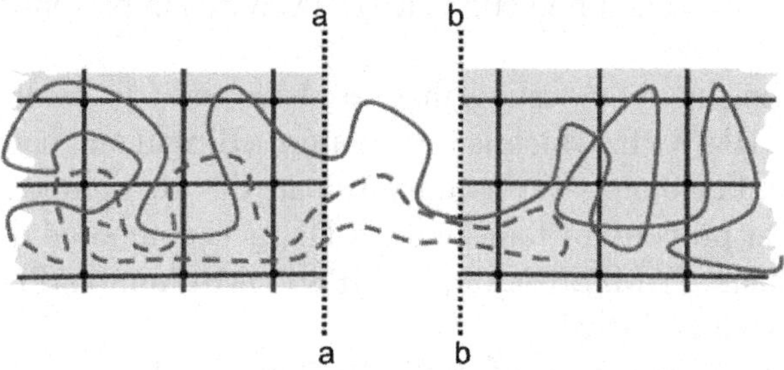

Figure 12. Schematic of a crack inside a DN-gel. The fracture surfaces a–a and b–b of PAMPS network are bridged by PAAm chains depicted as red lines. The *solid red line* denotes a tie chain along which the local stress can be transferred effectively, whereas the *dashed line* denotes a chain that is shallowly buried on one side of a crack and hence is not an effective stress-bearing chain across the crack. (Reproduced from [37], with permission.)

5. Biomedical applications

Normal cartilage tissues are a kind of multi-functional hydrogel with high toughness. They have a friction coefficient in the range 0.001–0.03, remarkably low even for hydrodynamically lubricated journal bearings [43]. It is not yet understood why the cartilage friction of the joints is so low even in conditions where the pressure between the bone surfaces reaches values as high as 3–18 MPa. Some low friction conditions on synthetic hydrogels had already been found [1,44]. For example, the surface frictional coefficient of PAMPS gel synthesized on a polystyrene plate are as low as 10^{-4}, which is at least two orders lower in magnitude than that of the gel synthesized on glass. The frictional coefficient reached that of natural cartilage surfaces. The decreased frictional coefficient of the gel prepared on the hydrophobic substrate is attributed to the presence of brush-like dangling chains on the gel surface. This is supported by the result obtained for the gels containing free linear polymer chains prepared on a glass plate, which showed similar low friction coefficients [44]. Next we need to do is to make sure of their wearing property, biodegradation, and biocompatibility in terms of the application point of view.

5.1. WEARING PROPERTY

Recently, four types of unique DN-gels were developed as potential materials for artificial cartilage: The first gel is PAMPS/PDMAAm DN-gel, which consists of PAMPS and poly(N,N′-dimetyl acrylamide). The second gel is PAMPS/PAAm DN-gel. The third gel is Cellulose/PDMAAm DN-gel, which is composed of bacterial cellulose and poly-dimetyl-acrylamide. The fourth gel is Cellulose/Gelatin DN-gel, which consists of bacterial cellulose (BC) and gelatin. Among them, the PAMPS/PDMAAm DN-gel has, for a hydrogel, an amazing wear resistance property that is comparable to that of the ultra-higher molecular weight polyethylene (UHMWPE) in pin-on-flat-type wear testing [45].

The pin-on-flat wear testing that has been used to evaluate the wear property of UHMWPE, which is one established rigid and hard biomaterial used in the artificial joint, was used to evaluate the wear property of DN-gels [45]. The wear property of four kinds of DN-gels composed of synthetic or natural polymers, PAMPS/PAAm, PAMPS/PDMAAm, BC/PDMAAm, and BC/Gelatin were evaluated.

Under one million times of cyclic friction, which was equivalent to 50 km friction, the maximum wear depth of the PAMPS/PAAm, PAMPS/PDMAAm, BC/PDMAAm, and BC/Gelatin gel was 9.5, 3.2, 7.8, and 1302.4 µm, respectively. It is amazing that the maximum wear depth of PAMPS/PDMAAm gel is similar to the value for UHMWPE (3.33 µm). In addition, although the maximum wear depth of PAMPS/PAAm gel and BC/PDMAAm gel is about 2–3 times higher than that of UHMWPE, these gels could bear the one million times of cyclic friction. The results demonstrate that PAMPS/PAAm, PAMPS/PDMAAm and BC/PDMAAm gels are resistant to wear to a greater degree than conventional hydrogels, and PAMPS/PDMAAm gel is potentially useable as replacement material for artificial cartilage. On the other hand, BC/Gelatin gel, that is composed of natural materials, shows extremely poor wear properties compared to other DN-gels. Recent developments in the synthesis of mechanically strong hydrogel break through the conventional gel conception, and open a new era of soft and wet materials as substitutes for articular cartilage and other tissues.

5.2. BIODEGRADATION

Biodegradation properties of the above DN-gels were also evaluated [46]. Concerning the DN-gels, a total of 12 specimens were prepared, and 6 of the 12 specimens were examined to determine the mechanical properties without any treatments. The remaining 6 specimens were implanted into the subcutaneous tissue of mature rabbits. At 6 weeks after implantation, the mechanical properties and the water content of the implanted specimens were measured. In the PAMPS/PDMAAm gel the ultimate stress and the tangent modulus had significantly increased from 3.10 and 0.20 MPa, respectively, to 5.40 and 0.37 MPa, with a significant reduction of the water content after implantation (94–91%). In the PAMPS/PAAm gel and the cellulose/PDMAAm gel, the stress (11.4 and 1.9 MPa, respectively) and the modulus (0.30 and 1.70 MPa, respectively) or the water content rarely changed after implantation (90% and 85%, respectively). In the BC/Gelatin, the ultimate stress was dramatically reduced from 4.30 to 1.98 MPa with a significant increase of the water content after implantation (78–86%). This study suggested that each unique DN-gels tested in this study has its own degradation properties within the living body.

5.3. BIOCOMPATIBILITY

Evaluation must be made as to what kind of tissue reaction is induced by implantation of these gel materials in the living body and how strong the degree of the reaction is, in order to consider application of these DN-gels to clinical implants. No studies on biological tissue reaction of the implanted DN-gels had been reported as of yet.

To answer the above question, the pellet implantation test into the muscle and the subcutaneous implantation test with a massive material, are recommended to evaluate the local reaction in the muscle or in the subcutaneous tissue around the implanted material [47]. Therefore, we have conducted macroscopic observations on subcutaneous implantation test with a massive material and histological examinations carried out in the pellet implantation test into the muscle of rabbits.

The width of the inflammatory zone around the PAMPS/PAAm gel and the Cellulose/PDMAAm gel was significantly greater than that around the positive control (high-density polyethylene) at 1 week (Figure 13). At 4 and 6 weeks, the width around the PAMPS/PDMAAm gel was the same as that of the negative control (polyurethane containing 0.75% zinc diethyldithio-carbamate), while the width was the same around the Cellulose/PDMAAm gel as that of the negative control and significantly greater around the PAMPS/PAAm gel than that of the negative control. The reactive inflammation around the implanted Cellulose/Gelatin gel was obviously different from the area surrounding the other three gels. Figure 13 shows that the inflammatory zone width around the Cellulose/Gelatin gel was significantly less than the positive control at 1 week, while there was no significant difference from the negative control. At 4 and 6 weeks, conversely, the width became significantly greater than the negative control, while there was no significant difference from the positive control.

Because of its characteristics discussed above we believe that the PAMPS/PDMAAm DN gel shows possibilities to be applied as artificial cartilage in the future. However, to verify this, many other factors needed for artificial cartilage repair, such as porosity, cell nutrition, changes of water content and compressive strength, capsule formation in the cartilage tissue, and so on, should be evaluated in future studies.

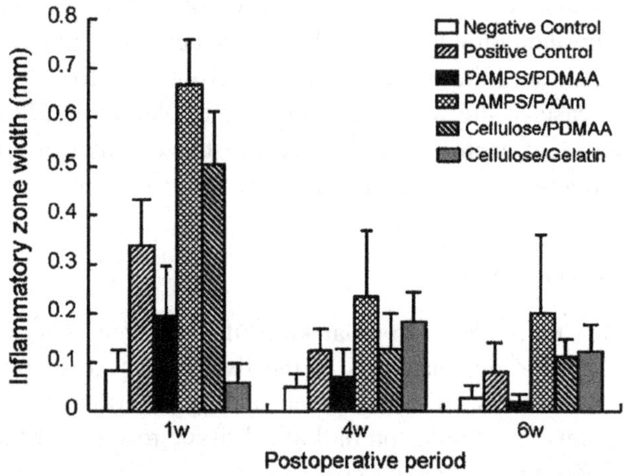

Figure 13. Comparisons of the inflammatory zone width at 1, 4, and 6 weeks. Positive control: high-density polyethylene; negative control: polyurethane containing 0.75% zinc diethyldithio-carbamate pellets of same shape. (Reproduced from [47], with permission.)

Acknowledgement

The authors thank Dr. M. Shibayama for dynamic light scattering studies, and Dr. W. Wu for small angle neutron scattering studies.

References

1. J. P. Gong, T. Kurokawa, T. Narita, G. Kagata, Y. Osada, G. Nishimura and M. Kinjo, *J. Am. Chem. Soc.* **123**, 5582 (2001).
2. J. P. Gong, M. Higa, Y. Iwasaki, Y. Katsuyama and Y. Osada, *J. Phys. Chem. B* **101**, 5487 (1997).
3. J. P. Gong and Y. Osada, *J. Chem. Phys.* **109**, 8062 (1998).
4. J. P. Gong, Y. Iwasaki, Y. Osada, K. Kurihara and Y. Hamai, *J. Phys. Chem. B* **103**, 6001 (1999).
5. J. P. Gong, Y. Iwasaki and Y. Osada, *J. Phys. Chem. B* **104**, 3423 (2000).
6. Y. Osada and J. P. Gong, *Adv. Mater.* **15**, 1155 (2003).
7. Y. Tanaka, J. P. Gong and Y. Osada, *Prog. Polym. Sci.* **30**, 1 (2005).
8. Y. M. Chen, J. P. Gong and Y. Osada, in *Macromolecular Engineering: Precise Synthesis Materials Properties, Applications* (K. Matyjaszewski, Y. Gnanou, L. Leibler, Eds., Wiley-VCH, Weinheim, 2007), pp. 2689–2717.
9. Y. Tanaka, R. Kuwabara, Y-H. Na, T. Kurokawa, J. P. Gong and Y. Osada, *J. Phys. Chem. B* **109**, 11559 (2005).
10. Y-H. Na, Y. Katsuyama, R. Kuwabara, T. Kurokawa, Y. Osada, M. Shibayama and J. P. Gong, *e-J. Surf. Sci. Nanotech.* **3**, 8 (2005).
11. Y-H. Na, T. Kurokawa, Y. Katsuyama, H. Tsukeshiba, J. P. Gong, Y. Osada, S. Okabe and M. Shibayama, *Macromolecules* **37**, 5370 (2004).
12. M. Huang, H. Furukawa, Y. Tanakana, T. Nakajima, Y. Osada and J. P. Gong, *Macromolecules* **40**, 6658 (2007).
13. T. Norisuye, M. Takeda and M. Shibayama, *Macromolecules* **31**, 5316 (1998).
14. M. Takeda, T. Norisuye and M. Shibayama, *Macromolecules* **33**, 2909 (2000).
15. G. J. Lake and A. G. Thomas, *Proc. R. Soc. London* **300**, 108 (1967).
16. P. G. de Gennes, *Langmuir* **12**, 4497 (1996).
17. J. Bastide and L. Leibler, *Macromolecules* **21**, 2407 (1988).
18. H. Tsukeshiba, M. Huang, N-H. Na, T. Kurokawa, R. Kuwabara, Y. Tanaka, H. Furukawa, Y. Osada and J. P. Gong, *J. Phys. Chem. B* **109**, 16304 (2005).
19. W. M. Kulicke, R. Kniewske and J. Klein, *Prog. Polym. Sci.* **8**, 373 (1982).
20. Y-H, Na, Y. Tanaka, Y. Kawauchi, H. Furukawa, T. Sumiyoshi, J. P. Gong and Y. Osada, *Macromolecules* **39**, 4641 (2006).
21. H. Brown, *Macromolecules* **40**, 3815 (2007).
22. Y. Tanaka, *Europhys. Lett.* **78**, 56005 (2007).
23. C. J. Glinka, J. G. Barker, B. Hammouda, S. Krueger, J. Moyer and W. J. Orts, *J. Appl. Cryst.* **31**, 430 (1998).
24. J. G. Barker, C. J. Glinka, J. J. Moyer, M-H. Kim, A. R. Drews and M. Agamalian. *J Appl. Cryst.* **38**, 1004 (2005).
25. T. Tominaga, V. R. Tirumala, E. K. Lin, J. P. Gong, H. Furukawa, Y. Osada and W-L. Wu, *Polymer* **48**, 7449 (2007).

26. S. Sukhishvili, E. Kharlempieva and V. Izumrudov, *Macromolecules* **39**, 8873 (2006).

27. T. Tominaga, V. R. Tirumala, S. Lee, E. K. Lin, J. P. Gong and W-L. Wu, *J. Phys. Chem. B* **112**, 3903 (2008).

28. M. Zeghal and L. Auvray, *Europhys. Lett.* **45**, 482 (1999).

29. Y. D. Zaroslov, V. I. Gordeliy, A. I. Kuklin, A. H. Islamov, O. E. Philippova, A. R. Khokhlov and G. Wegner, *Macromolecules* **35**, 4466 (2002).

30. O. E. Philippova, Y. D. Zaroslov, A. R. Khokhlov and G. Wegner, *Macromol. Symp.* **200**, 45 (2003).

31. S. Durmaz and O. Okay, *Polymer* **41**, 3693 (2000).

32. O. Okay and S. Durmaz, *Polymer* **43**, 1215 (2002).

33. A. Ozdogan and O. Okay, *Polym. Bull.* **54**, 435 (2005).

34. K. B. Zeldovich and A. R. Khokhlov, *Macromolecules* **32**, 3488 (1999).

35. C. W. Frank et al., International Patent Application 116137 (2006).

36. X. Yu, A. Tanaka, K. Tanaka and T. Tanaka, *J. Chem. Phys.* **97**, 7805 (1992).

37. V. R. Tirumala, T. Tominaga, S. Lee, P. D. Butler, E. K. Lin, J. P. Gong and W-L. Wu, *J. Phys. Chem. B* **112**, 8024 (2008).

38. Y. Okumura and K. Ito, *Adv. Mater.* **13**, 485 (2001).

39. K. Haraguchi and T. Takehisa, *Adv. Mater.* **14**, 1120 (2002).

40. S. Miyazaki, T. Karino, H. Endo, K. Haraguchi and M. Shibayama, *Macromolecules* **39**, 8112 (2006); Y. Shinohara, K. Kayashima, Y. Okumura, C. Zhao, K. Ito and Y. Amemiya, *Macromolecules* **39**, 7386 (2006).

41. A. Kelly and W. R. Tyson, *J. Mech. Phys. Solids* **13**, 329 (1965).

42. A. T. DiBenedeto, *Pure Appl. Chem.* **57**, 1659 (1985).

43. W. A. Hodge, R. S. Fijian, K. L. Carlson, R. G. Burgess, W. H. Harris and R. W. Mann, *Proc. Natl. Acad. Sci. USA* **83**, 2879 (1986)

44. J. P. Gong, *Soft Matter* **7**, 544 (2006).

45. K. Yasuda, J. P. Gong, Y. Katsuyama, A. Nakayama, Y. Tanabe, E. Kondo, M. Ueno and Y. Osada, *Biomaterials* **26**, 4468 (2005).

46. C. Azuma, K. Yasuda, Y. Tanabe, H. Taniguro, F. Kanaya, A. Nakayama, Y. M. Chen, J. P. Gong and Y. Osada, *J. Biomed. Mat. Res A* **81**, 373 (2006).

47. Y. Tanabe, K. Yasuda, C. Azuma, H. Taniguro, S. Onodera, A. Suzuki, Y. M. Chen, J. P. Gong and Y. Osada, *J. Mat. Sci. Mat. Med.* **19**, 1379 (2008).

CHEMOMECHANICAL DYNAMICS OF RESPONSIVE GELS

Stéphane Métens (`stephane.metens@univ-paris-diderot.fr`),
Sébastien Villain
Matière et Systèmes Complexes UMR 7057 CNRS,
Université Paris 7-Paris Diderot.
10, rue Alice Domon et Léonie Duquet
75205 Paris cedex 13, France.

Pierre Borckmans
Unité de Chimie-Physique nonlinéaire,
Université Libre de Bruxelles.
CP 231, Bd du Triomphe,
B1050 Bruxelles, Belgique.

Abstract. In this contribution we present a formalism to describe the spatio-temporalevolution of a responsive gel submitted to some autocatalytic chemical reaction. This theory is based on an hydrodynamical multi-diffusional approach of a gel, which is plunged in a chemically active mixture. Emergent volume self-oscillation dynamics of the gel result from the nonlinear coupling of the elastic deformation, the chemical kinetics and the transport phenomenon, that take place in the system. We apply our formalism to the case of the Belouzov-Zhabotinsky oscillatory chemical reaction, for which Yoshida et al. (see in this volume) have obtained many experimental results. In particular we discuss some possible coupling between the gel and the chemical reaction.

Keywords: Chemomechanics, responsive gels, autocatalytic chemical reaction, nonlinear elastic deformation, diffusion, multi-component mixture, swelling, chemical oscillations.

1. Introduction

The control and formulation of soft responsive or "smart" materials is a fast growing field of material science, specially in the area of polymer networks, due to their broad field of potential applications in bio-science, chemical sensors, intelligent microfluidic devices and drugs delivery systems [1–6]. Most of the systems are composed of gels, that consist in a cross-linked three-dimensional polymer network embedded in a fluid phase. This phase may be a pure fluid or a mixture of chemical species. We call the solvent the majority component. Our main concern here will be the case when the solute chemical species may react among themselves or with chemical groups attached to the

P. Borckmans et al. (eds.), *Chemomechanical Instabilities in Responsive Materials,*
© Springer Science + Business Media B.V. 2009

monomers that make up the backbone of the gel matrix. These soft materials present rigidity properties that are characteristic of both the solid and liquid states.They are well known to exhibit volume variations, that may be quite large and possibly exhibit a true phase transition [7], in response to a variety of stimuli such as temperature or pH modifications, the imposing of electric field, the shining of light, ... [8]. The theoretical framework to describe such complex systems finds its foundations in the interplay of polymer science, elasticity of large deformation and nonequilibrium physics. Many theoretical approaches, based on multi-phasic mixture theory and macroscopic trans-port equations have been developed [9–11]. Alternative modeling may also prove very useful [12–14]. In particular, the formalism of Sekimoto, based on hydrodynamics of systems with broken symmetry, describes the gel as an interpenetration of a fluid and an elastic medium interpreted as a two-fluid system [15–18].

Many applications of gel designed systems are based on some repetitive tasks, on-off switching of the stimuli has to be applied. More recently, a novel class of dynamical response of active gels has been achieved through the cou-pling of the gel with a chemical reaction. In these systems chemical energy is converted into mechanical work that essentially gives rise to volume de-formation of the system. We are, in this contribution, especially interested in the volume response to chemical stimuli resulting from a modification of the solvent composition because of the chemical reactions that take place in its midst. In this way some form of autonomous property is endowed to develop self-oscillating gels. This was realized by allowing a chemical reaction, the concentrations of which vary periodically in time, to take place inside a re-sponsive gel [19–29]. In these remarkable experiments, that could eventually lead to novel biomimetic intelligent materials exhibiting rhythmical action, the volume changes are nevertheless slaved to the chemical oscillations. Other experiments are in the making to take profit from the two decades research on gel unsensitive (inert) to their chemical environment. There the gel lay at the core of open reactors, to study the asymptotic properties of autocatalytic reaction-diffusion patterns. Their main role was to shield these phenomena from any disturbance of hydrodynamic flow nature. Foremost in these works lies the discovery of Turing patterns and the study of their properties (see [30] and references therein).

We propose a formalism that extends the Sekimoto approach to the case of a gel plunged into a multi-component reactive mixture. Our modeling will be applied to an oscillatory autocatalytic chemical reaction in relation with the above mentioned experiments. In the first section we summarize the broad lines of the Sekimoto approach. We begin with some basic ingredients of the formalism (definitions and application range of the considered system), followed by some thermodynamical considerations. The hydrodynamical component of the theory is then presented with an explicit expression for the entropy production from which we express the resulting elastodynamics evolution equations. In the next section we extend the formalism to a multi-component reactive mixture. We only present the problematic of the reaction-diffusion systems very briefly, as this is considered in two other chapters. Then follows a discussion on the coupling between chemistry and mechanics. We finally put all the pieces of the jigsaw together and obtain the chemoelastodynamical evolution equations of the system. In the last section we solve these equations for the case of the Belousov-Zhabotinsky oscillatory chemical reaction in view to compare our numerical simulations with experiments by Yoshida et al.

2. Hydrodynamics of responsive gels in a multi-component non reactive mixture

2.1. INGREDIENTS OF THE FORMALISM

The system: definitions, assumptions and framework   As we already mentioned, we consider a system composed of a chemical gel plunged in a multi-component fluid reactive mixture. Because they do not play an important role in the experiments we wish to describe, we will neglect the effect of electrical charges usually present in the gels. The case of ionic gels is, for instance, treated explicitly in the contribution of Boissonade in this volume [31] or in other approaches [11, 32]). The system is also assumed to be isothermal.

The basis of the formalism is the combination of macroscopic transport balance equations for mixtures, elasticity theory and the local equilibrium assumption. The latter allows to make use of all the thermodynamical relations[1]. Therefore, the natural framework, proposed by Sekimoto and extended here to a multi-component system, is a theory of the hydrodynamics of systems with spontaneously broken symmetry [16, 17]. The free energy is composed by a mixing and an elastic contribution. As an often used first approximation, the mixing contribution is given by the Flory-Huggins theory[2] [33], and a Gaussian chain approximation is applied to model the elastic contribution [36]. To close these balance equations, one resorts, as usual, to Onsager's linear approximation to express the thermodynamic forces-fluxes relations.

---

[1]Elasticity is of application even for very large deformation of gels, since plastic flow does not take place in the experiments considered here.

[2]Even if most of the works devoted to the thermodynamics of gels use this formalism, it is certainly not completely adapted to describe polymer solutions. Many extensions have been proposed in the literature [8, 34, 35].

2.2. THERMODYNAMICS OF GELS

As already noted above, the local equilibrium approximation plays a crucial role in the derivation of closed hydrodynamic equations. Therefore, we start from the differential of the free energy of the polymeric matrix system, with solvent and solute species. We adopt Einstein's repeated indices convention (except for m and s).

$$d\widetilde{F} = \mu_p^m dM_p + \mu_s^m dM_s + \sum_{\alpha=1}^{N} \mu_\alpha^m dM_\alpha + VT_v^\lambda \left(F^{-1}\right)_\lambda^p dF_p^v \tag{1}$$

where μ_s^m is the chemical potential per unit mass of solvent, μ_α^m, that by unit mass of species α, and μ_p^m is the "chemical potential" of the monomers. M_p, M_s and M_α are thermodynamical variables that respectively represent the mass of the monomers, the solvent and the species α in the system. F_p^v is the deformation gradient tensor and T_v^λ the Cauchy stress tensor of the polymeric network, V is the volume of the gel. The tensor F_p^v is a thermodynamical variable defined by [37]

$$F_p^v = \frac{\partial x^v}{\partial X^p}$$

where $\mathbf{X}$ are the coordinates of a particle of gel in a relaxed reference state, and $\mathbf{x}$ the coordinates in the deformed state. The deformation gradient tensor $\mathbf{F}(\mathbf{X}, t)$ characterizes the motion behavior in the vicinity of a point as represented on Figure 1.

The relation between the volume of a deformed gel and its value in the reference state is given by

$$V = V^0 J = \frac{M_p J}{\rho_p^0}$$

Where $J = det(F_p^v)$ and $\rho_p^0 = M_P/V^0$ is the density of the monomers chains in the reference state. We assume in this work that the monomers, the solvent and the solute species are incompressible. This means that the specific volume of each constituent does not vary in the presence of the others. More precisely, this expresses that the total volume of the system is the sum of each individual volume of its constituents

$$V = M_p \overline{V_p} + M_s \overline{V_s} + \sum_{\alpha=1}^{N} M_\alpha \overline{V_\alpha} \tag{2}$$

Where $\overline{V_\alpha} = V_\alpha/M_\alpha$ is the specific volume of the α species. $\overline{V_p} = V/M_p$ and $\overline{V_s} = V_s/M_s$ correspond respectively to the specific volume of the polymer

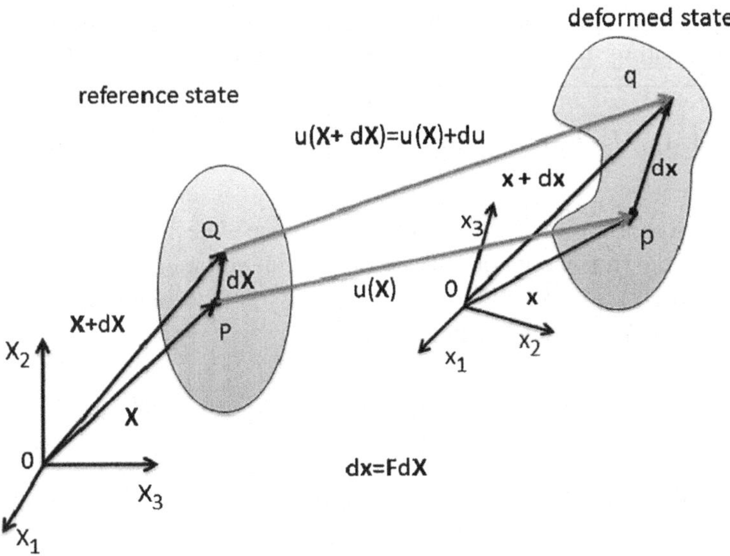

Figure 1.   Deformation gradient tensor **F** which expresses the relation between the infinitesimal length $d\mathbf{x}$ and $d\mathbf{X}$ respectively in the reference and deformed state.

and the solvent. In terms of the densities or volume fractions the incompressibility is expressed as

$$\sum_{\alpha=1}^{N} \rho_\alpha \overline{V_\alpha} + \rho_p \overline{V_p} + \rho_s \overline{V_s} = 1$$

$$\sum_{\alpha=1}^{N} \Phi_\alpha + \Phi_s + \Phi_p = 1 \tag{3}$$

Where $\rho_\alpha, \rho_s, \rho_p$ and $\Phi_\alpha, \Phi_s, \Phi_p$ are the density and the volume fractions respectively of the solute species, the solvent and the monomers. Various ways exist to take this constraint into account[16]. Putting $\mu_s^m = 0$, we have to substitute μ_α^m and μ_p^m by the corresponding exchange chemical potential $\widehat{\mu_\alpha^m}$ and $\widehat{\mu_p^m}$ [35]. In place of the Cauchy stress tensor, we then have to use the osmotic stress tensor [3], that will be denoted by Π_α^β. Equivalently a Lagrange multiplier, p, can take care of this constraint. It induces a "constraint" force on the gel

---

[3] As a gel consists of a network of polymer chains, its surface may be assimilated to a semi-permeable membrane, i.e. solvent (and other solutes) may be exchanged with the surroundings, whereas the polymer chains are constrained to stay inside the system. Whenever a "dry" gel is plunged in a solvent, an osmotic phenomenon takes place. The solvent tends to invade the gel that responds by developing an osmotic pressure, noted by Π_{mix}. It results from the difference of chemical potentials values inside and outside of the gel. Nevertheless, the network is trying to spread out as far as possible by absorbing appropriate quantities of

given by $-\nabla p$. This leads to an additional term $-p\delta_v^\lambda$ into the momentum balance equation. In consequence, relation (1) has to be replaced by

$$d\overline{F}\left(M_p; M_\alpha; F_p^v\right) = \widehat{\mu_p^m}dM_p + \sum_{\alpha=1}^N \widehat{\mu_\alpha^m}dM_\alpha + V\,(\Pi)_v^\lambda \left(F^{-1}\right)_\lambda^P dF_p^v \qquad (4)$$

Dividing expression (4) by the volume reference state V^0, we obtain the differential of the free energy density by unit initial volume. Taking into account that $1/V^0 d(M_p) \sim 1/\overline{V_p}d(\Phi_p^0) = 0$, where Φ_p^0 is the volume fraction of the monomers in the reference state, we have

$$d\widetilde{f}\left(\Phi_\alpha; F_p^v\right) = \sum_{\alpha=1}^N \frac{\widehat{\mu_\alpha^m}}{V_\alpha}d\,(J\Phi_\alpha) + J\,(\Pi)_v^\lambda \left(F^{-1}\right)_\lambda^P dF_p^v \qquad (5)$$

It turns out to be easier to write relation (5) in terms of the Piola-Kirchoff tensor $(\Pi^R)_p^v$ related to the osmotic stress tensor by $(\Pi^R)_p^v = J(\Pi)_v^\lambda(F^{-1})_\lambda^P \cdot (\Pi^R)_v^\lambda$ [34, 37]

$$d\widetilde{f}\left(\Phi_\alpha; F_p^v\right) = \sum_{\alpha=1}^N \frac{\widehat{\mu_\alpha^m}}{V_\alpha}d\,(J\Phi_\alpha) + (\Pi^R)_p^v dF_p^v \qquad (6)$$

The state equations of the system, i.e. the osmotic stress tensor $(\Pi)_v^\lambda$ and the solutes exchange chemical potentials $\widehat{\mu_\alpha^m}$ are directly obtained from relations (5) and (6)

$$(\Pi)_v^\lambda = \frac{F_p^\lambda}{J}\frac{\partial \widetilde{f}}{\partial F_p^v}|_{T;J\Phi_\alpha} \qquad (7)$$

$$\frac{\widehat{\mu_\alpha^m}}{V_\alpha} = \frac{\partial \widetilde{f}}{\partial(J\Phi_\alpha)}|_{T;F_p^v;J\Phi_{\beta\neq\alpha}} \qquad (8)$$

Free energy of gels   The free energy density may be expressed as the sum of an elastic and a mixing contribution; $\widetilde{f} = \widetilde{f}_{mix} + \widetilde{f}_{elastic}$ [8]. The *mixing free energy* is obtained from the Flory-Huggins theory [33],

$$\widetilde{f}_{mix} = \frac{Jk_BT}{v_1}\left(\sum_{i=1}^{N+s}\chi_{ip}\Phi_i\Phi_p + \frac{1}{2}\sum_{i\neq j}^{N+s}\chi_{ij}\Phi_i\Phi_j + \sum_{k=1}^{N+s}\frac{\Phi_k}{N_k}\ln\Phi_k\right) \qquad (9)$$

---

solvent. This dilation entails deformation of the network chains, which calls in the chain elasticity. The elastic reaction is expressed as $\Pi_{elastic}$ and acts against swelling. In the case of polyelectrolyte gels, not considered here, the swelling pressure includes a new contribution reflecting the osmotic pressure of the counter-ions in the gel.

where v_1 denotes the lattice site volume, i.e. the volume of a monomer or of a solute molecule; k_B is the Boltzmann constant, T the temperature, χ_{ip} and χ_{ij} are the Flory-Huggins energy parameters respectively between the polymer and the i molecules, and between the molecules i and j; N_k are the numbers of lattice sites occupied by molecules k and there are s solvent molecules on the lattice.

From relation (7) and (9) and using the identity $(\partial J/\partial F_p^v)|_{T;\Phi_\alpha} = J(F^{-1})_v^p$, we obtain the mixing free energy contribution to the osmotic stress tensor

$$
\begin{aligned}
(\Pi_{mix})_v^\lambda &= \delta_v^\lambda \frac{k_B T}{v_1} \Bigg(\frac{\ln \Phi_s}{N_s} + \frac{\Phi_p}{N_s} + \chi_{sp}\Phi_p^2 + \sum_{k=1}^{N} \Phi_k \Big(\frac{1}{N_s} - \frac{1}{N_k} \Big) + \sum_{k=1}^{N} \chi_{sp}\Phi_k\Phi_p \\
&\quad + \sum_{i=1}^{N} \chi_{is}\Phi_i\Phi_p + \sum_{k=1}^{N}\sum_{i=1}^{N} \chi_{is}\Phi_i\Phi_k - \sum_{i=1}^{N} \chi_{kp}\Phi_k\Phi_p \Bigg)
\end{aligned}
\tag{10}
$$

where δ_v^λ is the Kronecker symbol, while N_s and N_k are the numbers of lattice sites occupied respectively by solvent and solute molecules. The free species exchange chemical potential is expressed from (8) and (9) as

$$
\begin{aligned}
\frac{\widehat{\mu_{\alpha,int}^m}}{V_\alpha} &= \frac{k_B T}{v_1} \Big(\chi_{\alpha p}\Phi_p + \sum_{j\neq\alpha}^{N} \chi_{\alpha j}\Phi_j + \frac{1}{N_\alpha}\ln \Phi_\alpha + \frac{1}{N_\alpha} \Big) \\
&\quad - \frac{k_B T}{v_1} \Big(\chi_{sp}\Phi_p + \sum_{j=1}^{N} \chi_{sj}\Phi_j - \chi_{s\alpha}\Phi_s + \frac{1}{N_s}\ln \Phi_s + \frac{1}{N_s} \Big)
\end{aligned}
\tag{11}
$$

The second term corresponds to the *elastic free energy density* of the lattice obtained in the Gaussian chain approximation [36]

$$
\widetilde{f}_{elastic} = \frac{v_e k_B T}{2} \Big(\mathrm{Tr}\left(\mathbf{FF^T}\right) - 3 - 2\ln\left(\det \mathbf{F}\right) \Big)
\tag{12}
$$

where v_e is the number of partial chains by unit of reference volume, i.e. the number of monomers between two cross-links. The elastic part of the osmotic stress tensor is derived from (8) and (12) as

$$
(\Pi_{elastic})_v^\lambda = \frac{\Phi_p}{\Phi_p^0} v_e k_B T (F_p^\lambda F_p^v - \delta_v^\lambda)
\tag{13}
$$

The osmotic stress tensor is the sum of the mixing and elastic part, $\mathbf{\Pi} = \mathbf{\Pi}_{mix} + \mathbf{\Pi}_{elastic}$.

In the free energy density, we also have to take into account the contribution due to the fluid surrounding the gel

$$
\widetilde{f}_{mix,ext} = \frac{k_B T}{v_1} \Bigg(\frac{1}{2} \sum_{i\neq j}^{N+s} \chi_{ij}\Phi_{i,ext}\Phi_{j,ext} + \sum_{k=1}^{N+s} \frac{\Phi_{k,ext}}{N_k} \ln \Phi_{k,ext} \Bigg)
\tag{14}
$$

where $\Phi_{k,ext}$ is the external volume fraction of species k outside the gel. The solute exchange chemical potential $\widehat{\mu^m_{\alpha,ext}}/\overline{V_\alpha}$ of surrounding fluid is therefore given by

$$\frac{\widehat{\mu^m_{\alpha,ext}}}{\overline{V_\alpha}} = \frac{k_B T}{v_1}\left(\sum_{j\neq\alpha}^{N}\chi_{\alpha j}\Phi_{j,ext} + \frac{1}{N_\alpha}\ln\Phi_{\alpha,ext} + \frac{1}{N_\alpha}\right)$$

$$- \frac{k_B T}{v_1}\left(\sum_{j=1}^{N}\chi_{sj}\Phi_{j,ext} - \chi_{s\alpha}\Phi_{s,ext} + \frac{1}{N_s}\ln\Phi_{s,ext} + \frac{1}{N_s}\right) \quad (15)$$

The *boundary conditions* that have to be imposed at the surface of the gel are given by [17]

$$\mathbf{\Pi} \cdot \mathbf{n} = -\Pi_0\mathbf{n} \quad (16)$$

$$\widehat{\mu^m_{\alpha,int}} = \widehat{\mu^m_{\alpha,ext}} \quad (17)$$

where we have defined $\Pi_0 = -(\widetilde{f}_{mix,ext} - \sum_{i=1}^{N}\Phi_{\alpha,ext}\,\widehat{\mu^m_{\alpha,ext}}/\overline{V_\alpha})$. Relation (17) expresses the osmotic equilibrium condition (i.e. the external composition of the fluid that can be controlled), while relation (16) implies that there are neither external forces acting directly on the surface of the network, nor irreversible surface stress that accompanies the transfer of the solvent across the surface [38].

2.3.   HYDRODYNAMICS OF GELS

The starting point of the formalism is the conservation laws of mass, momentum and energy of the different constituents of the system. In a simple non reactive fluid, we known that relaxation to equilibrium of conserved quantities are described by five hydrodynamics modes corresponding respectively to two damped propagative and three dissipative modes. Here, we further have to take into account the deformation of the gel, which does not result from a symmetry breaking process[4]. The situation differs somewhat from the solid case, where the strain satisfies a local conservation law that expresses the broken translational symmetry [39, 40]. In gels, the permanent nature of crosslinks prevents the permeation flow of monomers in the gel under a fixed deformation, i.e. permeation of fluid inside the gel occurs only when deformation take place. As plastic flow is not taken into account in

---

[4]The chemical hydrogels are not obtained through an equilibrium phase transition, but by a chemical process. Therefore the situation is not completely comparable to the case of a liquid-solid equilibrium phase transition.

the description, we have to impose a "coherency constraint" that reduces the number of acting hydrodynamics modes in the system. Indeed, plastic flow, were it present, induces a vacancy-diffusion mode that relaxes shear stress in the equilibrium states [41]. Therefore, in that case, shear deformation must be taken as an additional thermodynamic variable. The expression of balance equations, closed within the framework of Onsager theory, will give rise to a two-fluid model that describes the gel evolution. More recently, such an approach has also been successfully developed for nematics gels [42].

To satisfy the incompressibility condition, the boundary of the gel has to be considered as a semi-permeable membrane (see [17] for more details). The mass-averaged velocity $\mathbf{v}_{int}$ of the gel is defined by

$$\rho \mathbf{v}_{int} = \rho_p \dot{\mathbf{x}} + \rho_s \mathbf{u}_s + \sum_{k=1}^{N} \rho_k \mathbf{u}_k \tag{18}$$

where ρ is the total mass density of the system, $\dot{\mathbf{x}}$ is the center of mass velocity of the monomers defined by

$$\dot{x}^{\mu} = \frac{\partial x^{\mu}(\mathbf{X}, t)}{\partial t} \tag{19}$$

$\mathbf{u}_k$ and $\mathbf{u}_s$ are respectively the center of mass velocities of the species k and the solvent.

As already mentioned, we assume local equilibrium in the system, then the local balance equations for the total mass of the system, the mass of species k, the mass of polymer forming the network, the total momentum and the total energy are given by

$$\partial_t \rho = -\partial_{\mu}(\rho v^{\mu}) \tag{20}$$

$$\partial_t \Phi_k = -\partial_{\mu}(\Phi_k u_k^{\mu}) \tag{21}$$

$$\partial \Phi_p = -\partial_{\mu}(\Phi_p \dot{x}^{\mu}) \tag{22}$$

$$\partial_t(\rho v_{int}^{\nu}) = -\partial_{\mu}\left(\rho v_{int}^{\mu} v_{int}^{\nu} - (\sigma^{tot})^{\mu\nu}\right) \tag{23}$$

$$\partial_t\left(\rho(e + \frac{v_{int}^2}{2})\right) = -\partial_{\mu}\left(\rho v_{int}^{\mu}(e + \frac{v_{int}^2}{2}) - v_{int}^{\nu}(\sigma^{tot})^{\mu\nu} + j_E^{D\mu}\right) \tag{24}$$

where e, $(\sigma^{tot})^{\mu\nu}$ and $j_E^{D\mu}$ refer respectively to the energy mass density of gel, the total stress tensor and the dissipative part of the energy current; v_{int}^{μ} is the velocity of the local center of mass. In addition to those laws, we also have a relation of geometrical nature that expresses the fact that the more we act on the gel, the more it is deformed

$$\partial_t\left(\frac{F_p^{\nu}}{J}\right) = \partial_{\mu}\left(\frac{1}{J}(F_p^{\mu}\dot{x}^{\nu}) - F_p^{\nu}\dot{x}^{\mu})\right) \tag{25}$$

The incompressibility condition can be written as

$$\partial_\nu \left(\Phi_p \dot{x}^\nu + \Phi_s u_s^\nu + \sum_{k=1}^{N} \Phi_k u_k^\nu \right) = 0 \tag{26}$$

To close the above set of balance equations, we have to apply the Onsager linear theory of irreversible process, to express the "flux-force" relations. Therefore we need to obtain the entropy production. As we consider an isothermal system, and assume negligible viscous effects, we have

$$T\sigma_s = -\partial_\mu \left(\overline{V_s^m} p \right) j_s^{D\mu} - \sum_{k=1}^{N} \partial_\mu \left(\widehat{\mu_k^m} + \overline{V_k^m} p \right) j_k^{D\mu}$$
$$- \left(\widehat{\mu_p^m} \mathbf{I} - \frac{\mathbf{\Pi}}{\rho_p} + \overline{V_p^m} p \mathbf{I} \right)_{\mu;\lambda}^{\lambda} j_p^{D\mu} \tag{27}$$

where we use the covariant derivative, defined by

$$A_{\nu;\sigma}^\mu \equiv \partial_\sigma A_\nu^\mu + \Gamma_{\lambda\sigma}^\mu A_\nu^\lambda - \Gamma_{\nu\sigma}^\lambda A_\lambda^\mu$$

with Christofel's symbol $\Gamma_{\mu\lambda}^\nu$ given by

$$\Gamma_{\mu\lambda}^\nu \equiv \frac{\partial^2 X^p}{\partial x^\mu \partial x^\lambda} \frac{\partial x^\nu}{\partial X^p}$$

In relation (27), the various j^D are the mass density fluxes of the different components of the system expressed with respect to the local center of mass; p is the Lagrange multiplier introduce to take into account the incompressibility condition. $j_s^{D\mu}$ and $j_p^{D\mu}$ are the flux of mass diffusion, with respect to local center of mass, respectively of solvent and of monomers. The corresponding Gibbs-Duhem relation

$$0 = \rho_s \partial_\mu (\overline{V_s^m} p) + \sum_{k=1}^{N} \rho_k \partial_\mu (\widehat{\mu_k^m} + \overline{V_k^m} p) + \rho_p (\widehat{\mu_p^m} \mathbf{I} - \frac{\mathbf{\Pi}}{\rho_p} + \overline{V_p^m} p \mathbf{I})_{\mu;\lambda}^\lambda$$
$$+ \partial_\lambda (\mathbf{\Pi} - p\mathbf{I})_\mu^\lambda \tag{28}$$

is used to eliminate the covariant derivative in (27). As the relevant experimental quantities are the permeation and free species diffusion coefficients, we have to introduce, as in [16], new fluxes given by:

$$j_k^\mu = \Phi_k \left(u_k^\mu - \overline{u}^\mu \right) \tag{29}$$

$$j_{perm}^\mu = \left(\sum_{k=1}^{N} \Phi_k + \Phi_s \right) (\overline{u}^\mu - \dot{x}^\mu) \tag{30}$$

$$j^\mu_{tot} = \sum_{k=1}^{N} \Phi_k u^\mu_k + \Phi_s u^\mu_s + \Phi_p \dot{x}^\mu \tag{31}$$

where j^μ_k is the flux of species k with respect to the mean velocity of the fluid, $\overline{u^\mu}$, defined as

$$\overline{u^\mu} = \left(\sum_{k=1}^{N} \Phi_k u^\mu_k + \Phi_s u^\mu_s \right) \Bigg/ \left(\sum_{k=1}^{N} \Phi_k + \Phi_s \right) \tag{32}$$

and with $\mathbf{j}_{perm}$ the permeation flux of the fluid in the gel. Also using the total flux $\mathbf{j}_{tot}$, the entropy production (27) may be rewritten as

$$T\sigma_s = -\sum_{k=1}^{N} \partial_\mu \left(\widehat{\frac{\mu^m_k}{V^m_k}} \right) j^\mu_k - \left(\sum_{l=1}^{N} \frac{\Phi_l}{1-\Phi_p} \partial_\mu \left(\widehat{\frac{\mu^m_l}{V^m_l}} \right) + \partial_\lambda (\Pi)^\lambda_\mu \right) j^\mu_{perm}$$
$$- \partial_\lambda (\Pi - p\mathbf{I})^\lambda_\mu \left(v^\mu_{int} - j^\mu_{tot} \right) \tag{33}$$

If we suppose mechanical equilibrium then $\partial_\lambda (\Pi - p\mathbf{I})^\lambda_\mu = 0$. Imposing a linear relation between fluxes and the conjugate thermodynamic forces, we have

$$j^\mu_k = -\frac{L_k}{T} \partial_\mu \left(\widehat{\frac{\mu^m_k}{V^m_k}} \right) \tag{34}$$

$$j^\mu_{perm} = -\frac{L_{perm}}{T} \left(\sum_{l=1}^{N} \frac{\Phi_l}{1-\Phi_p} \partial_\mu \left(\widehat{\frac{\mu^m_l}{V^m_l}} \right) + \partial_\lambda (\Pi)^\lambda_\mu \right) \tag{35}$$

From the resulting flux relations, we are able to write down a closed set of evolution equations for a non reactive system

$$\partial_t \Phi_k = -\partial_\eta \left(j^\eta_k + j^\eta_{perm} \left(\frac{\Phi_k}{1-\Phi_p} \right) + \Phi_k \dot{x}^\eta \right) \tag{36}$$

$$j^\mu_{tot} = j^\mu_{perm} + \dot{x}^\mu \tag{37}$$

$$0 = \partial_\mu (j^\mu_{tot}) \tag{38}$$

$$1 = \sum_{k=1}^{N} \Phi_k + \Phi_s + \Phi_p \tag{39}$$

The elastodynamics evolution equations of the system have to be solved with the appropriate boundary and initial conditions discussed previously. Here we will not present the application of those equations to the case of the relaxation kinetics of a piece of gel submitted to a stimulus that induce a swelling or shrinking process. There are many works, using various approaches, that have tackled the subject [9, 18, 43–53]

3.   Chemoelastodynamics of a gel in a reactive multi-component system

3.1.   REACTION-DIFFUSION SYSTEMS

The problematic of reaction-diffusion systems is considered in Chapters 1 [54] and 3 [55] of this volume. We briefly present here some kinetic results on the oscillatory Belousov-Zhabotinsky (B-Z) chemical reaction in solution and that has been used in most of the experiments where it is made to interact with hydrogels.

The most standard B-Z reaction is the oxidation of an organic species, as malonic acid, by the acidic bromate ion catalyzed by a metal ion that may alter its oxidation state [56, 80]. Mostly, two different catalysts are used: the ferroin-ferriin $(Fe(Phen)_3^{3+}/Fe(phen)_3^{2+})$ and the ruthenium tris(2, 2'-bipyridine) $(Ru(bpy)_3^{3+}/Ru(bpy)_3^{2+})$ couples. The chemical oscillations in time are revealed through a drastic changes in the colour of the solution corresponding to the transition from one state of the catalyst to the other. More quantitative measurement of the oscillations have made using spectrophotometry or specific electrodes [58].

Field, Körös and Noyes derived a three variables model to describe the B-Z kinetics [59]

$$\frac{dX}{dt} = k_1 H^2 AY - k_2 HXY + k_3 HAX - 2k_4 X^2 \tag{40}$$

$$\frac{dY}{dt} = -k_1 H^2 AY - k_2 HXY + fk_5 BZ \tag{41}$$

$$\frac{dZ}{dt} = 2k_3 HAX - k_5 BZ \tag{42}$$

where $X = [HBrO_2]$, $Y = [Br^-]$, $Z = [Ru(bpy)_3^{3+}]$, $A = [BrO_3^-]$, $B =$ [organic oxidisable species], $H = [H^+]$, and the k_i denote kinetics constants. Concentrations A, B and H can be taken as constant as they are consumed very slowly (pool chemical approximation), therefore only the intermediate species X, Y and Z oscillate in time. Because of the slowness of the depletion of A, B and H, the B-Z reaction may be used, with hardly any drift in properties, in a closed reactor. This is an uncommon property of this reaction that explains its success (besides its historical importance). This characteristic is not possessed by other time oscillating reactions, that therefore have to be run in open reactors. It is also worth noting that with the ruthenium containing catalyst, the reaction becomes sensitive to light. As the second step in the reduced kinetics mechanism (40-42) is much faster than the others, the quasi-stationarity of Y, i.e. $dY/dt = 0$, may be assumed and therefore

$$Y = \frac{1}{2} \frac{f k_5 B Z}{k_1 H^2 A + k_2 H X} \tag{43}$$

According to this assumption, the so-called two variables Oregonator model is obtained [61]

$$\frac{du}{dt} = \frac{1}{2} f k_5 B v \frac{k_1 H^2 A - k_2 H u}{k_1 H^2 A + k_2 H u} + k_3 H A u - 2 k_4 u^2 \tag{44}$$

$$\frac{dv}{dt} = 2 k_3 H A u - k_5 B v \tag{45}$$

where $u = X$ and $v = Z$. This model has been extensively studied in the literature to describe either the well-mixed reaction or its coupling with diffusion; among its characteristics are bistability between homogeneous steady states emerging from saddle-node bifurcations, temporal oscillations corresponding to Hopf bifurcations, excitability and many other complex behaviors [56, 62–66, 80]. The coupling with diffusion allows for various wave phenomena that have been studied in thin layers of liquid solutions or in inert gel reactors (see [54] and references therein).

3.2.   PHENOMENOLOGY OF THE CHEMOELASTODYNAMICS

As mentioned in the introduction, in the experiments we are concerned with, responsive gels are plunged into a autocatalytic chemical reaction. Until now, two types of situations have been considered. First, self-organized response of a cylindrical gel in a C.S.T.R. has been observed in the experiments of De Kepper et al. that use the chlorite-tetrathionate (C-T) reaction pH clock reaction. With this reaction volume variations of the cylinder and chemical excitation waves may be induced to travel along the gel, that then responds by developing a deformation that propagates in correlation [68–70]. In fact the C-T reaction allows chemical spatial bistability in the gel ([54, 55]). Theory predicts that the coupling between chemistry and elasticity may then induce *emergent* autonomous oscillations, that exist neither in the gel nor the chemical reaction alone [31, 53, 71, 72, 81]. However the characteristics of the gel and chemical concentrations range used in the experiment do not allow this to occur. Preliminary results obtained with the thiourea iodate-sulfate reaction [82] do exhibit a few damped oscillations [67]. This illustrates the difficulties arising from the determination of a good adequation between the choice of a gel and the choosing of an adequately interacting chemical reaction. The number of free parameters also calls for an extensive search in phase space.

Secondly, and we will focus on this case here, is the use of oscillatory chemical reactions to induce volume oscillations of the gel through a chemo-mechanical coupling. Two situations were investigated. In the first, the gel is plunged in a CSTR where an oscillatory pH reaction of the Landolt-type occurs and provides for the coupling. Therefore the reacting solute concentrations oscillations occur both in the bath (outside the gel) and inside the gel [20, 21, 29]. The C.S.T.R. imposes a temporal *forcing* on the gel dynamics through the periodic pH variations. In the second, use is made of the B-Z reaction, discussed in the preceeding section. The ingeneous trick is to incorporate the ruthenium tris$(2, 2'$-bipyridine) catalyst of the reaction in the gel matrix. When such gel is dipped in the bath of the BZ reactants, they invade the gel and the reaction oscillates only inside the gel, while the surroundings remain quiescent. Because of the property of the B-Z reaction, alluded to previously, the use of a C.S.T.R. is not necessary [19, 22–27]. The grafting of the catalyst on its matrix endows the gel with *autonomous* time-periodic volume variations.

To describe the situation, an intuitive interpretation of the observed volume oscillations has been proposed by Yoshida. Remember the catalyst exhibits an oxidized $[Ru(byp)_3^{3+}]$ and a reduced state $[Ru(byp)_3^{2+}]$ respectively. On the other hand, the poly-NIPAAm hydrogel used exhibits a sharp volume transition around 32^oC. The key is that, at fixed temperature, the swelling rate associated with this transition, is larger for the oxidized state than for the reduced state of the catalyst, while the two states have the same temperature dependence as represented on Figure 2.

In the course of the chemical reaction, the catalyst oscillates between its two oxidation states. Therefore the gel is able to cycle between its two corresponding possible swelling states. This mechanism is apparent for a miniature cubic gel (size 0.5 mm); indeed, in this case the chemical wavelength is larger than the system size and oscillations are homogeneous in space [26, 27]. Nevertheless, in the case of a larger rectangular gel [22] ($1\,mm\ x\ 20\,mm$), clamped on the shorter side, the propagation of a longitudinal chemical wave generates a deformation wave of the gel visible along the longer edge. The free lateral boundaries respectively expand and shrink as the oxidized and reduced part of the wave come by. The situation is there somewhat more complex: the gel is in a prestressed state imposed by its clamping which induces an anisotropy in the system. Therefore, a chemical wave is generated in this direction, oscillations are no more homogeneous, and phase opposition may take place.

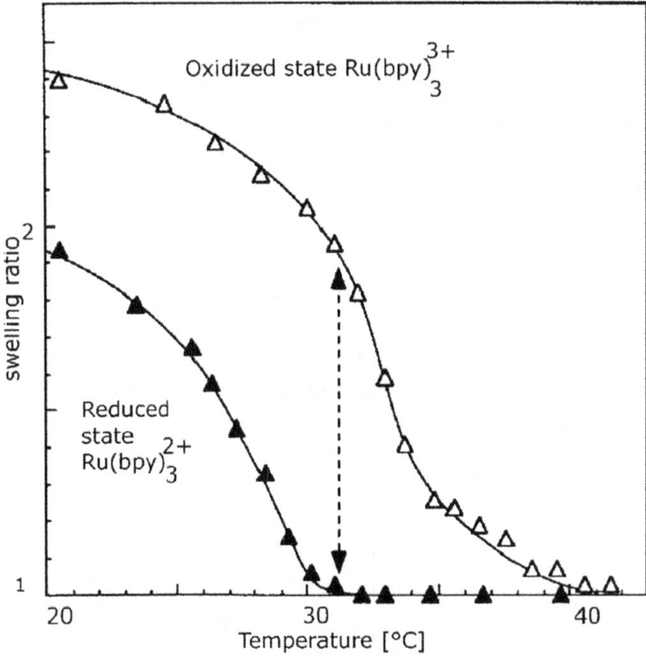

Figure 2.   Rate swelling at equilibrium of gel poly[NIPAAm-co-Ru(bpy)$_3$] in function of temperature respectively for the reduced state $Ru(bpy)_3^{2+}$ of the catalyst, in a $Ce(SO_4)_2$ solution, and for the oxided state $Ru(bpy)_3^{3+}$, in a $Ce(SO_4)_3$ solution (adapted from R. Yoshida in [22]).

3.3.   CHEMOELASTODYNAMICS EQUATIONS OF EVOLUTION

From hereon we focus on the case of a *spherical miniature bead of gel* that has the catalyst of the B-Z reaction grafted on its matrix. Each point of the sphere is defined by a triplet of variables (r, θ, ϕ) and we further assume that the spherical symmetry is maintained during the volume oscillation, i.e. deformations are purely radial as in [18, 71, 72].

Evolution equation for the gel   In the following we consider situations, as in the experiments, where the *solute chemical reacting species* are *diluted* so that $\Phi_i \ll \Phi_p$. In addition we have $\Phi_p < \Phi_s$ as in the experiments of [26, 69]. This restricts the validity of the presentation to small volume variations. Therefore, diffusion of free species in the solvent inside the gel are not affected by variations of Φ_p. Accordingly, relation (34) for the diffusion fluxes of the free (from the matrix) reactive species are given by

$$j_k^\mu = -\frac{L_k}{T}\partial_\mu\left(\frac{\widehat{\mu_k^m}}{\overline{V_k^m}}\right) \equiv D_k\partial_\mu\Phi_k \qquad (46)$$

where D_k are their diffusion coefficients. Relation (35) for the permeation flux is

$$j^\mu_{perm} \approx -\frac{L_{perm}}{T}\partial_\lambda\,(\Pi)^\lambda_\mu \equiv -\frac{D_{perm}}{k_B T v_e}\partial_\lambda\,(\Pi)^\lambda_\mu \tag{47}$$

where we have neglected

$$\sum_{l=1}^{N}\frac{\Phi_l}{1-\Phi_p}\partial_\mu\left(\frac{\widehat{\mu^m_l}}{\overline{V^m_l}}\right)$$

because $\Phi_k/(1-\Phi_p) \sim \Phi_k/\Phi_s \ll 1$.

Otherwise, as we assume a dilute media, relation (30) for the permeation flux can be simplified as

$$\mathbf{j}_{perm} = (1-\Phi_p)(\overline{\mathbf{u}} - \dot{\mathbf{r}}(R,t)) \tag{48}$$

$$= (1-\Phi_p)\left(\frac{\sum_{k=1}^{N}\Phi_k\mathbf{u}_k + \Phi_s\mathbf{u}_s}{\sum_{k=1}^{N}\Phi_k + \Phi_s} - \dot{r}(R,t)\right) \tag{49}$$

$$= (1-\Phi_p)(\mathbf{u}_s - \dot{\mathbf{r}}(R,t)) \tag{50}$$

where $\dot{\mathbf{r}}(R,t)$ is the radial velocity of a *gel particle*, labelled R, the movement of which we follow. To obtain (50) from (49), we have invoked $\Phi_k \ll \Phi_p < \Phi_s$ to neglect $\sum_{k=1}^{N}\Phi_k\mathbf{u}_k$ and $\sum_{k=1}^{N}\Phi_k$. As D_{perm} depends on Φ_p, we introduce the friction coefficient $\zeta(\Phi_p)$ between the matrix and the solvent as [73]

$$\zeta(\Phi_p) = (T/L_{perm})(1-\Phi_p) \tag{51}$$

Equaling relation (50) and (47), we obtain

$$\nabla \cdot \Pi + \zeta(\Phi_p)(\mathbf{u}_s - \dot{\mathbf{r}}(R,t)) = 0 \tag{52}$$

This equation is very similar to those obtains by other approaches [31]. For some specific geometries, there exist steady points, i.e. points where simultaneously, the solvent, free species and polymer matrix velocities are equal to zero. In the case of a sphere, its center is such a point. As $\nabla \cdot \mathbf{j}_{tot} = 0$, we have

$$\nabla \cdot \mathbf{u}_s = \nabla \cdot (\Phi_p[\mathbf{u}_s - \dot{\mathbf{r}}(R,t)]) \tag{53}$$

Integrating (53) over the sphere, taking into account relation (52) and the center steady point, we have

$$\mathbf{u}_s = (\Phi_p[\mathbf{u}_s - \dot{\mathbf{r}}(R,t)]) = -\frac{\Phi_p}{\zeta(\Phi_p)}\nabla \cdot \Pi \tag{54}$$

Substituting relation (54) into equation (52), we obtain the evolution equation for the gel

$$\dot{r}(R,t) \; = \; \frac{1 - \Phi_p}{\zeta(\Phi_p)} \left(\frac{\partial \Pi_r^r}{\partial r} + \frac{1}{r}(2\Pi_r^r - \Pi_\theta^\theta - \Pi_\phi^\Phi) \right) \tag{55}$$

In the following we have to make explicit the dependence of ζ on Φ_p. Many different models exist in the literature, as the Ogston model [9], based on molecular diffusion in porous media, or scaling laws theories verified in some experiments [73]. We follow the model proposed in [18]

$$\zeta(\Phi_p) \; = \; \zeta_0 h(\Phi_p) = \zeta_0 \frac{\Phi_p^{(3/2)}}{(\Phi_p^0)^{(3/2)}(1 - \Phi_p)^3} \tag{56}$$

Starting from relations (47) and (51), we identify $\zeta_0 \; = \; k_B T v_e / D_{perm}$. Therefore we have

$$\dot{r}(R,t) \; = \; \frac{D_{perm}}{k_B T v_e} \frac{1 - \Phi_p}{h(\Phi_p)} \left(\frac{\partial \Pi_r^r}{\partial r} + \frac{1}{r}(2\Pi_r^r - \Pi_\theta^\theta - \Pi_\phi^\phi) \right) \tag{57}$$

Reactives species evolution equations. We first express relation (36) in terms of $\mathbf{r}(R,t)$, then write it in terms of species concentrations (rather than volume fractions), and finally add the chemical kinetics as a source term. Starting from $\nabla \cdot \mathbf{j}_{tot} = 0$ and integrating on the sphere volume, we obtain, as the spherical symmetry is preserved,

$$\int_V \nabla \cdot \mathbf{j}_{tot} dV \; = \; \int_V \nabla \cdot (\mathbf{j}_{perm} + \dot{\mathbf{r}}(R,t)) dV$$

$$= \; \int_S (\mathbf{j}_{perm} + \dot{\mathbf{r}}(R,t)) \cdot d\mathbf{S}$$

$$= \; \int_S (j_{perm}^r + \dot{r}(R,t)) 4\pi r^2 dr = 0$$

$$\Rightarrow \; \dot{r}(R,t) = -j_{perm}^r \tag{58}$$

where the steady point of the sphere ensure the flux to be zero at its center. This allows us to write relation (36) as

$$\partial_t \Phi_k \; = \; -\nabla \cdot (\mathbf{j_k}) - \frac{\Phi_p}{1 - \Phi_p} \mathbf{j_{perm}} \cdot \nabla \Phi_k - \frac{\Phi_k \Phi_p}{1 - \Phi_p} \nabla \cdot \mathbf{j_{perm}}$$

$$- \frac{\Phi_k}{\left(1 - \Phi_p\right)^2} \mathbf{j_{perm}} \cdot \nabla \Phi_p \tag{59}$$

As equation (59) is linear in the Φ_k, we express it in terms of the corresponding concentrations c_k, so that we may add the reaction velocity term as a source term

$$\partial_t c_k = D_k \Delta c_k + f_k\{c_k\} + \frac{\Phi_p}{1-\Phi_p} j^r_{perm} \frac{\partial c_k}{\partial r} + c_k \frac{\Phi_p}{1-\Phi_p} \left(\frac{\partial j^r_{perm}}{\partial r} + \frac{2}{r} j^r_{perm} \right)$$
$$+ c_k \frac{1}{(1-\Phi_p)^2} j^r_{perm} \frac{\partial \Phi_p}{\partial r} \tag{60}$$

where $f_k\{c_k\}$ corresponds to the kinetics of the species k. As usual, because the reactant concentrations are small, we have considered the D_k as constants. This assumption is experimentally verified as long as the solvent is in excess and the mixture is ideal(see discussion in the next section). To give a physical meaning to the different contributions in relation (60), we express the evolution in terms of material derivatives $D_t = \partial_t + \dot{r}\partial_r$. Therefore, we have

$$\frac{D\Phi_k}{Dt} = -\partial_r j^r_k - \partial_r \left(J^r_{perm} \frac{\Phi_k}{\Phi_s + \sum\limits_{\alpha=1}^{N} \Phi_\alpha} \right) + \frac{\Phi_k}{\Phi_p} \frac{D\Phi_p}{Dt} \tag{61}$$

The first term expresses the variation of Φ_k due to the flux of Φk with respect to the solvent. The second one depicts the effect of the permeation flux and the last one represents the contribution due to the volume variation of the gel particle, the motion of which we follow. Indeed the term $(1/\Phi_p)D\Phi_p/Dt$ is similar to a concentration-dilution term introduced to describe variable-volume chemical reactors [74, 75]. Returning to Eq.(60), when the chemical reactions do not affect the mechanical properties of the gel, the osmotic tensor is zero and we recover the standard reaction-diffusion equations. One should however be aware that the three supplementary terms could also arise from stresses resulting from other sources such as the clamping of the gel in the reactor, or swelling restrictions due to the boundaries of the reactor, or from other physical stimuli applied to the gel, such as imprinting with light beams,... In summary, the evolution of the system made of a spherical bead of gel plunged in a chemically reactive mixing is given by equations (57) and (60) with the appropriate initial and boundary conditions.

4. Autonomous volume oscillations induced by the BZ reaction

Chemomechanical coupling of a responsive gel   If the chemical processes are to influence the gel, some coupling with a property of the gel must be introduced. This is not a generic feature and will depend on the particular nature of the gel composition and chemical kinetics at work. In view to specify the mechanisms for the coupling, we have to give explicit expressions, for the chemical potential and the osmotic stress tensors. We first give those expressions in the case of a two chemical reaction species denoted by u and v. Relations (11) and (15) for the chemical potential respectively inside and outside the gels are given by

$$\frac{\widehat{\mu^m_{u,int}}}{\overline{V^m_u}} = \frac{k_B T}{v_1}\left[\left(\chi_{up}\Phi_p + \chi_{uv}\Phi_v + \frac{1}{N_u}\ln\Phi_u + \frac{1}{N_u}\right) - \Upsilon^a_{int}\right] \tag{62}$$

$$\frac{\widehat{\mu^m_{v,int}}}{\overline{V^m_v}} = \frac{k_B T}{v_1}\left[\left(\chi_{vp}\Phi_p + \chi_{vu}\Phi_u + \frac{1}{N_v}\ln\Phi_v + \frac{1}{N_v}\right) - \Upsilon^b_{int}\right] \tag{63}$$

where Υ^a_{int} and Υ^b_{int} denote

$$\Upsilon^a_{int} = \left(\chi_{sp}\Phi_p + \chi_{su}\Phi_u + \chi_{sv}\Phi_v - \chi_{su}\Phi_s + \frac{1}{N_s}\ln\Phi_s + \frac{1}{N_s}\right)$$

and Υ^b_{int} is similar to Υ^a_{int} where $\chi_{su}\Phi_s$ is replaced by $\chi_{sv}\Phi_s$
and

$$\frac{\widehat{\mu^m_{u,ext}}}{\overline{V^m_u}} = \frac{k_B T}{v_1}\left[\left(\chi_{uv}\Phi_{v,ext} + \frac{1}{N_u}\ln\Phi_{u,ext} + \frac{1}{N_u}\right) - \Upsilon^a_{ext}\right] \tag{64}$$

$$\frac{\widehat{\mu^m_{v,ext}}}{\overline{V^m_v}} = \frac{k_B T}{v_1}\left[\left(\chi_{vu}\Phi_{u,ext} + \frac{1}{N_u}\ln\Phi_{v,ext} + \frac{1}{N_v}\right) - \Upsilon^b_{ext}\right] \tag{65}$$

where Υ^a_{ext} denotes

$$\Upsilon^a_{ext} = \left(\chi_{su}\Phi_{u,ext} + \chi_{sv}\Phi_{v,ext} - \chi_{su}\Phi_{s,ext} + \frac{1}{N_s}\ln\Phi_{s,ext} + \frac{1}{N_s}\right)$$

and Υ^b_{ext} is similar to Υ^a_{ext} where $\chi_{su}\Phi_{s,ext}$ is replaced by $\chi_{sv}\Phi_{s,ext}$.

The osmotic stress tensor components given by (10) and (13) take the explicit form

$$\Pi^r_r = \frac{\Phi_p}{\Phi^0_p}v_e k_B T((F^r_r)^2 - 1) + \Xi \tag{66}$$

$$\Pi^\theta_\theta = \frac{\Phi_p}{\Phi^0_p}v_e k_B T((F^\theta_\theta)^2 - 1) + \Xi \tag{67}$$

where the following notation is introduced

$$\Xi = \frac{k_B T}{v_1}\left[\frac{\ln\Phi_s}{N_s} + \frac{\Phi_p}{N_s} + \chi_{sp}\Phi_u\Phi_p + \chi_{sp}\Phi_v\Phi_p + \chi_{sp}\Phi^2_p + \Phi_u\left(\frac{1}{N_s} - \frac{1}{N_u}\right)\right.$$

$$+ \; \Phi_v \left(\frac{1}{N_s} - \frac{1}{N_v} \right) + \chi_{us}\Phi_u\Phi_p + \chi_{vs}\Phi_v\Phi_p + + \chi_{us}\Phi_u^2 + \chi_{us}\Phi_u\Phi_v + \chi_{vs}\Phi_v^2$$

$$+ \; \chi_{vs}\Phi_v\Phi_u - \chi_{up}\Phi_u\Phi_p - \chi_{vp}\Phi_v\Phi_p \Big] \tag{68}$$

The numerous contributions, when only two reactants are considered, enlighten our caveat at the beginning of this section. One should bear in mind that we are already working with a free energy density in the given frame of approximation that, as detailed in section 2.4 of reference [76], while it allows the description of the essential features of the volume phase transition, does not lead to a quantitative agreement between theory and experiments. These authors also describe the typical interactions among the monomers. Because chemical reactions stand at the center of the stage we wish to describe, we will introduce their couplings, with the monomers (and among themselves if the reactive mixture were not ideal) through the dependence of the Flory interaction parameter χ_{sp} on the concentrations of the reactive species. This way of proceeding is not new. It was first invoked in [77] to permit the description of the volume phase transition of a neutral gel by an expansion of χ_{sp} in terms of the volume fraction of monomers such that $\chi_{sp} = \chi_1 + \chi_2\Phi_p + \cdots$. This procedure was generalized [78] to include a dependence on the solute species. In some experiments [79], such explicit dependence of χ_{sp} as a function of the volume fraction has even been measured. More examples may be found in the Polymer-Solvent Interaction section of reference [10]. We have also adopted such an approximation in [53, 71], but here we add an interaction energy with parameter χ_{vp} dependent on the concentration v.

In this paper, as we mentioned earlier, we restrict our attention to the Oregonator autocatalytic model, for which the kinetics is given by equations (44,45) because we assume the pool chemical approximation (chemostat), such that species A, B and H are constant in the course of the reaction [80]. The chemomechanical coupling is introduced through χ_{vp} as $v = [Ru(bpy)_3^{3+}]$ is grafted on the polymer matrix. Therefore, its diffusion coefficient D_v is equal to zero (this species moves around only because of the matrix permeation), whereas $D_u = 10^{-5} \; cm^2 s^{-1}$. The explicit form of the chemomechanical equations (60) for the Oregonator model is given by

$$\partial_t u = D_u \Delta u + \frac{\Phi_p}{1 - \Phi_p} \left(j_{perm}^r \frac{\partial u}{\partial r} + u \left(\frac{\partial j_{perm}^r}{\partial r} + \frac{2}{r} j_{perm}^r \right) \right)$$

$$+ \; \frac{u j_{perm}^r}{(1 - \Phi_p)^2} \frac{\partial \Phi_p}{\partial r} + f k_5 B v \frac{k_1 H^2 A - k_2 H u}{2(k_1 H^2 A + k_2 H u)} + k_3 H A u - 2k_4 u^2 \tag{69}$$

$$\partial_t v = \frac{\Phi_p}{1 - \Phi_p} \left(j_{perm}^r \frac{\partial v}{\partial r} + v \left(\frac{\partial j_{perm}^r}{\partial r} + \frac{2}{r} j_{perm}^r \right) \right)$$

$$+ \; \frac{v \, j^r_{perm}}{(1 - \Phi_p)^2} \frac{\partial \Phi_p}{\partial r} + 2k_3 H A u - k_5 B v \tag{70}$$

In the following, we assume the quasi-ideality of the system, i.e. $\chi_{up} = \chi_{us} = \chi_{vs} = 0$. The only source of non-ideality is due to the $[Ru(bpy)_3^{3+}]$-polymer interaction described by $\chi_{vp} \neq 0$, whereas, the solution of the other reactants may be considered as ideal. Because the catalyst consists in a large molecular group, we may assume that $\Phi_v \gg \Phi_u$ and then neglect $\chi_{sp}\Phi_u\Phi_p$ in comparison with $\chi_{sp}\Phi_v\Phi_p$. We have further assumed that each molecule occupies only one site in the network, i.e. $N_s = N_i = 1$

According to these hypothesis, the osmotic stress tensor components are given by

$$\Pi_r^r = \frac{k_B T}{v_1}\left(\Phi_p\left\{\chi_{sp}\Phi_p + 1 + \beta v + \frac{v_e v_1}{\Phi_p^0}[(F_r^r)^2 - 1]\right\} + \ln(1 - \Phi_p)\right) \tag{71}$$

$$\Pi_\theta^\theta = \frac{k_B T}{v_1}\left(\Phi_p\left\{\chi_{sp}\Phi_p + 1 + \beta v + \frac{v_e v_1}{\Phi_p^0}[(F_\theta^\theta)^2 - 1]\right\} + \ln(1 - \Phi_p)\right) \tag{72}$$

Where $\beta = \overline{V_v}(\chi_{sp} - \chi_{vp})$, plays the role of the control parameters in the numerical simulations, and $\overline{V_v}$ denotes the molar volume of species v. The parameter χ_{vp} models the hydrating effect the of the catalyst in its oxidized form at the origin of the swelling of the gel. We expressed the chemical species in terms of the their concentration, so that $\Phi_v = v\overline{V_v}$. In the absence of experimental determination of the χ parameters, there is a great deal of arbitrariness in the choice of the values to be attributed. Therefore, we also tested other kinds of couplings and showed the relative independence of the obtained results [53, 81]. The chemical potential of v is written as

$$\frac{\widehat{\mu_{v,int}}}{\overline{V_v^m}} = \frac{k_B T}{v_1}\left(\ln \Phi_{v,int} - \chi_{sp}\Phi_p - \chi_{vp}\Phi_p\right) \tag{73}$$

As we have assumed a dilute mixture, we neglect $\chi_{sp}\Phi_p$ and $\chi_{vp}\Phi_p$ with respect to $|\ln \Phi_{v,int}|$, indeed the volume fraction of $\Phi_{v,int}$ is close to zero. Therefore the continuity of the chemical potential at the gel-embedding fluid interface gives

$$\Phi_{u,int} = \Phi_{u,ext} \tag{74}$$

$$\Phi_{v,int} = \Phi_{v,ext} \tag{75}$$

that ensure also continuity for the concentrations.

The condition at the gel surface for the osmotic stress tensor is written as

$$\mathbf{\Pi} \cdot \mathbf{n} = -\pi_0 \mathbf{n} \tag{76}$$

with

$$\pi_0 = -\left(f_{mix}^{ext} - \sum_{\alpha=1}^{N} \Phi_{\alpha,ext} \frac{\widehat{\mu_{\alpha,ext}^m}}{V_\alpha}\right)$$

$$= -\frac{k_B T}{v_1} \ln \Phi_{s,ext} \approx 0 \tag{77}$$

because the volume fraction of the solvent is close to unity. consequently, the osmotic stress tensor will be taken equal to zero at the interface

$$\Pi_r^r = 0 \tag{78}$$

Numerical simulations of the chemoelastodynamics for the BZ reaction Equations (57) - (69,70) have been integrated numerically for a spherical bead of gel. [5]

In the simulations, some parameters values were kept fixed to:

$$B = 0.0035M \qquad H = 0.3M \qquad k_1 = 2M^{-3}s^{-1} \quad k_2 = 310^6 M^{-2}s^{-1}$$

$$k_3 = 42M^{-2}s^{-1} \quad k_4 = 310^3 M^{-1}s^{-1} \quad k_5 = 6M^{-1}s^{-1} \quad f = 0.7$$

corresponding to a region of parameter space where temporal concentrations oscillations take place inside an inert gel. Switching to a sensitive gel, the temporal evolution of the sphere radius R is represented on Figure 3 for different values of our coupling parameter β.

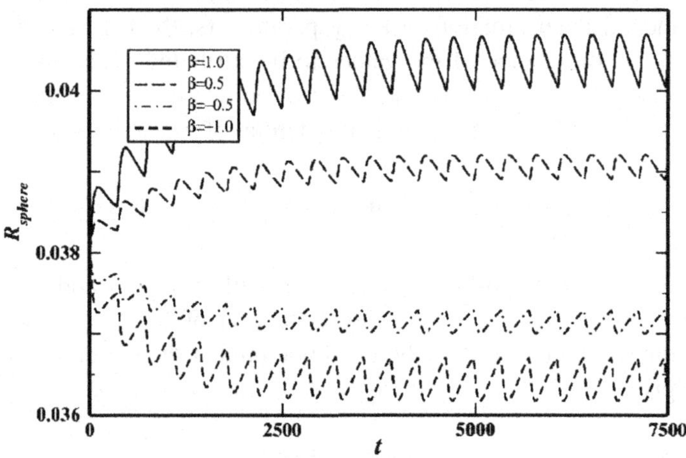

Figure 3. Evolution with time of the sphere radius for different values of the interaction parameter β, with $\chi_{sp} = 0.536052, \Phi_0 = 0.1, \Phi_p(t = 0) = 0.1, R_{sphere}(t = 0) = 0.038\,cm$, $A = 0.085\,M$.

---

[5]The present formalism has also successfully been tested to generate oscillations [53,81], using a clock reaction with a kinetics analogous to that of the C-T reaction, in the situation discussed above and in [31].

Initially the gel is in some chosen equilibrium (reference) state, $\Phi_p(t = 0)$. It corresponds to the sphere plunged in the reactants solution where the coupling β is held off. When, at $t = 0$, the coupling β is now turned on, the reaction sets in and the concentrations start oscillating. The radius of gel follows the oscillation. But there is first, for $\beta > 0$, a transient mean swelling of the gel (on which the oscillations are superposed) - the opposite behavior is true for $\beta < 0$ - as the mean radius adapts to the new nonequilibrium conditions. After this transient regime, the radius R of the sphere, during a period of oscillation, varies over an interval denoted by $\Delta R = R_{max} - R_{min}$ as a function of β (Figure 4). This behavior is easy to understand, as we might expect that, not only the mean new volume of the gel but also its variations during the oscillations are in direct relation with β, that models the intensity of the interaction between species v and the polymer network. We have however noticed no dependence of the oscillation period T with β.

Another parameter, that has been varied is the concentration of the substrate $A = [BrO_3^-]$. A decrease of A induces an increase of the oscillatory period T represented in Figure 5. This effect is in agreement with experiments. In the case of microgels an experimental dependence of period T as a function of malonic acid, sodium bromate and nitric acid concentrations can be captured by the relation $T \propto [MA]^{-0.506}[NaBrO_3]^{-0.667}[HNO_3]^{-0.478}$ [27]. Other simulations results (not given here) show that the period T is also a decreasing function of concentrations B and H.

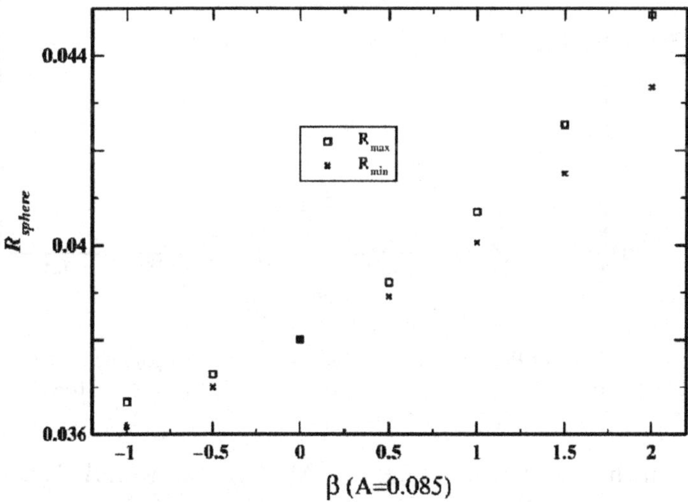

Figure 4.   Maximum and minimum gel sphere radius during a period of oscillation as a function of the interaction parameter β, with $\chi_{sp} = 0.536052$, $\Phi_0 = 0.1$, $\Phi_p(t = 0) = 0.1$, $R_{sphere}(t = 0) = 0.038\ cm$, $A = 0.085M$.

                    S. MÉTENS, S. VILLAIN AND P. BORCKMANS

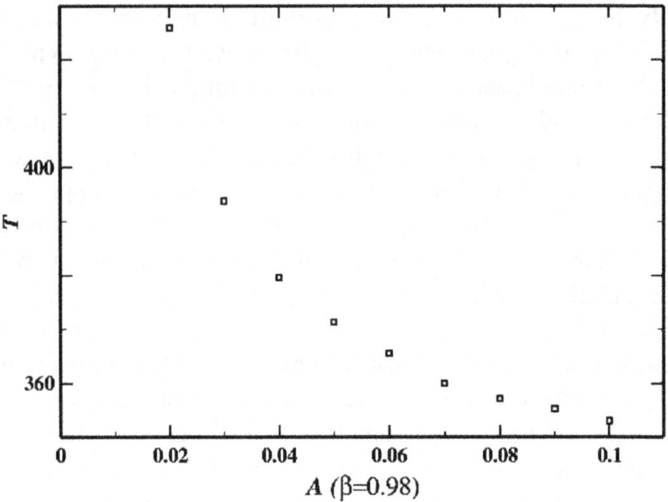

Figure 5.   Oscillatory period of the sphere radius as a function of parameter A, with $\chi_{sp} = 0.536052$, $\Phi_0 = 0.1$, $\Phi_p(t = 0) = 0.1$, $R_{sphere}(t = 0) = 0.038\ cm$, $\beta = 0.98$.

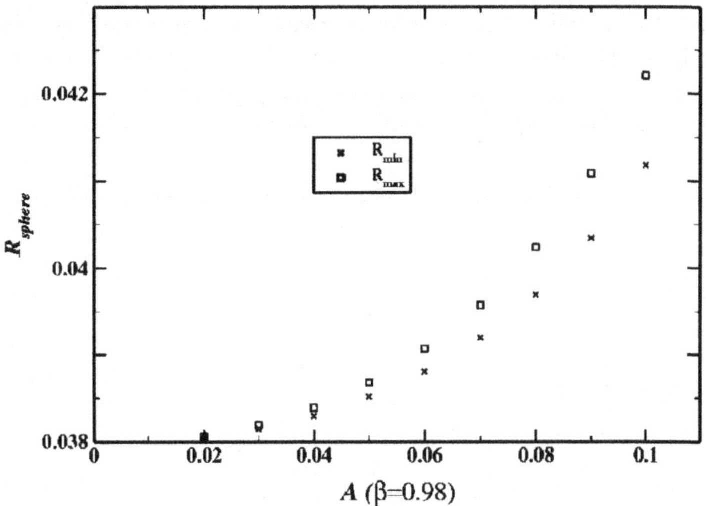

Figure 6.   Dependence of R_{max} and R_{min} as a function of A; other parameters have been fixed to $\beta = 0.98$, $\chi_{sp} = 0.536052$, $\Phi_0 = 0.1$, $\Phi_p(t = 0) = 0.1$, $R_{sphere}(t = 0) = 0.0380cm$.

In the meanwhile as shown in Figure 6, ΔR decreases with A. This behavior may be explained through the relation that exist between v_{max} and A (Figure 7), where v_{max} is the largest value taken by v inside the sphere during an oscillation and that occurs at the center, as we show further on. Now v_{max} is a monotonously increasing function of A. Therefore a variation

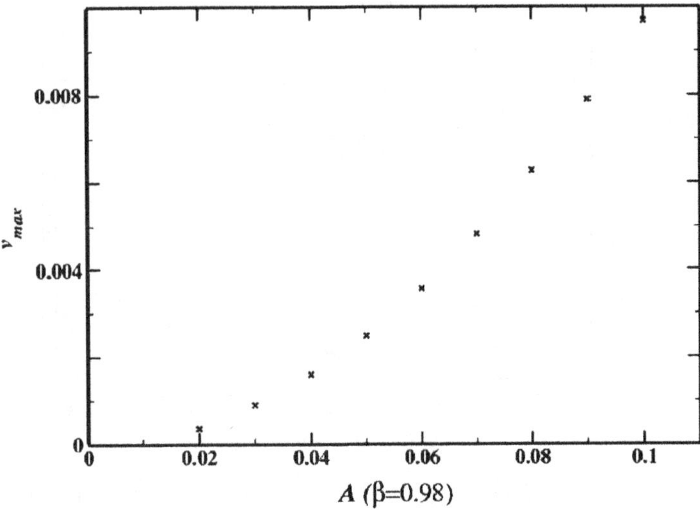

Figure 7. v_{max} as a function of parameter A, with $\beta = 0.98$, $\chi_{sp} = 0.536052$, $\Phi_0 = 0.1$, $\Phi_p(t = 0) = 0.1$, $R_{sphere}(t = 0) = 0.0380$.

of A induces a change in the interaction strength $\beta v \Phi_p$ that controls the gel volume. If we interpret the results presented on Figures 5-6 in terms of the oscillation period, we observe a smaller ΔR for a larger T.

This is a priori a counter-intuitive behavior. Indeed, in simple, naïve, thinking, the longer the period, the longer chemistry is able to act to make the gel swell in the first part of the period, before it recalls it to shrink in the second part (or vice versa), the larger ΔR would be. The existence of an opposite behavior probably finds its origin in the new time and space scales that emerge from the coupling between chemistry and mechanics. The simple picture based on a single relaxation time to equilibrium, as is the case of the response to a small step stimulus, does not apply anymore. Indeed the characteristic time and space scales involved in Eqs. (57) - (69,70) are related to the inverse of the eigenvalues of the corresponding linearized problem. Now the fact that the period T increases sharply when A decreases, as shown on Figure 5, might signal the existence of an infinite period bifurcation at a critical value A_c. In the case of an inert gel, we have found this to be the case with a critical value close to $A_c \sim 0.0105$ [6]. There the corresponding concentrations profiles for u and v in the sphere become closer and closer to the trivial uniform solution ($u = 0, v = 0$) when A reaches its critical value where the oscillations suddenly disappear. As the amplitude of the profile

---

[6]We have found no bifurcation study of the two variables FKN model in a sphere in the literature and only a few experimental results have been obtained by Yoshida et al. for other geometries.

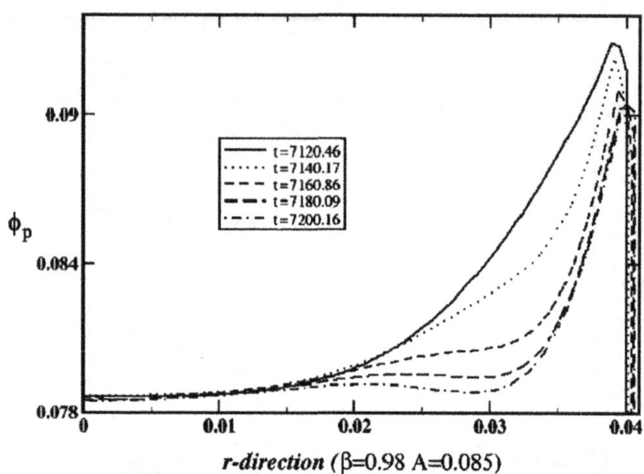

Figure 8. Temporal evolution of the monomers volume fraction Φ_p during the swelling part of the oscillation; $\chi_{sp} = 0.54, \Phi_0 = 0.1, \Phi_p(t = 0) = 0.1, R_{sphere}(t = 0) = 0.038\,cm, A = 0.085\,M, \beta = 0.98$.

decrease, the corresponding ΔR of the sphere decrease as β is proportional to v as observed on Figure 6. Usually infinite period bifurcations occur when some measure of the extension of the limit cycle characterizing the oscillations in phase space increases and collides with some separatrix. Here a similar behavior seems to occur when the extension of the limit cycle decreases. The literature however gives other examples of such global bifurcation behavior [62–64].

The evolution of $\Phi_p(r, t)$ and $v(r, t)$ during a swelling phase and a deswelling phase are represented respectively on (Figures 8 and 10) and (Figures 9 and 11). We observe, in the core of the gel, weak spatial and temporal variations of $\Phi_p(r, t)$. Those variations are important in the region closer to the boundary of the gel where the spatial gradient of $v(r, t)$ is large. Indeed one of the driving terms in the radius equation (57) is proportional to $\partial_r \Pi_r^r$, i.e. also of $\nabla v(r, t)$. The position at which $\Phi_p(r, t)$ takes its maximal value does not coincide with the corresponding value for $v(r, t)$ as suggested by the simple picture based on equilibrium properties of the gel. This offset is related to the above mentioned difference in the scales, corresponding respectively to the evolution dynamics and to the relaxation dynamics to a non equilibrium state.

On Figures 12 and 13 we represent the temporal evolution of the concentration profile $u(r, t)$ respectively during a swelling and a deswelling phase. Concentrations profiles for u and v are only weakly affected by the coupling with the gel. Such a result might however be different in the case of large gel volume variations or for another choice of the coupling constants.

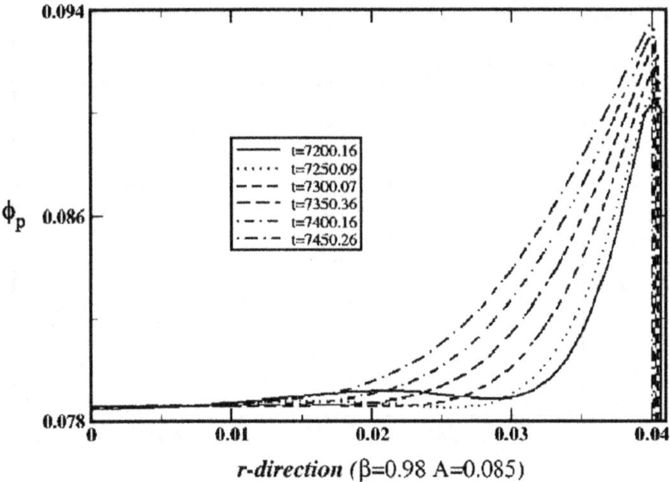

Figure 9.   Temporal evolution of the monomers volume fraction Φ_p during the deswelling part of the oscillation; $\chi_{sp} = 0.54, \Phi_0 = 0.1, \Phi_p(t = 0) = 0.1, R_{sphere}(t = 0) = 0.038\,cm,$ $A = 0.085\,M, \beta = 0.98.$

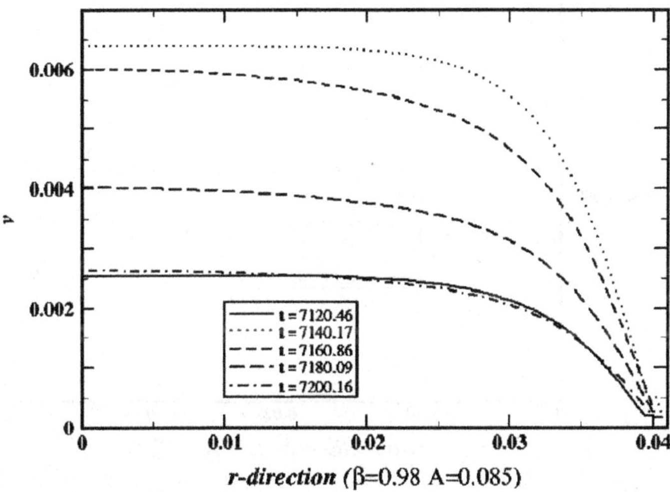

Figure 10.   Temporal evolution of $v(r,t)$ during the swelling phase; with $\chi_{sp} = 0.54,$ $\Phi_0 = 0.1, \Phi_p(t = 0) = 0.1, R_{sphere}(t = 0) = 0.038\,cm, A = 0.085\,M, \beta = 0.98.$

The Belousov-Zhabotinsky chemical reaction provides relaxation oscillations for u and v (Figures 14 and 15). The swelling phase comes out to be shorter than that of shrinking in agreement with the experiments.

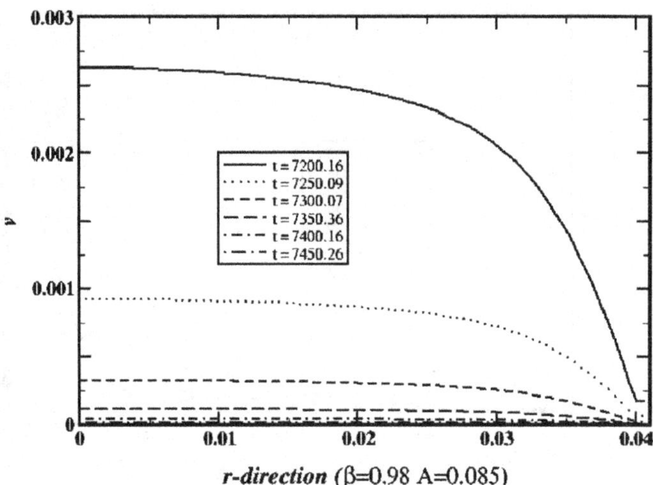

Figure 11. Temporal evolution of $v(r,t)$ during the deswelling phase; with $\chi_{sp} = 0.54$, $\Phi_0 = 0.1$, $\Phi_p(t = 0) = 0.1$, $R_{sphere}(t = 0) = 0.038\ cm$, $A = 0.085\ M$, $\beta = 0.98$.

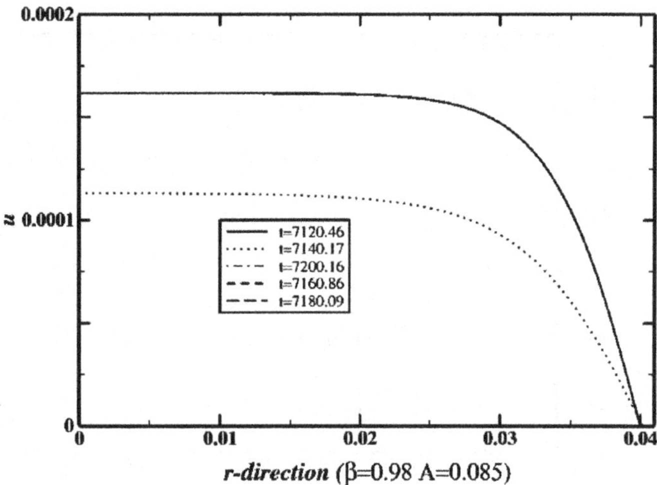

Figure 12. Temporal evolution of $u(0,t)$ during the swelling phase, with $\chi_{sp} = 0.54, \Phi_0 = 0.1, \Phi_p(t = 0) = 0.1, R_{sphere}(t = 0) = 0.038cm$, $A = 0.085\ M$, $\beta = 0.98$.

The cyclic evolution of the sphere radius over many time periods is represented on Figure 16.

The corresponding transient dynamics is represented in the phase space (v, R), where we observe the relaxation to the limit cycle corresponding to the non equilibrium asymptotic state (Figure 17). In agreement with the

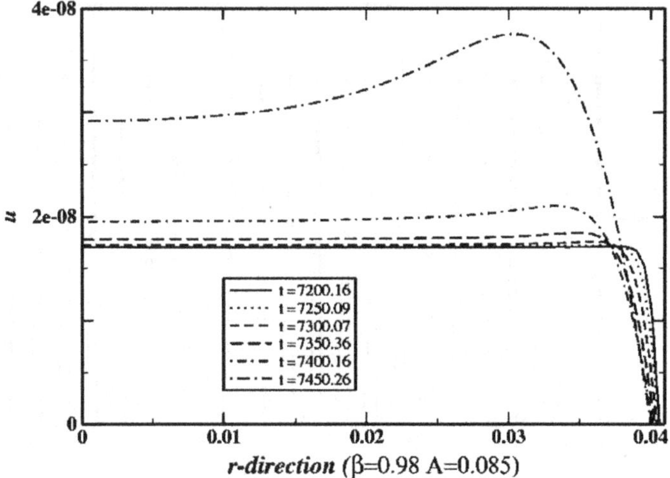

Figure 13.   Temporal evolution of $u(0, t)$ during a deswelling phase, with χ_{sp} $= 0.54$, $\Phi_0 = 0.1, \Phi_p(t = 0) = 0.1, R_{sphere}(t = 0) = 0.038\,cm, A = 0.085\,M, \beta = 0.98$.

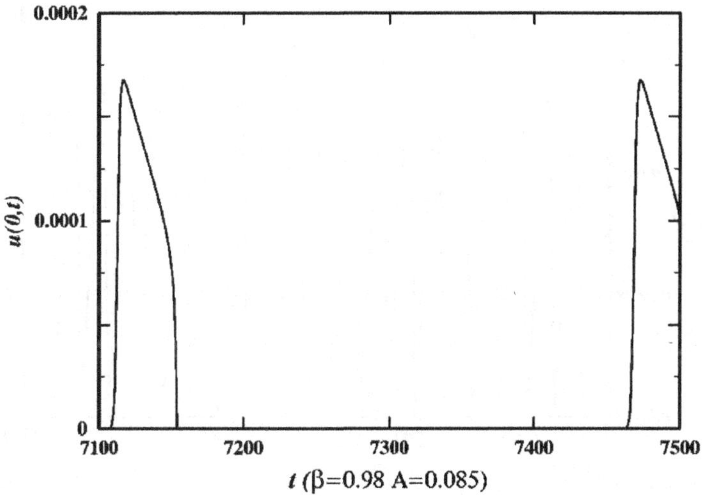

Figure 14.   Oscillation of $u(0, t)$ during one period, with χ_{sp} $= 0.54, \Phi_0 = 0.1$, $\Phi_p(t = 0) = 0.1, R_{sphere}(t = 0) = 0.038\,cm, A = 0.085\,M, \beta = 0.98$.

experiments, for a micro bead of gel, oscillations of size are nearly in phase with concentration oscillations of u and v. We represent on Figure 18 the oscillation of sphere radius R_{sphere} and $v(0, t)$ on one period of time to show this.

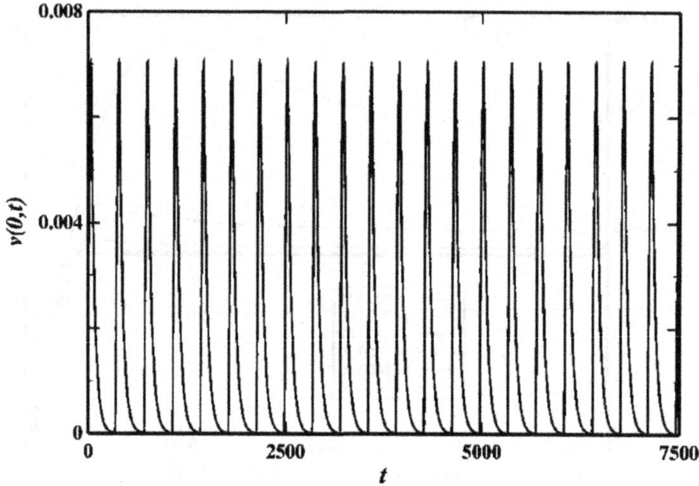

Figure 15.   Temporal oscillation of $v(0, t)$; (χ_{sp} = 0.54, Φ_0 = 0.1, $Phi_p(t = 0)$ = 0.1, $R_{sphere}(t = 0)$ = 0.038, A = 0.085, β = 0.98).

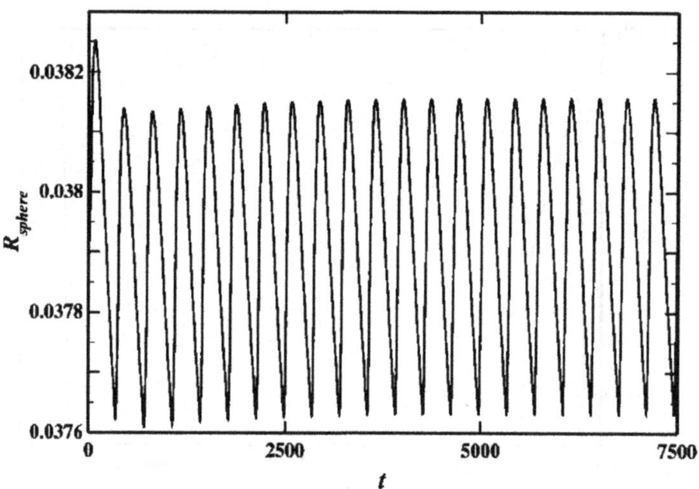

Figure 16.   Temporal oscillation of the sphere radius (χ_{sp} = 0.54, Φ_0 = 0.1, $\Phi_p(t = 0)$ = 0.1, $R_{sphere}(t = 0)$ = 0.038 cm, A = 0.085 M, β = 0.98).

5.   Conclusion and perspectives

We have presented the hydrodynamical multi-diffusional approach of gel dynamics proposed by Sekimoto extended to n-components and also to take into account the effects of autocatalytic chemical reactions affecting the volume of the gel. We have then applied this to the coupling of the gel with a kinetic

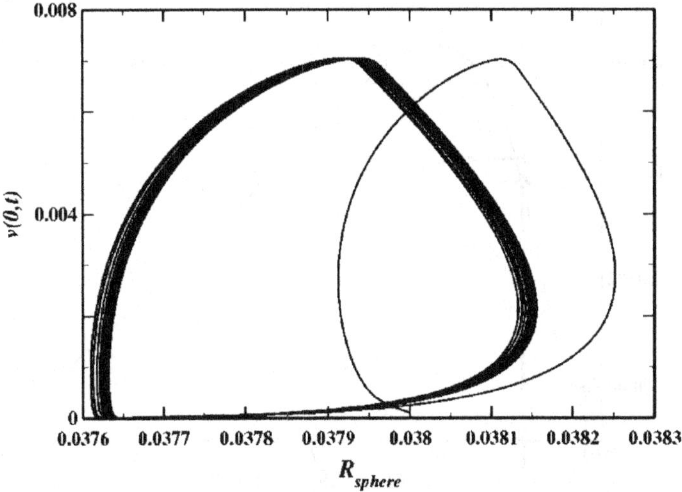

Figure 17. Relaxation from an initial condition to the limit cycle in the $(v(0, t), R_{sphere})$ phase space.

model for the BZ reaction when the catalyst is grafted to the polymer matrix, as in the experiments carried out by Yoshida's group.

These first results, however encouraging because we qualitatively recover some experimental dynamic behaviors, are limited to a spherical bead of gel, the volume variations of which remain small. To consider other geometries finite elements numerical codes will prove a necessity. Usually larger swelling ratios are the result of ionic effects, that we have here purposedly neglected, because it is an hydrating effect that arises in the experiments. The formalism described here could be generalized to take charges into account as has been done by Boissonade [31]. However when the volume variations become larger, one may possibly enter the region where the volume phase transition takes place. The introduction of the phase transition dynamics will definitely make matters much more complex [8]. Nevertheless, owing to the large number of approximations introduced and coupling parameters to be considered and measured, qualitative comparison with experiments based on trends, is all that seems possible. A bifurcation analysis remains a far cry.

It is also often mentioned, as a shortcoming, that gels tend to incur cracks when subjected to many swelling-deswelling cycles and are therefore not entirely suitable for repetitive tasks. Recently however double-network gels have been created and studied, that may help alleviate this difficulty [83–85]. The formalism presented here could also be extended to study the interaction of such materials with autocatalytic chemistry.

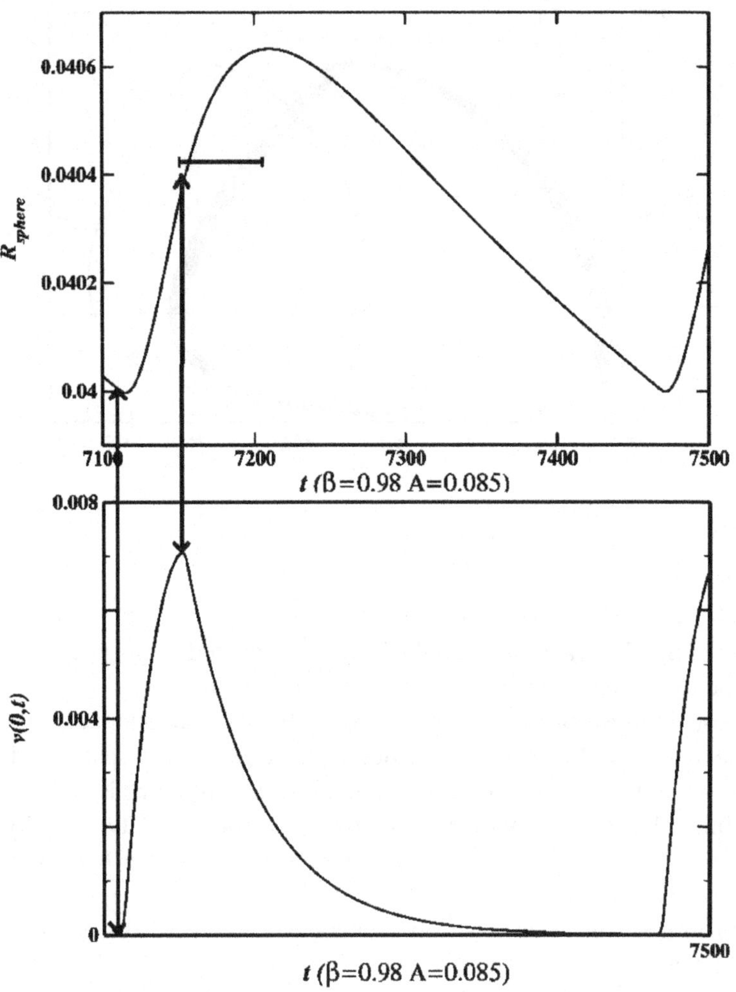

Figure 18.   Oscillation of sphere radius and of $v(0,t)$ over a period of time $(\chi_{sp} = 0.53602, \Phi_0 = 0.1, \Phi_p(t = 0) = 0.1, R_{sphere}(t = 0) = 0.038\,cm, A = 0.085\,M, \beta = 0.98)$.

Acknowledgements

We thank J. Boissonade, P. De Kepper, and F. Gauffre for numerous discussions, and K. Sekimoto for clarifying some of the intricacies of gel dynamics. P.B. was supported by the Fonds National de la Recherche Scientifique (Belgium) and S.V. by a DEFINO(HPRN-CT-2002-00198) grant.

References

1. B. Ziaie, A. Baldi, M. Lei, Y. Gu, R. A. Siegel, *Adv. Drug Deliv. Revs.* **56**, 145 (2004).

2. J. P. Gong, Y. Osada, in *Polymer Sensor and Actuators* (eds. Y. Osada, D. E. De Rossi, Eds, Springer Verlag, Berlin, 2000).

3. D. J. Beebe, J. S. Moore, J. M. Bauer, Q. Yu, R. H. Liu, C. Devadoss, B-H. Jo, *Nature* **404**, 588 (2000).

4. T. Shiga, *Adv. Polymer Sci.* **134**, 131 (1997).

5. H. Andersson, A. van den Berg, *Lab Chip* **4**, 98 (2004).

6. R. Siegel in this volume.

7. T. Tanaka, *Phys.Rev. Lett.* **40**, 820 (1978).

8. K. Dusek (Ed.), *Responsive Gels: Volume Transitions, Adv. Polymer Sci.* **109** and **110** (Springer, Berlin, 1993).

9. M. Bisschops, K. Luyben, L. van der Wielen, *Ind. Eng. Chem. Res.* **37**, 3312 (1998).

10. S. Wu, H. Li, J. P. Chen, K. Y. Lam, *Macromol. Theory Simul* **13**, 13 (2004).

11. E. C. Achilleos, K. N. Christodoulou, I. G. Kevrekidis, *Comput. Theor. Polym. Sci.* **1**, 63 (2001).

12. V. V Yashin, A. C. Balazs, *Science* **314**, 798 (2006).

13. V. V Yashin, A. C. Balazs, *Macromolecules* **39**, 2024 (2006).

14. O. Kuksenok, V. V Yashin, A. C. Balazs, *Phys. Rev. E* **78**, 041406 (2008).

15. M. Doi, 2008, Modeling of Gels (slides for IMA Tutorial). http://www.ima.umn.edu/matter/fall/t1.html.

16. K. Sekimoto, *J. Phys.II (Fr)* **1**, 19 (1991).

17. K. Sekimoto, *J. Phys.II (Fr)* **2**, 1755 (1992).

18. B. Barrière, L. Leibler, *J. Polym. Sci.* B, **41**, 166 (2003).

19. R. Yoshida, in this volume.

20. R. Yoshida, H. Ichijo, T. Hakuta, T. Yamaguchi, *Macromol. Rapid. Commun.* **16**, 305 (1995).

21. R. Yoshida, T. Yamagushi, H. Ichijo, *Material Science and Engineering.* **4**, 107 (1996).

22. R. Yoshida, E. Kokufuta, T. Yamagushi, *Chaos* **9**, 260 (1999).

23. K. Miyakawa, F. Sakamoto, R. Yoshida, E. Kokufuta, T. Yamaguchi, *Phys. Rev. E* **62**, 793 (2000).

24. R. Yoshida, T. Takahashi, T. Yamaguchi, H. Ichijo, *J. Am. Chem. Soc.* **118**, 5134 (1996).

25. R. Yoshida, T. Takahashi, T. Yamaguchi, H. Ichijo, *Adv. Mater.* **9**, 175 (1997).

26. R. Yoshida, T. Sakai, O. Tambata, T. Yamaguchi, *Sci. and Tech. Adv. Mat.* **3**, 95 (2002).

27. T. Sakai, R. Yoshida, *Langmuir* **20**, 1036 (2004).

28. R. Yoshida, T. Onodera, T. Yamaguchi, E. Kokufuta, *J. Phys. Chem.* **103**, 8573 (1999).

29. C. Crook, A. Smith, R. Jones, A. Ryan, *Phys. Chem. Chem. Phys.* **4**, 1367 (2002).

30. P. Borckmans, G. Dewel, A. De Wit, E. Dulos, J. Boissonade, F. Gauffre, P. De Kepper, *Int. J. Bif. and Chaos* **12**, 2307 (2002).

31. J. Boissonnade and P. De Kepper, in this volume.

32. U. P. Schröder, W. Opperman, W. in *The physical properties of polymeric gels*, ed. J. P. Cohen-Addad, John Wiley and Sons Ltd., (1996).

33. P. J. Flory, *Statistical Mechanics of Chain Molecules*, Wiley-Intersciences (1969).

34. G. Strobl, *The Physics of Polymers*, Third Edition, Springer (2007).

35. M. Rubinstein, R. Colby, *Polymer Physics*, Oxford University Press (2003).

36. J. J. Hermanns, *J. Poly. Sci.*, **59**, 191 (1962).

37. W. M. Lai, D. Rubin, E. Krempl, *Introduction to Continuum Mechanics*, Third Ed., Butterworth Heinemann (1999).

38. K. Yoshimura, K. Sekimoto, *J. Chem. Phys* **101**, 4407 (1994).

39. P. C. Parodi, O. Parodi, P. S. Pershan, *Phys. Rev. A* **6**, 2401 (1972).

40. P. D. Fleming, C. Cohen, *Phys. Rev. B* **13**, 500 (1976).

41. D. L. Johnson, *J. Chem. Phys* **77**, 1531 (1982).

42. T. C. Lubensky, R. Mukhopadhyay L. Radzihovsky, X. Xing, *Phys. Rev. E* **66**, 66011702 (2002).

43. T. Tanaka, J. Filmore, *J. Chem. Phys.* **70**, 1214 (1979).

44. Y. Li, T. Tanaka, *J. Chem. Phys* **92**, 1365, (1990).

45. E. Geissler, A. M. Hecht, *J. Chem. Phys* **77**, 1548 (1982).

46. E. S. Matsuo, T. Tanaka, *J. Chem. Phys* **90**, 5161 (1989).

47. A. Peters, S. J. Candau, *Macromolecules* **21**, 2278 (1988).

48. C. J. Durning, K. N. Morman, *J. Chem. Phys* **98**, 4275 (1992).

49. T. Tomari, M. Doi, *Macromolecules* **28**, 8334 (1995).

50. T. Tomari, M. Doi, *J. Phys. Soc. Japan* **63**, 2093 (1994).

51. K. De Sudipto, N. R. Aluru, B. Johnson, W. C. Crone, D. J. Beebe, J. Moore *J. of Microelectromechanical Syst.* **11**, 544 (2002).

52. G. Rossi, A. Mazich, *Phys. Rev. E* **48**, 1182 (1992).

53. S. Métens, S. Villain, P. Borckmans to appear in *Physica D.* doi:10.1016/j.physd. 2009.06.002 (2009).

54. P. De Kepper, J. Boissonade, I. Szalai, in this volume.

55. P. Borckmans, S. Métens, in this volume.

56. P. Gray, S. K. Scott, *Chemical oscillations and Instabilities*, Clarendon Press - Oxford (1990).

57. I. R. Epstein, J. Pojman, *An introduction to Nonlinear Chemical Dynamics*, Oxford University Press, (1998).

58. Z. Nagy Ungvarai, B. Hess, *Physica D* **49**, 33 (1991).

59. R. J. Field, E. Körös, R. M. Noyes, *J. Am. Chem Soc.*, **94** 8649 (1972).

60. L. Kuhnert, K. I. Agladze, and V. I. Krinsky, *Nature* **337**, 244 (1989).

61. J. Tyson, *J. Phys. Chem.* **86**, 3006 (1982).

62. P. De Kepper, K. Bar Eli, R. Noyes, *J. Phys. Chem* **87**, 480 (1983).

63. K. Bar Eli, R. Noyes, *J. Chem. Phys* **86**, 1927 (1987).

64. K. Bar Eli, M. Brons, *J. Phys. Chem* **94**, 7170 (1990).

65. J. Tyson, *J. Chem. Phys* **66**, 905 (1977).

66. J. Tyson, *J. Chem. Phys* **67**, 4297 (1977).

67. P. De Kepper, private communication.

68. F. Gauffre, V. Labrot, J. Boissonade, and P. De Kepper in *Nonlinear Dynamics in Polymeric Systems* (J. A. Pojman and Q. Tran-Cong-Miyata, Eds), ACS Symposium Series **869**, 80 2003.

69. V. Labrot, P. De Kepper, J. Boissonade, I. Szalai, and F. Gauffre, *J. Phys. Chem. B* **109**, 21476 (2005).

70. J. Boissonade, P. De Kepper, F. Gauffre and I. Szalai, *Chaos* **16**, 037110 (2006).

71. J. Boissonade, *Phys. Rev. Lett.* **90**, 188302 (2003).

72. J. Boissonade, *Chaos* **15**, 023703 (2005).

73. M. Tokita, T. Tanaka, *J. Chem. Phys.* **95**, 4613 (1991).

74. O. Levenspiel, *Chemical Reaction Engineering*, 3rd Ed. Wiley (1998).

75. P. Borckmans, K. Benyaich, A. De Wit, G. Dewel *in Nonlinear Dynamics in Polymeric Systems* (J. A. Pojman, Qui Tan-Cong-Miyata, Eds), ACS Symposium Series **869**, 58 (2003).

76. Shibayama and Tanaka, *Adv. Polymer Sci.* **109**, 1 (1993).

77. B. Erman, P. J. Flory, *Macromolecules* **19**, 2342 (1986).

78. T. A. Orofino, P. J. Flory, *J. Chem. Phys.* **26**, 1067 (1957).

79. H. Chiu, Y. Lin, Y. Hsu, *Biomaterials* **23**, 1103 (2002).

80. I. R. Epstein, J. Pojman, *An introduction to Nonlinear Chemical Dynamics, Oxford University Press*, (1998).

81. S. Villain, *Comportement mécanique de gels soumis à des réactions auto-catalytiques* (in French). Ph.D. Thesis, Université Paris 7 (2007).

82. J. Horvath, I. Szalai, and P. De Kepper, *Science* **324**, 772 (2009).

83. J. P. Gong, Y. Katsuyama, T. Kurokawa, Y. Osada, *Adv. Mater.* **14**, 15 (2003).

84. V. R. Tirumala, T. Tominaga, S. Lee, P. D. Butler, E. K. Lin, J. P. Gong, W-l. Wu, *J. Phys. Chem. B* **112**, 8024 (2008).

85. T. Tominaga, Y. Osada, and J. P. Gong, in this volume.

AUTONOMOUS RHYTHMIC DRUG DELIVERY SYSTEMS BASED ON CHEMICAL AND BIOCHEMOMECHANICAL OSCILLATORS

Ronald A. Siegel[*]
*Departments of Pharmaceutics and Biomedical Engineering
University of Minnesota, Minneapolis, MN 55455, USA*

Abstract. While many drug delivery systems target constant, or zero-order drug release, certain drugs and hormones must be delivered in rhythmic pulses in order to achieve their optimal effect. Here we describe studies with two model autonomous rhythmic delivery systems. The first system is driven by a pH oscillator that modulates the ionization state of a model drug, benzoic acid, which can permeate through a lipophilic membrane when the drug is uncharged. The second system is based on a nonlinear negative feedback instability that arises from coupling of swelling of a hydrogel membrane to an enzymatic reaction, with the hydrogel controlling access of substrate to the enzyme, and the enzyme's product controlling the hydrogel's swelling state. The latter system, whose autonomous oscillations are driven by glucose at constant external activity, is shown to deliver gonadotropin releasing hormone (GnRH) in rhythmic pulses, with periodicity of the same order as observed in sexually mature adult humans. Relevant experimental results and some mathematical models are reviewed.

Keywords: drug delivery, rhythmic, hydrogels, glucose oxidase, nonlinear feedback

1. Introduction and context

Introduction of therapeutic agents such as drugs and hormones into the clinic requires that adequate delivery systems be developed. A major goal of drug delivery research is to develop systems that release drug into the body at the proper place and at proper times, in order to optimize drug effect and minimize toxic side effects.

---

[*] To whom correspondence should be addressed. e-mail: siege017@umn.edu

P. Borckmans et al. (eds.), *Chemomechanical Instabilities in Responsive Materials*,
© Springer Science + Business Media B.V. 2009

The simplest and commonest notion is that the drug effect, or therapeutic endpoint, is directly related to the present unbound concentration of drug in blood. In this case, the most logical strategy is to administer the drug at a rate leading to concentrations within a so-called therapeutic range, TR, as shown in Figure 1a. At concentrations below TR, drug is unable to produce its desired effect, while above TR, toxicity occurs. While the bounds of TR are somewhat fuzzy and may vary between patients, they are useful in determining the desired dosing regimen. Ideally, drug is administered at a constant, or zero-order rate, but more intermittent regimens, e.g. once or several times daily oral admininstration, are often preferred so long as the resulting swings in plasma concentration do not take it out of TR. Recently developed oral controlled release formulations are intended to reduce fluctuations in plasma concentration.

While the forgoing philosophy holds for most drugs, important exceptions exist. Insulin is a hormone that regulates glucose homeostasis. A normally functioning pancreas secretes insulin in response to rise in glucose just before or during a meal, as illustrated in Figure 1b. Insulin then mobilizes transport of glucose into the liver and other tissues for storage or utilization. As blood glucose level falls, insulin secretion is reduced until a basal level is

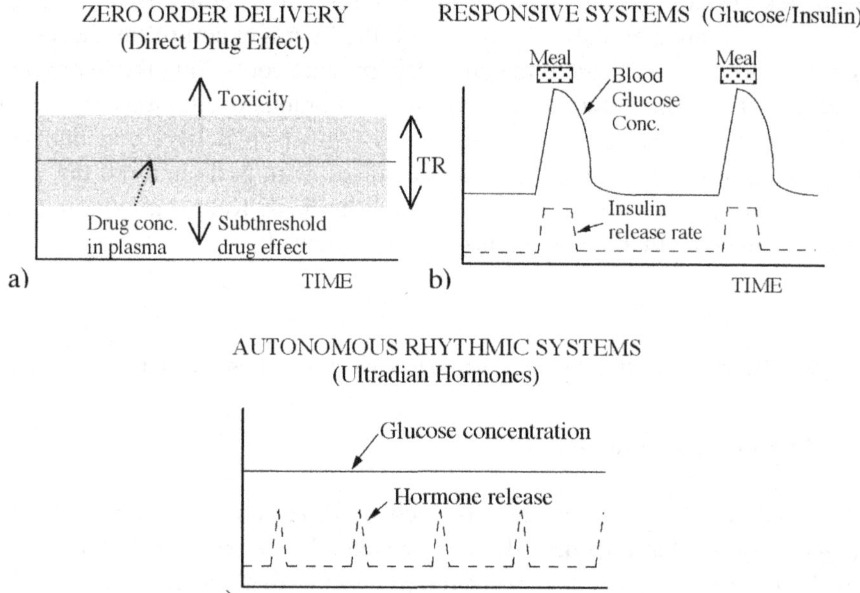

Figure 1. Various drug delivery strategies. Ordinates are release rates or concentrations in blood. (a) Zero order release; (b) "Closed loop" response to a physiologic variable, e.g. in response to glucose; (c) Autonomous rhythmic release fueled by a constant blood glucose concentration. (Panels (b) and (c) reproduced from [58], with permission from Elsevier.)

re-established. In type I diabetes, the pancreas fails to secrete insulin and does not exert the required "closed loop" control of blood glucose level, so insulin must be delivered exogenously. Considerable effort is being invested in developing pancreas mimics that respond properly to blood glucose levels [1–4]. Such systems must rapidly switch insulin delivery on and off, and hence are not of the zero-order type.

In recent decades, it has been established that many other hormones are endogenously secreted in episodic pulses, rather than continuously, as illustrated in Figure 1c [5]. Pulses are typically separated by tens of minutes or by hours, and hence the hormonal rhythms are termed "ultradian" [6]. A list of ultradian hormones, along with their observed pulse frequencies, is shown in Table 1 [7]. This list includes hormones involved in growth, reproduction, fluid balance, response to stress, etc. The reader will notice multiple entries in the "# Pulse/Day" column. Since hormones are assayed in the blood, and not at the site of secretion, washout kinetics must be factored into estimations of secretion rate, and a deconvolution procedure is typically used. Such procedures can be prone to noise artifacts, and different procedures may lead to varied identifications of secretion pulses [5,8].

TABLE 1. Frequencies of secretory hormones, as summarized in [7]. Multiple entries correspond to estimates from different primary sources.

Hormone	#Pulses/day	Hormone	#Pulses/day
Growth hormone	9–16, 29	Parathyroid hormone	24–129, 23
Prolactin	4–9, 7–22	Insulin	108–144, 120
Thyroid stimulating hormone	15, 54	Pancreatic polypeptide	96
Gonadotropin releasing hormone	7–18	Somatostatin	72
Leuteinizing hormone	7–15	Glucagon	103, 144
Follicle stimulating hormone	4–16, 19	Estradiol	8–16, 8–19
β-endorphin	13	Progesterone	8–16, 6–16
Melatonin	18–24, 12–20	Testosterone	13, 8–12
Vasopressin	12–18	Aldosterone	6, 9–12
Renin	6,8–12	Cortisol	15, 39

It is interesting to note that insulin is contained in Table 1. In the normally functioning pancreas, insulin is secreted in discrete pulses separated by 11–13 min, and experimental delivery systems that administer insulin in this pattern require less insulin than systems that deliver it continuously [9,10]. In view of the mitogenic side affects of insulin, which appear to be reduced with pulsatile insulin, future practical delivery systems might be programmed to take this factor into account.

Of the hormones displayed in Table 1, gonadotropin releasing hormone (GnRH) is probably the most important for the present discussion. GnRH, also known as leuteinizing hormone releasing hormone (LHRH), is a master hormone that is secreted in regular pulses by the hypothalamus in both males and females [11–14]. GnRH travels directly to the anterior pituitary gland, where it stimulates secretion of leuteinizing hormone (LH) and follicle stimulating hormone (FSH). These hormones circulate and stimulate secretion of other peptide and steroidal reproductive hormones in the gonads. In females, these hormones feed back on the secretion rate of GnRH, causing an alteration of LH and FSH balance, thus controlling the reproductive/ menstrual cycle [15]. In males the secretion period of GnRH is much more regular. In normal males and females, rhythmic secretion develops during sexual maturation, and failure to develop this pattern results in arrested sexual and reproductive development. Patients with hypogonadotropic hypo- gonadism (HH), in which secretion of GnRH is attenuated or nonexistent, can be treated with rythmic pulsatile intravenous (IV) GnRH, and most symptoms are usually reversed [16,17]. However, this treatment may be chronic in severe cases, and the dangers (e.g. infection) and inconvenience associated with long term IV treatments are such that fully implantable systems should be considered for GnRH delivery.

Another class of agents for which intermittent delivery is desired is drugs that exhibit tolerance, or reduced effect following sustained exposure. Opium derivatives, cocaine and nicotine are perhaps the best known drugs of this kind because of their association with abuse and addiction, but there are also purely therapeutic compounds that exhibit tolerance. For example, nitroglycerin (glyceryl trinitrate, GTN), a vasodilator that acts through a nitric oxide (NO) pathway, develops tolerance after several hours of con- tinuous administration. Early GTN skin patches, which were expected to pro- vide continuous 24-h prevention against angina, failed because this tolerance was not recognized as an important aspect. Presently, skin patches contain only a 12-h dose of GTN, and the patient must replace the patch only once daily. During the intermittent GTN-free period, the patient regenerates his or her response to GTN [18].

A number of mechanisms may underly drug tolerance, including physio- logical counter-regulation and receptor down-regulation [19]. In either case, the periods of drug exposure and drug removal should be determined by the kinetics associated with these mechanisms [20]. Hormonal periodicity is believed to be matched to the kinetics of down-regulation and recovery of receptors in target cells, and it has been further hypothesized that changes in period of hormonal secretion may permit a single hormone to act on different tissues at different times [7,19].

To complete this brief survey of periodicity in drug and hormone requirements, we recall perhaps the most common periodic rhythm, the 24-h circadian rhythm. This rhythm, which is shared by diverse life forms, adapts the organism to changes it experiences during day and night [21]. Pharmaco-kinetic factors such as blood flow to various tissues, and pharmacodynamic susceptibilities of these tissues to positive and negative drug effects, vary with time of day [21–25]. The need to take circadian rhythms into account in drug therapy is becoming more evident in the treatment of cardiovascular problems, asthma, and cancer.

In this contribution we summarize our own attempts to harness chemical and biochemomechanical oscillators to provide rhythmic delivery of drugs and hormones. In the first example, a pH oscillator drives an acidic drug between a charged and uncharged state. The drug is only permeable through a lipophilic membrane when it is uncharged, so pH oscillations drive oscil-lations in drug flux. In the second example, a hydrophobic polyelectrolyte gel separates a glucose reservoir from a chamber containing the enzyme glucose oxidase. The membrane modulates glucose flux to the enzyme, and the enzyme reaction produces hydrogen ions, which in turn modulate hydrogel swelling and permeability. A negative feedback instability leads to oscillations in membrane permeability to both glucose and GnRH, which is released in rhythmic pulses.

2. pH oscillator driven drug delivery

Numerous drugs are essentially hydrophobic in structure, but contain ioniz-able acid or base groups. When ionized, these drugs become more soluble in water but less soluble in lipophilic media. Flux of a drug through a lipophilic membrane from a high concentration drug source to a low con-centration drug sink can therefore be controlled by the pH of the source medium. Oscillations in pH can, under proper circumstances, be converted to oscillations in flux of drug across the membrane. In this section, we consider a model system, in which a bromate-sulfite-marble pH oscillator medium containing benzoic acid (BzA) is placed next to a lipophilic ethylene vinyl acetate copolymer (EVAc, 28 wt % vinyl acetate) membrane. The pH oscillations modulate the fraction of BzA that is uncharged, and available to partition into the membrane and diffuse across it.

The partition coefficient of uncharged drug between a lipophilic mem-brane and water is independent of pH, and a saturated solid suspension of BzA would provide a constant level of uncharged BzA, even while pH is oscillating. For the pH oscillator scheme to work, BzA should be unsaturated at all encountered pH values.

Many pH oscillators are of the relaxation type [26–28], featuring relatively long, quiescent periods with small pH changes, punctuated by rapid pH "pulses," leading in turn to pulses in uncharged drug concentration and entry into the membrane [29–31]. Diffusion of uncharged drug through the membrane will delay and smooth these pulses at the point of exit. To evaluate the extent of these effects as a function of pulse width and interpulse interval, we solve the one-dimensional diffusion equation in the membrane [32,33], subject to the boundary conditions that concentration is zero at the (downstream) exit point (perfect sink), and that the compartment upstream of the membrane is well stirred, with concentration given by a pulse of amplitude C_0 and width $\Delta << \tau = h^2 / \pi^2 D$. Here h is the membrane thickness and D is the diffusion coefficient of drug across the membrane. The parameter τ is a useful characteristic time for diffusion across the membrane. Denoting the uncharged drug's membrane/water partition coefficient by K, the flux of drug into the receptor, $J_\Delta(t)$, following the pulse is calculated as (see Appendix for details)

$$J_\Delta(t) = -2PC_0\Delta\sum_{n=1}^{\infty}(-1)^n n^2 \tau e^{-n^2 t/\tau}$$

where $P = KD/h$ is the permeability of the membrane to uncharged drug under steady state conditions. The time integral of J_Δ, $Q_\Delta(t)$, is proportional to the accumulated transfer of drug resulting the pulse, and is given by

$$Q_\Delta(t) = PC_0\Delta\left[1 + 2\sum_{n=1}^{\infty}(-1)^n e^{-n^2 t/\tau}\right]$$

The functions J_Δ and Q_Δ, normalized by $PC_0\Delta$, are plotted *versus* t/τ^*, where $\tau^* = h^2/D = \pi^2\tau$, in Figure 2. These plots show that the effects of a single pulse die out well before time $t = \tau^*$; this characterizes the degree of pulse "smearing" by the membrane. Pulses separated in time by at least τ^* will not interfere with each other. The peak of J occurs near $t = \tau$, which may be taken as an effective delay of drug appearance downstream following a pulse delivered upstream.

While numerous pH oscillators are available, they are not necessarily suitable, either due to toxicity of their components, or their short periods. For our studies we selected the bromate-sulfite-marble pH oscillator, which is relatively mild in toxicity and relatively long in period. This oscillator covers the pH range 2–7, which broadly brackets the range of ionization of BzA, whose pKa = 4.2.

The experimental setup is illustrated in Figure 3 [31]. pH oscillations occur in a continuous stirred tank reactor (CSTR) that is fed by one flask containing H_2SO_4 (6mM), Na_2SO_3 (96 mM) and BzA (0.5 mM), and a second flask containing $NaBrO_3$ (50 mM). Reactor volume is maintained constant by aspiration at the fluid-air interface into a waste collection flask. The reactor contains small chips of marble of determined total surface area.

The contents of the reactor are circulated rapidly into and out of the donor side of a diffusion cell, returning to the reactor. The donor phase of the cell is separated from the receptor phase (distilled water) by an EVAc membrane of thickness 2 mm, through which uncharged BzA diffuses. The ionic species cannot enter the membrane due to the latter's lipophilicity. A pH electrode monitors the donor phase, while accumulation of BzA in the receptor phase is monitored spectrophotometrically ($\lambda = 226$ nm).

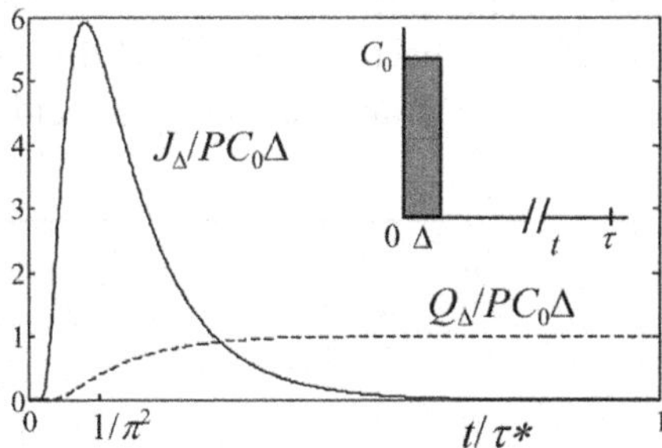

Figure 2. Normalized flux and accumulated flux of drug across a membrane following application of a narrow pulse (*inset*) of drug on the upstream side of the membrane. See text for symbol definitions.

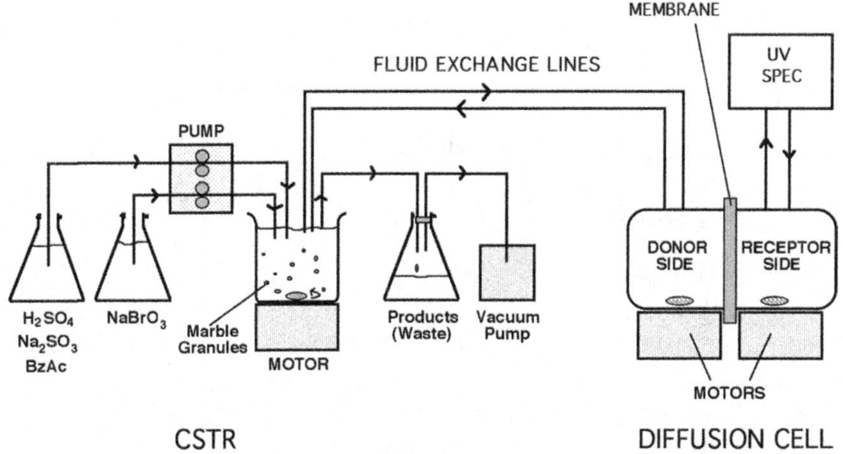

Figure 3. Setup for pH oscillator-driven rhythmic permeation of benzoic acid (BzA) across a lipophilic EVAc membrane. (Reproduced from [31], with permission from Elsevier.)

It is believed [34] that the pH oscillator functions according to a mechanism in which HSO_3^- and H_2SO_3, produced initially from H_2SO_4 and Na_2SO_3, react with BrO_3^-, releasing Br^-, SO_4^{2-}, and H^+. The H^+ product binds to other SO_3^{2-} and HSO_3^-, producing more HSO_3^- and H_2SO_3, respectively, which further react with BrO_3^-. This "autocatalytic" cycle ultimately exhausts itself when sulfite species is depleted, and a precipitous drop in pH occurs due to the remaining H^+ ions. The H^+ is eventually removed primarily by heterogeneous reaction with the marble ($CaCO_3$). The cycle is primed again by inflow of reactants into the CSTR.

When the H_2SO_4, Na_2SO_3, and BzA are mixed in the feeder flask at the prescribed concentrations, the resulting pH is close to 7.0 due to the strong buffer capacity of SO_3^{2-} near that pH. BzA is therefore completely dissociated, and remains that way until the rapid pH drop occurs. BzA is only capable of crossing the EVAc membrane during an acid pulse.

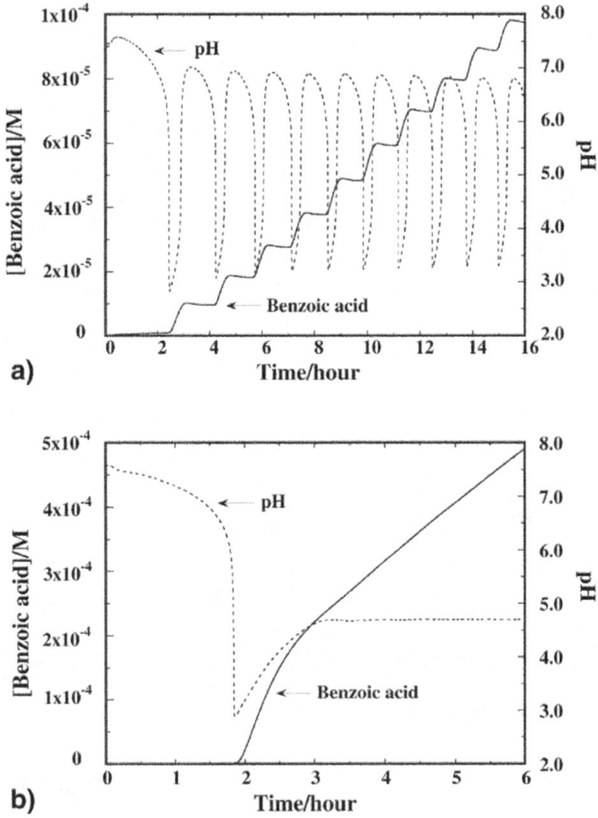

Figure 4. Time courses of pH in donor cell and benzoic acid accumulation in receptor cell. (a) 0.5 mM BzA in reaction; (b) 7.5 mM BzA in reaction. (Reproduced from [31], with permission from Elsevier.)

Figure 4a demonstrates sharp, pulsed oscillations in pH. Acidic pulses are separated by quiescent interpulse periods during which pH is close to 7.0. The interpulse interval decreases from ~2 to ~1.5 h over the period of observation due to a gradual acidic drift in the CSTR, perhaps due to CO_2 buildup in the reactor headspace. It is clear from Figure 4a that BzA accumulates in the receptor cell in a stepwise manner, with steps closely following the acid pulses. The shapes of these steps resemble $Q(t)$ in Figure 2, although there is a minor but consistent "retreat" during the quiescent phase, possibly due to backflux of BzA.

In Figure 4b it is shown that when BzA concentration is increased to 7.5 mM, the pH oscillation is quenched after an initial acid pulse. Rather than returning to near pH 7.0, the system settles at a steady state with pH 4.8, which is in the buffering range of BzA. Apparently, during the initial acid pulse, the BzA in the reactor becomes almost fully protonated. Any subsequent attempt by the system to rid itself of H^+ through the marble is compensated by release of H^+ from BzA. A constant flux of BzA across the EVAc membrane is recorded after steady state is reached.

To further study the buffering effect of drug on pH oscillations, the CSTR was run without circulation into the diffusion cell, runs being performed at several concentrations of drug, and for three drugs with different acid dissociation constants: BzA (pKa = 4.2), salicylic acid (SA: pKa = 2.96), and acetic acid (AA: pKa = 4.72) [30]. Oscillations were sped up by using higher reactant concentrations than before: 22 mM H_2SO_4, 194 mM Na_2SO_3, and 200 mM $NaBrO_3$. Results are summarized in Figure 5. At low drug concentrations, oscillations are sustained. At high concentrations, all drugs ultimately quench oscillations. At intermediate concentrations, oscillations are attenuated and may exhibit a "spindle" shape, ultimately decaying to steady state [30]. These shape factors may be due in part to drift in marble activity as the experiment proceeds. At high concentrations, the steady state pH value seems correlated with pKa of the drug. Of particular interest is the strong effect of drug pKa on the system's behavior. For example, oscillations are robust with 5 mM SA, but are quenched with 5 mM BzA, and AA quenches pH oscillations at and above 2 mM.

A mathematical model of the pH oscillator, without drug, was proposed by Rábai and Hanazaki (RH), who also presented the initial experimental studies of that system [34]. The RH model assumes three slow, irreversible reactions, with accompanying velocity expressions:

$$3HSO_3^- + BrO_3^- \rightarrow 3SO_4^{2-} + Br^- + 3H^+; \quad v_1 = k_1[BrO_3^-][HSO_3^-] \qquad (R1)$$

$$3H_2SO_3 + BrO_3^- \rightarrow 3SO_4^{2-} + Br^- + 6H^+; \quad v_2 = k_2[BrO_3^-][H_2SO_3] \qquad (R2)$$

$$H^+ + CaCO_3 \rightarrow Ca^{2+} + HCO_3^-; \quad v_3 = k_3[H^+] \qquad (R3)$$

Where $k_1 = 2.71 \times 10^{-2}$ M^{-1}sec^{-1}, $k_2 = 6$ M^{-1}sec^{-1}, and $k_3 =$ is a heterogeneous rate constant which depends on exposed marble surface area, stirring conditions, and chamber volume. In addition, the reaction of sulfite and bisulfite ions with H$^+$ is assumed to be reversible and fast, and is represented by equilibrium constants:

$$H^+ + SO_3^{2-} \leftrightarrow HSO_3^-; \quad K_4 = 10^{-7} \text{ M} \tag{R4}$$

$$H^+ + HSO_3^{2-} \leftrightarrow H_2SO_3; \quad K_5 = 1.8 \times 10^{-2} \text{ M} \tag{R5}$$

The effect of ionizable drug, A, is to introduce one more acid–base reaction into the mechanism [30]:

$$H^+ + A^- \leftrightarrow HA; \quad K_6 = 10^{-pKa} \text{ M} \tag{R6}$$

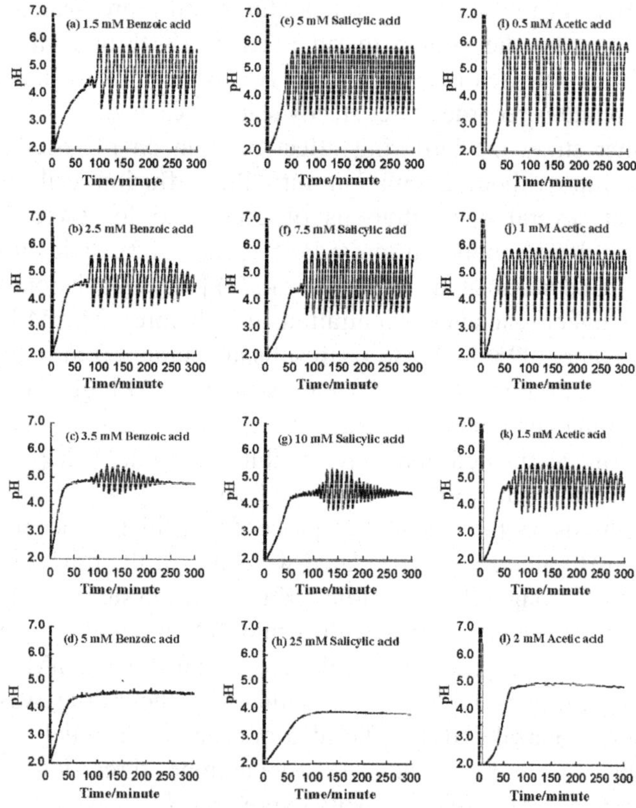

Figure 5. pH oscillations in CSTR with BzA (a–d), SA (e–h), and AA (i–l) in feed. (Reproduced from [30], with permission from Wiley.)

In the CSTR setup, reactants are introduced into and are removed from the reactor as part of a total volume flow, F. Denoting the volume of the fluid in the reactor by V, the ratio V/F is the residence time of fluid in the reactor, and its inverse, $k_0 = F/V$, is the fluid turnover rate. Differential equations representing conservation of total SO_3 containing species (**S**), acidic hydrogen (**H**) and bromate (**B**) based on the reaction mechanism are:

$$d\mathbf{H}/dt = -k_3[\mathrm{H}^+] + k_0(2[\mathrm{H_2SO_4}]_f + \alpha[\mathrm{A}]_f - \mathbf{H})$$

$$d\mathbf{S}/dt = -3k_1\mathbf{B}[\mathrm{HSO_3}^-] - 3k_2\mathbf{B}[\mathrm{H_2SO_3}] + k_0([\mathrm{Na_2SO_3}]_f - \mathbf{S})$$

$$d\mathbf{B}/dt = -k_1\mathbf{B}[\mathrm{HSO_3}^-] - k_2\mathbf{B}[\mathrm{H_2SO_3}] + k_0([\mathrm{NaBrO_3}]_f - \mathbf{B})$$

where the subscript "f" denotes concentration in the feed stream (half the concentrations in the two source flasks, due to mixing) and the coefficient α represents the fraction of drug added to the reactants in its acid (HA) form. In the experiments reported here, $\alpha = 1$.

Integration of the differential equations can be carried out, and forms of the pH curves *versus* time can be shown as a function of controllable parameters such as k_0, k_3 and feed concentrations of reactants. Instead of displaying such curves, we present a phase diagram (Figure 6) representing qualitative behaviors over the space of drug concentrations in the feed, $[\mathrm{A}]_f$ and marble reactivities, k_3, with all other parameters held constant. For each drug, a phase boundary (Hopf bifurcation) curve surrounds a region of (k_3, $[\mathrm{A}]_f$) space in which pH oscillates indefinitely. Outside the curve the system reaches a steady state, characterized by a constant pH, either by a smooth

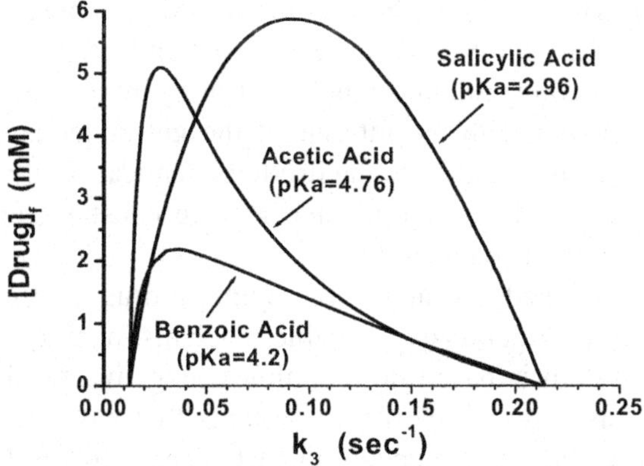

Figure 6. Phase diagrams representing presence or absence of sustained oscillations in reactions with varying drug concentration and marble reactivity. (Reproduced from [30], with permission from Wiley.)

decay, possibly following an overshoot, or by an exponentially decaying oscillation. Drug pKa has a salient impact on the phase boundary, but the relative positions of the curves are not readily explained.

To close this section, we remark that the Landolt pH oscillator has also been studied as a candidate for rhythmic drug delivery [29]. Unfortunately, only one transient pH swing was observed when BzA or nicotine were added, and the resulting flux of these agents across an EVAc membrane failed to oscillate more than once, probably due to buffering of the pH oscillator by drug. The modulating role of mono- and dihydrophosphate buffers, which serve as acidic proton acceptors and donors, respectively, on the performance of a mixed Landolt oscillator, have also been studied in detail [35,36]. Combining these results with the work summarized here, it is apparent that pH oscillators provide rather limited opportunities for rhythmic drug delivery. Even dilute concentrations of drug alter or compromise the behavior of the oscillators, and once delivered, drug will be further diluted in its volume of distribution in the body. Topical drugs may be an exception. Further, reduction of the experimental CSTR system to practice will introduce difficulties. Finally, all pH oscillators based on inorganic redox reactions are potentially toxic. Therefore, it is fruitful to seek alternative mechanisms for rhythmic drug delivery.

3. Hydrogel/enzyme oscillator for rhythmic GnRH delivery

In the previous section, the membrane's permeability to drug was considered constant, and oscillations in delivery were due to modulation of concentration of drug available for permeation by changes in pH. Alternatively, one could seek systems in which the membrane's permeability is pH sensitive and use pH oscillations to drive rhythmic drug delivery. Indeed, pH oscillators have been shown to generate undulatory behavior in hydrogels [36–42] and oscillations can be restricted to the interior of the gel under proper conditions [43–51]. However, in most cases the driving chemical oscillators are based, as before, on toxic redox reactions and require constant replenishment of reactants, limiting their practicality.

We have introduced a strategy for rhythmic drug or hormone delivery that utilizes only endogenously available reactants, namely glucose and O_2 [52–61]. This system is based on a nonlinear negative feedback instability principle, and has some similarity to systems that have been investigated both theoretically and experimentally in past decades [62–67]. The rhythmic delivery device, illustrated in Figure 7, consists of a chamber that contains the drug, and the enzymes glucose oxidase (*gluox*), catalase (*cat*), and gluconolactonase (*glulac*). The chamber is bounded by a thin hydrogel membrane, composed of poly(N-isopropylacrylamide-*co*-methacrylic acid)

(NIPA-*co*-MAA), lightly crosslinked with methylene bisacrylamide. The membrane separates the chamber from the external medium, which contains glucose at constant concentration and is held at physiological pH. The hydrogel shrinks and decreases its permeability to glucose and drug (for hydrophilic drugs which diffuse through aqueous channels in the hydrogel) with increasing local H^+ concentration, and swells, with increasing permeability, with decreasing local H^+ concentration. As will be shown below, swelling and shrinking occur with hysteresis.

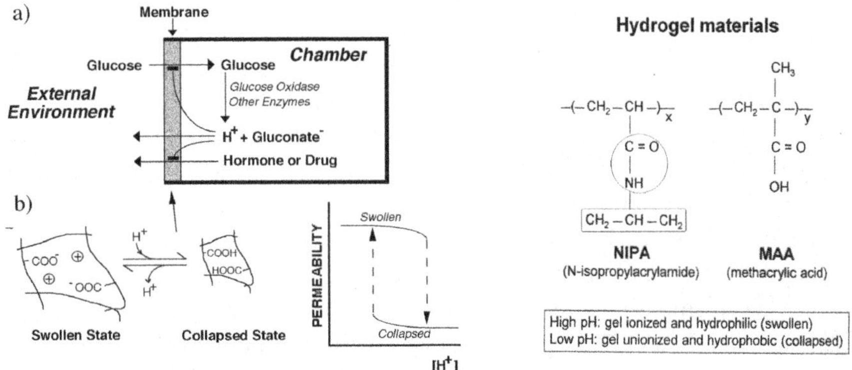

Figure 7. (a) Schematic of rhythmic drug/hormone delivery device, showing chamber containing drug and enzymes, bounded by membrane. (b) Characteristics of polyacid hydrogel membrane, which shrinks (decreases permeability) with increasing H^+ concentration and swells (increases permeability) with decreasing H^+ concentration, with hysteresis. c) Chemical structure of poly (NIPA-co-MAA) hydrogel membrane, highlighting regions of hydrophilic (*circle*) and hydrophobic (*rectangle*) hydration of NIPA. (Panel (a) reproduced from [57], with permission from Taylor & Francis.)

Here is how the rhythmic delivery device works. Starting with the membrane in the swollen state, glucose and O_2 diffuse through the hydrogel into the chamber, where the enzymes convert these reactants into an acidic proton and gluconate⁻ ion, according to the combined reaction

$$\text{glucose} + O_2 \xrightarrow{\;\text{enzymes}\;} H^+ + \text{gluconate}^- + \tfrac{1}{2}O_2$$

The released H^+ lowers the pH in the reaction chamber, and also diffuses into the membrane proximal to the chamber, binding to carboxylate⁻ (MAA) sidechains and neutralizing them. As the proximal hydrogel is neutralized, it forms a collapsed skin layer. Layers of the membrane close to the external medium are always swollen since the medium is at physiological pH, well above the pKa of the MAA sidechains, which is about 4.7. When the skin layer consolidates, it blocks glucose transport, and production of H^+ in the chamber is shut down. Eventually, the acidic protons bound to the MAA in the hydrogel are released, and diffuse into the external medium, where they are diluted and removed, e.g. by circulation. Meanwhile, pH in the chamber rises. Finally, when enough of the MAA is re-ionized, the collapsed skin swells again, reestablishing its permeability to glucose and drug. The system has now returned to its initial state, and is poised to enter another cycle of collapse and reswelling of the membrane skin. As the skin collapses and reswells repeatedly, drug or hormone is released rhythmically [58].

The description in the last paragraph is purely heuristic, and more conditions must be met for the system in Figure 7 to function as an oscillator. For example, a simple decrease in membrane permeability with increasing H^+ production (lowered pH in the chamber) is not sufficient to guarantee oscillations. A weak relation between pH and glucose permeability produces a stationary state, in which an intermediate glucose flux leads to a steady production of H^+ by the enzymes, and just sufficient protonation of the membrane to maintain glucose permeability at a point that enables the former intermediate flux of glucose to proceed. In engineering terms, there is a negative feedback loop between H^+ production and the glucose flux that enables the former. This feedback will arrive after a delay corresponding to diffusion of H^+ into the membrane and swelling kinetics of the gel in response to change in protonation state. Low negative feedback gains are known to stabilize systems at stationary values. High feedback gains accompanied by delay, on the other hand, can drive a system into oscillations [6,68,69]. Returning to the present context, this means that a minimal degree of sharpness in the pH-glucose permeability characteristic is needed to produce sustained oscillations [53].

Unfortunately, the sharper the permeability characteristic, the narrower the pH range over which oscillations can occur, and the system's behavior becomes extremely sensitive to small perturbations due, for example, to physiological fluctuations or drift in membrane parameters as the system ages [53]. A more reliable oscillator is based on a membrane whose glucose permeability is bistable over a range of pH values. The permeability characteristic of such a membrane, illustrated in Figure 7b, exhibits hysteresis, meaning that the pH at which the membrane "flips" from its high permeability state to its low permeability state, lies below the pH at which the membrane flips from low to high permeability. Within the pH bistability band, the membrane retains "memory" of its most recent history, hence the term "hysteresis."

Bistability/hysteresis is a common attribute of systems that undergo first order phase transitions. For example, liquid water can be superheated, remaining in the liquid state even though temperature exceeds 100°C. Conversely, water can be supercooled below 100°C. Both the superheated and supercooled states are metastable, meaning that they do not represent the lowest free energy state. With time, supercooled steam will condense and superheated water will evaporate. Similarly, at any pH within the swelling/permeability hysteresis band, one of the two swelling states will be more thermo-dynamically stable, but not necessarily accessible over the time frame of an oscillation. Persistence of a metastable state in a hydrogel may be enhanced by the interconnectedness of the polymer chains, which inhibits or at least retards local "nucleation" and "growth" of macroscopic domains [70].

As a candidate hydrogel that would undergo a pH-driven first order swelling/permeability phase transition, poly(NIPA-co-MAA), with 10 mol % MAA substitution was chosen based on considerable experience with similar hydrogels that undergo thermal phase transitions, i.e. with increasing and decreasing temperature. The chemical structure of randomly copoly-merized poly(NIPA-co-MAA) is illustrated in Figure 7c. The NIPA mono-mer is unusually structured insofar as its sidechain contains an amide group, which tends to be surrounded by "hydrophilic" water molecules, and an isopropyl group, which is "hydrophobically" hydrated. The two types of hydration waters are configurationally distinct, and it is believed that they produce a positive interfacial free energy where they meet [71–73]. The enthalpy and entropy of formation of hydrophobic waters around the iso-propyl group are negative, so hydrophobic hydration is favored below a lower critical temperature (LCT). Above LCT, hydrophobic hydration water escapes to the bulk, where the entropy is higher, and the isopropyl com-ponents of the chains aggregate, leading to hydrogel collapse.

Addition of MAA, even in small quantities, alters LCT [74,75]. When protonated, MAA is itself more hydrophobic than NIPA. Moreover, the protonated carboxylic acid group of PMAA can form a hydrogen bond with the amide group of NIPA, displacing hydrophilic hydration waters. As a result, LCT of neutral poly(NIPA-co-MAA) is lower than that of poly(NIPA). On the other hand, deprotonated MAA, with its negative charge, attracts water by ion-dipole forces, and breaks the hydrogen bond with NIPA. More-over, the charged MAA$^-$ sidechain requires the presence of a free, mobile counterion, usually Na$^+$, which contributes to an osmotic swelling pressure. The LCT of charged poly(NIPA-co-MAA) is therefore raised, according to degree of substitution of MAA, and the fraction of MAA units that are ionized. As the latter depends on pH, incorporating MMA into a hydrogel based on poly(NIPA) converts the thermal transition to a pH-driven phase transition.

To demonstrate that the poly(NIPA-co-MAA) hydrogel membrane exhi-bits the supposed bistability in permeability as a function of pH, a series of glucose transport experiments was carried out [55,76]. The membrane was mounted between two chambers of a diffusion cell (Figure 8a), with the donor phase (Cell I) held at pH 7.0 by an autotitrator, and pH in the receptor chamber (Cell II) was stepped down and up in 0.1 pH increments. Meanwhile, flux of ^{14}C-glucose across the membrane was monitored. Figure 8b shows the results of one experiment. Starting at pH 5.6 and decreasing by 0.1 pH every hour, flux was essentially constant until 4.9 was reached. At that point flux became strongly attenuated. Flux remained low as pH was stepped down to 4.5, and then brought back up to pH 5.2. At that point, the original glucose flux was restored, and persisted as pH was raised further. The difference in pH where flux shut off and the pH where flux was restored was suggestive of hysteresis and bistability. To demonstrate that this effect was not due to a kinetic delay, several other pH protocols were presented, and shutoff and restoration of glucose permeability always occurred at the same pH values (data not shown here). Bistability was unaffected by the presence of a background of cold (^{12}C) glucose.

Early attempts to base an oscillator on the pH-bistable membrane were unsuccessful [54]. Figure 9a is a schematic of the oscillator prototype. The same diffusion cell as was used in the flux experiments (see previous paragraph) was used, but instead of controlling pH in Cell II, a pH monitor electrode was installed and enzymes were dissolved in that cell. Enzyme concentrations were in excess to ensure that oscillations were not controlled by enzyme kinetics – if they were, then enzyme degradation with time

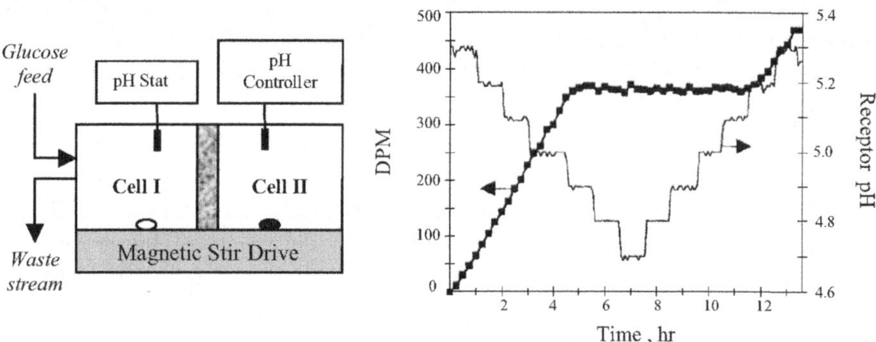

Figure 8. (a) Schematic of test cell used to measure transport of glucose across the poly(NIPA-*co*-MAA) hydrogel membrane as a function of pH in the receptor cell. Membrane separates Cell I and Cell II. The devise is water jacketed to provide constant temperature. (b) Transport of ^{14}C-glucose, recorded by liquid scintillation counting of receptor fluid across membrane (left ordinate) as a function of pH in Cell II (right ordinate), with static pH 7.0 in Cell I. Shutoff of glucose transport occurs around pH 4.9, and re-establishment of transport occurs around pH 5.2, indicating hysteresis. (Panel (b) reproduced from [76], with permission from Wiley.)

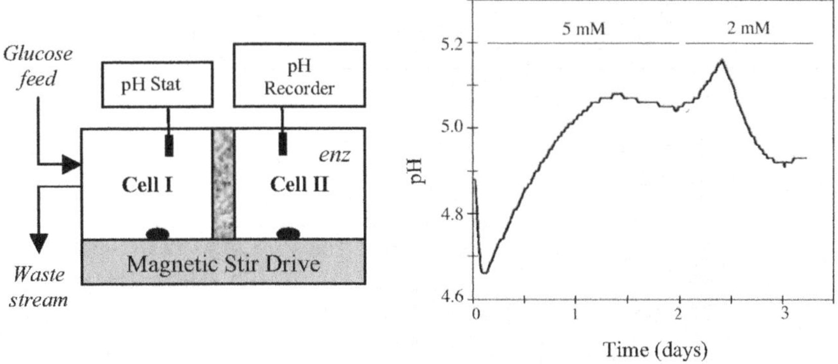

Figure 9. (a) Schematic of oscillator prototype, which is nearly identical to test cell in Figure 8, but with enzymes *gluox*, *cat*, and *glulac*, collectively denoted by *enz*, incorporated in receptor cell, and pH controller replaced by pH recorder, which digitally records and saves pH record on a PC. (b) Early attempt to elicit oscillations was unsuccessful, probably due to slow rates of pH change in receptor cell. (Panel (b) reproduced from [54], with permission from the American Institute of Physics.)

might affect performance. Figure 9b shows that a down- and upswing in pH occurred in Cell II, and the system slowly reached a steady state pH value. Restimulation of the receptor phase by changing glucose concentration flowing though Cell I led to another up/down transient in receptor phase pH, which again settled to a stationary point. In addition to the frustrating lack of sustained oscillations in Cell II, it was noticed that pH transients were extremely slow. Even if the oscillator worked, its rhythmic period would be too long.

A hypothesis was formed that the very slow swings in pH would permit the membrane to approach an intermediate steady state whose permeability lies somewhere between that of the collapsed state and that of the glucose-permeable state. It was further conjectured that accelerating the slow rate of pH change would force the membrane to choose between the high and low permeability states, bypassing the intermediate state. To this end marble, similar to that used in the experiments with the chemical pH-oscillator, was introduced into Cell II.

As already described, marble irreversibly reacts with hydrogen ion, and therefore acts as a "shunt" pathway for H^+ removal, increasing $|(d[H+]/dt)|/[H+] = 2.3|dpH/dt|$ [57]. Since this shunt will also reduce H^+ concentration in Cell II given a fixed glucose feed rate in Cell I, this feed rate must be increased to keep pH values in Cell II in a range that includes the region of bistability of the hydrogel membrane.

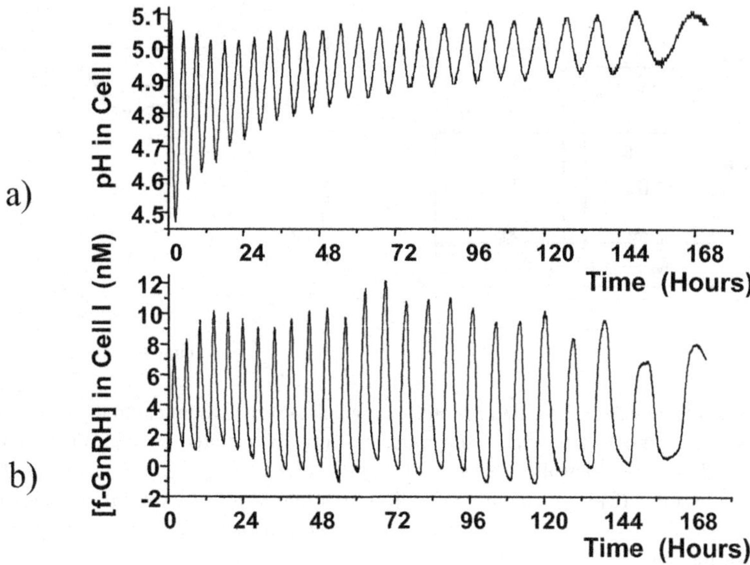

Figure 10. (a) pH oscillations in Cell II achieved when marble was introduced into it (cf. Figure 9), and glucose concentration entering Cell I was increased to 50 mM. (b) Oscillations in concentration of f-GnRH in Cell I, due to diffusion across the membrane during permeable phase, followed by removal by circulation of Cell I fluid. (Reproduced from [58], with permission from Elsevier.)

With proper selection of marble activity in the receptor chamber and glucose concentration in the donor chamber, pH swings were accelerated and sustained, robust pH oscillations were developed in the receptor chamber. pH oscillations, lasting a week, are illustrated in Figure 10a [57–59].

In the experiment leading to Figure 10a, fluorescein-labeled GnRH (f-GnRH) was initially loaded into Cell II. Figure 10b shows that f-GnRH was transported across the hydrogel membrane in nearly rhythmic pulses that were coherent with the pH oscillations. The graph in Figure 10b represents f-GnRH concentration in Cell I, and is a convolution of the delivery rate of f-GnRH across the membrane with the residency time distribution of f-GnRH in Cell I. Since that phase was well stirred, and solution was fed and removed at a constant rate, the residency distribution is that of a CSTR, i.e. monoexponentially declining with time constant equal to phase volume divided by feed flow rate. Deconvolution of this distribution from the raw data in Figure 10b provides an estimate of instantaneous permeation rate, and the resulting peaks are somewhat sharper than those in Figure 10b [58]. Nevertheless, the peaks in Figure 10b are well resolved since the residence time distribution of f-GnRH is much narrower than the interpeak interval.

While the one-week sustained oscillations can be regarded as a proof of principle, an ultimately practical implantable device will be required to function for a much longer durations, perhaps several years. The mechanism causing oscillations to cease after one week must therefore be identified and remediated. A look at the pH oscillations in Figure 10a provides potential clues. Oscillation amplitude initially decreases with time, primarily at the pH nadirs, but amplitude stabilizes after several oscillations. Also, the period between peaks increases with time, until oscillations finally cease. Reduction in amplitude in the early stages of the experiment may in fact be caused in part by increase in period, insofar as the membrane may not be able to respond immediately in its swelling to very rapid changes in pH. We therefore focus on mechanisms causing the slowing down of oscillations with time, and reasons why such slowing may lead to final cessation of oscillations.

When glucose is converted by *gluox* and *glulac*, the product is gluconic acid. This compound reversibly dissociates into H^+ and gluconate$^-$ ions, with pKa $\approx$ 3.5. The reaction of H^+ with $CaCO_3$ in marble leads to production of Ca^{2+} and bicarbonate$^-$ ions. Gluconate, bicarbonate, and calcium ions accumulate in Cell II and the hydrogel membrane, and may affect the kinetics of various processes. Both gluconate$^-$ and bicarbonate$^-$ act as pH buffers, which slow down pH changes in Cell II. Ca^{2+} may affect enzyme efficiency, or it may alter the relationship between pH in Cell II and hydrogel swelling, either by altering ionic strength or by specific interactions between this divalent ion and the carboxylate sidechains.

As of this writing, the strongest evidence implicates accumulation of gluconate in Cell II in the slowdown of oscillations [77]. Titration studies carried out on aliquots taken from Cell II at various stages of oscillations indicate a strong, increasing buffering presence of gluconate with time, which correlates well with increasing oscillation period. Bicarbonate ion appears to contribute very little to the titration properties of the system at any stage. Thus, if slowdown of pH changes is ultimately responsible for cessation of oscillations, then gluconate is the most likely culprit. It has also been shown that oscillations can be abruptly halted by spiking Cell II with a sufficiently high concentration of gluconic acid [77].

Bicarbonate's pKa $\approx$ 6.4, and it is therefore a much stronger base than gluconate. However, one gluconate equivalent is produced by the enzymes for each equivalent of glucose delivered into Cell II, while bicarbonate is produced more slowly, even in the presence of marble. Bicarbonate$^-$ (HCO_3^-) is also converted at the observed acidic pH values to H_2CO_3 and then CO_2. As all of these are small molecules compared to gluconate, they should be removed from Cell II very rapidly, either by diffusion across the hydrogel membrane or by evaporative loss in the case of CO_2.

Spiking Cell II with a high concentration of Ca^{2+} also halts oscillations abruptly, and oscillations are restarted by refreshing Cell II with calcium-free solution [77]. However, calcium accumulation studies have not been carried out yet, so the role of this ion is not clear, although it clearly has no role as a pH buffer. It is expected that Ca^{2+}, like bicarbonate species, will be produced more slowly than gluconate and will diffuse much more rapidly across the membrane.

Several auxiliary checks have been carried out [78]. Enzyme activity remains strong after one week, as expected since enzyme was incorporated in excess. Oscillations can be restarted by reconstituting Cell II with the "starter" solution, which is free of accumulated reaction products, and the oscillations strongly resemble those seen in Figure 10a, with initially decreasing amplitude and slowly growing period between peaks. Thus, cessation of oscillations need not be attributed to elimination of enzyme activity, or to irreversible damage to the membrane.

Combining the observations that oscillations are suppressed in the absence of marble, where pH changes are very slow, and that long-standing oscillations cease after a prelude of slowing oscillations, we decided to observe changes in membrane permeability in response to slow pH ramps in Cell II, with pH in Cell I held constant at pH 7.4 [79]. Figure 11 shows the time course of glucose flux across the membrane in response to three consecutive sets of down-then-up ramps of pH in Cell II. As expected, glucose flux was highly attenuated when pH was ramped below a certain value, and full flux was restored when pH was ramped above another value that was higher than the shutoff pH, indicating hysteresis.

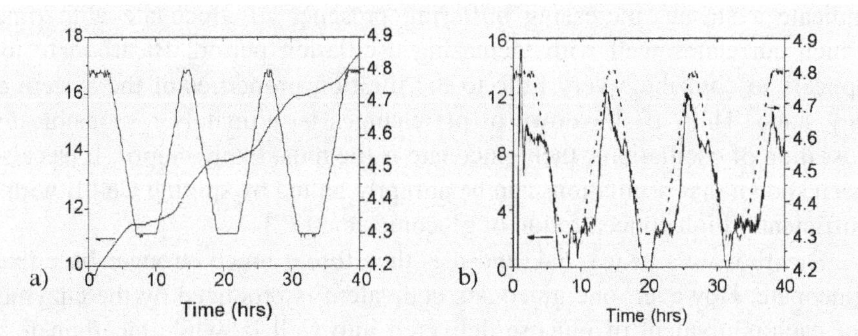

Figure 11. (a) Accumulated flux of glucose across membrane from Cell II to Cell I as function of pH in Cell II which is varied as ramps, as determined by pH stat in Cell I. (b) Time derivative of accumulation curve, reflecting glucose flux. Notice that initial shutoff is followed by transition to intermediate permeability, both occurring at pH 4.3. (Reproduced from [79], with permission from Wiley-VCH.)

A new feature however is the transition, after a certain period at low pH, of glucose flux from near zero to an intermediate value. This pattern repeated itself throughout the experiment. (Note: in these experiments glucose was loaded in Cell II, and the enzymes were in Cell I. Glucose flux, which was in the opposite direction than before, was determined by neutralizing acidic protons generated by the enzymes with NaOH from an autotitrator [79]).

We believe that the appearance of the intermediate glucose permeability state causes cessation of oscillations in the hydrogel/enzyme/marble system. At the intermediate permeability, a stable stationary attractor becomes available, in which glucose flux across the membrane from Cell I to Cell II exactly matches removal of H^+ in Cell II, a balance that is not sustained when the membrane flips between a highly permeable and a highly impermeable state.

Why does slowing of oscillations lead to an intermediate permeability state of the membrane? The lead hypothesis is derived from the observation that clamping the membrane with a pH gradient across it produces differential swelling stresses and strains in the membrane. In particular, the glucose permeation-controlling skin layer is under contractile stress when it collapses. With rapid changes between high and low pH in Cell II, the skin might collapse and swell uniformly. However, long periods of exposure to low pH in Cell II could lead to a thicker and denser skin, with increased lateral stress. At sufficiently high stress, these stresses may lead to mechanical phase separation and formation of channels permitting glucose permeaion through the skin, at the intermediate rate. This notion, illustrated in Figure 12, has been corroborated by photographic studies [79].

To summarize, sustained pH oscillations in Cell II, accompanied by coherent delivery pulses of GnRH, have been achieved in the glucose hydrogel/enzyme system, with the aid of marble. Slowing of oscillations over time appears to be due primarily to accumulation of gluconate in Cell II, which acts as a pH buffer. With sufficiently slow pH ramps, the membrane appears to revert to an intermediate permeability state, leading to stationary glucose flux and pH in Cell II and quenched oscillations.

This system is an early prototype of an implantable system that should function for months or years, so a number of refinements are needed. First, the system needs to be substantially reduced in size for implantation. Microfabrication approaches are being pursued towards this end. Second, the operating pH of the oscillator must be raised closer to physiological values and to function in the presence of physiological levels of bicarbonate and phosphate buffers. Third, it would be preferable for the system to operate without marble which, although not specifically toxic, adds mass to the system and is depleted as oscillations proceed.

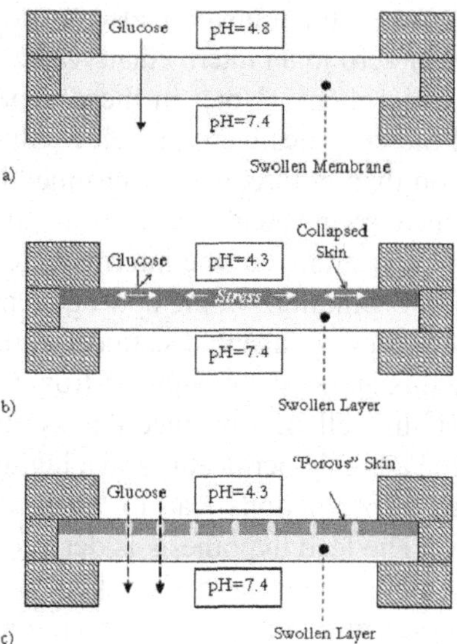

Figure 12. Behavior of clamped hydrogel membrane under pH gradient conditions, with "bottom" side fixed at pH 7.4. (a) At pH 4.8 on "top" side, the membrane is swollen throughout and glucose permeates at its maximal rate. (b) When top side is lowered to pH 4.3, the membrane forms a collapsed skin layer, initially blocking glucose permeation. Lateral stress builds up in the skin due to clamping. (c) Eventually stresses lead to later phase separation and development of "channels" through which glucose diffuses at an "intermediate" rate. (Adapted from [79], with permission from Wiley-VCH.)

Recent studies have shown that alkaline shifts in pH oscillation range can be accomplished by either decreasing the acidic monomer content in the hydrogel, or by increasing the hydrophobicity of the acidic comonomer, e.g. by substituting for MAA, n-alkylacrylic analogs of increasing alkyl chain length [59,61]. Figure 13 shows the results of substituting ethylacrylic acid (EAA) and butylacrylic acid (PAA). To explain these results, recall that swelling transitions occur when osmotic forces associated with hydrogel ionization overwhelm hydrophobic forces; hydrogel collapse occurs when the force dominance is reversed. With reduced concentration of acid groups in the hydrogel, a larger fraction of these groups must be ionized to produce a particular ion osmotic force, hence the requirement of a more alkaline pH.

With increasing hydrophobicity of the hydrogel, increased ionization is needed, also mandating higher pH. Since pH is determined by hydrogel ionisconcentration and the latter is determined by glucose flux, it turns out that glucose concentrations in Cell I needed to obtain oscillations are also reduced when the pH oscillation range is shifted in the alkaline direction.

An alternative path to shifting pH towards the physiologic range would be to utilize an acidic monomer whose intrinsic pKa lies close to pH 7. Bae et al. [80,81] have demonstrated that sulfonamide-based molecules can be functionalized with vinyl end groups and incorporated into hydrogels by standard free radical polymerization. Sulfonamides are noncarboxylic acids in which an electron withdrawing sulfonyl group acidifies the neighboring amine hydrogen. By configuring other electron withdrawing groups near the sulfonamide moiety, a range of pKa values can be achieved. Along these lines, we copolymerized acryloylated sulfamethoxypyridazine with NIPA and a small amount of MAA. As shown in Figure 14 this hydrogel, when mounted into the test cell, displays behavior suggestive of oscillatory potential at reduced glucose concentration (10 mM), in the presence of bicarbonate buffer (15 mM), and without addition of marble [78].

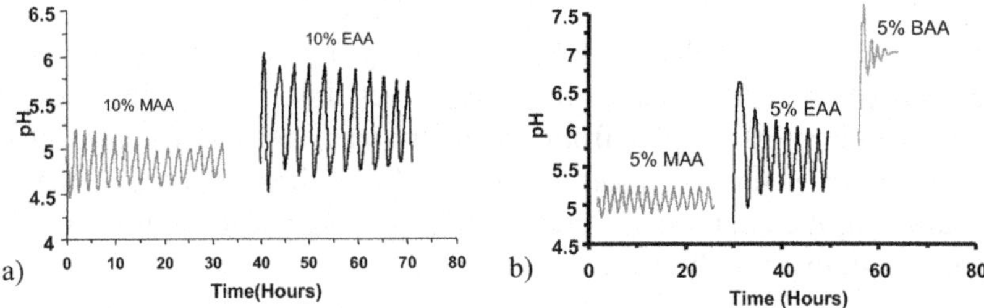

Figure 13. Alkaline shifts in pH oscillations in Cell II resulting from either increasing chain length of n-alkylacrylate comonomer (MAA = methacrylic acid, EAA = ethylacrylic acid, PAA = propylacrylic acid), or by decreasing n-alkylacrylate comonomer content (compare panels a and b). (Reproduced from [61], with permission from Wiley-VCH.)

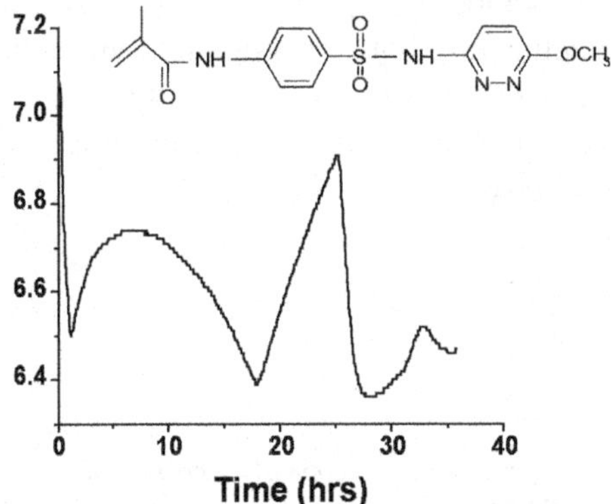

Figure 14. pH oscillations in Cell II when hydrogel membrane consists of crosslinked NIPA (97 mol %), MAA (1 mol %), and acryloylated sulfamethoxypyridazine (ASMP). The structure of ASMP is shown as an inset. Glucose concentration flowing through Cell I was 10 mM, and solutions in both cells were buffered by 15 mM bicarbonate. (Reproduced from [78].)

Thus, future work in improving oscillator performance should include both microfabrication and molecular design of the hydrogel, based on physico-chemical principles.

Acknowledgments

Various stages of this work were funded by Amgen, Inc., the National Science Foundation (CHE-961551), and the National Institutes of Health (HD-040366). Contributions by Drs. John P. Baker, Jean-Christophe Leroux, Xiaoqin Zou, Gauri P. Misra, Anish P. Dhanarajan, Amardeep S. Bhalla, Siddharthya K. Mujumdar and Eric Nuxoll are gratefully acknowledged.

Appendix: Derivation of expressions for J_Δ and Q_Δ

Our starting point is the well known theory for diffusional permeation across a membrane of thickness h into a perfect sink [32,33]. With permeant initially (time $t = 0$) absent from the membrane, and with permeant concentration in the source reservoir given by $C_0 H(t)$, where $H(t)$ is the Heaviside step function, the expression for accumulated drug permeation is, using the parameters defined in the text,

$$ Q_H(t) = PC_0 \left[t - \frac{\tau^*}{6} - 2\tau \sum_{n=1}^{\infty} \frac{(-1)^n}{n^2} e^{-n^2 t / \tau} \right] $$

The "pulse" of drug in the source reservoir, of concentration C_0 and duration Δ, can be represented by $C_0 H(t) - C_0 H(t - \Delta)$. Since the diffusion and partition coefficients are both independent of concentration, super-position gives $Q_\Delta(t) = Q_H(t) - Q_H(t - \Delta) \cong (dQ_H / dt)\Delta$, the latter pair approaching equality for $\Delta \ll \tau$. Differentiating once more in time, we obtain the expression for $J_\Delta(t)$.

References

1. T. G. J. Farmer, T. F. Edgar and N. A. Peppas, *J. Pharm. Pharmacol.* **60**, 1–13 (2008).
2. H. E. Koschwanez and W. M. Reichert, *Biomaterials* **28**, 3687–3703 (2007).
3. E. Renard, G. Costalat, H. Chevassus and J. Bringer, *Diabet. Metab.* **32**, 497–502 (2006).
4. G. M. Steil and K. Rebrin, *Expert Opin. Drug Deliv.* **2**, 353–362 (2006).
5. W. F. Crowley and J. G. Hofler, *The Episodic Secretion of Hormones* (Wiley, New York, 1987).
6. A. Goldbeter, *Biochemical Oscillations and Cellular Rhythms* (Cambridge University Press, Cambridge, U.K., 1996).

7.   G. Brabant, K. Prank and C. Schöfl, *Trends Endocrinol. Metab.* **3**, 183–190 (1992).

8.   J. D. Veldhuis, M. L. Johnson and W. K. Bolton, *Pediatr. Nephrol.* **5**, 522–528 (1991).

9.   C. H. Courtney, A. B. Atkinson, C. N. Ennis, B. Sheridan and P. M. Bell, *Metabolism* **52**, 1050–1055 (2003).

10. D. R. Matthews, B. A. Naylor, R. G. Jones, G. M. Ward and R. C. Turner, *Diabetes* **32**, 617–621 (1983).

11. P. E. Belchetz, T. M. Plant, Y. Nakai, E. J. Keogh, and E. Knobil, *Science* **202**, 631–633 (1978).

12. P. M. Conn and W. F. Crowley Jr., *Ann. Rev. Med.* **45**, 391–405 (1994).

13. N. Santoro, M. Filicori and W. F. Crowley Jr., *Endocr. Revs.* **7**, 11–23 (1986).

14. A. V. Schally, *Science* **202**, 18–28 (1978).

15. J. E. A. McIntosh and R. P. McIntosh, *J. Endocr.* **109**, 155–161 (1986).

16. G. L. Leyendecker, L. Wildt and M. Hansmann, *J. Clin. Endocrinol. Metab.* **51**, 1214–1216 (1980).

17. M. B. Southworth, A. M. Matsumoto, K. Gross, M. R. Soules and W. J. Bremner, *J. Clin. Endocrin. Metab.* **72**, 1286–1289 (1991).

18. J. O. Parker, M. H. Amies, R. W. Hawkinson, J. M. Heilman, A. J. Hougham, M. C. Vollmer and R. R. Wilson, *Circulation* **91**, 1368–1374 (1995).

19. D. Stanislaus, J. H. Pinter, J. Janovick and P. M. Conn, *Mol Cell Endocrinol* **144**, 1–10 (1998).

20. Y.-X. Li and A. Goldbeter, *Biophys. J.* **55**, 125–145 (1989).

21. M. H. Smolensky and N. Peppas, *Adv. Drug Del. Revs.* **59**, 828–851 (2007).

22. B. Lemmer, *J. Control. Release* **16**, 63–74 (1991).

23. A. E. Reinberg, *Ann. Rev. Pharmacol. Toxicol.* **32**, 51–66 (1992).

24. B. Bruguerolle, A. Boulamery and N. Simon, *J. Pharm. Sci.* **97**, 1099–1108 (2008).

25. M. Baraldo, *Expert. Opin. Drug Metabol. Toxicol.* **4**, 175–192 (2008).

26. I. R. Epstein and J. A. Pojman, *An Introduction to Nonlinear Chemical Dynamics*, Oxford, New York, 1998.

27. P. Gray and S. K. Scott, *Chemical Oscillations and Instabilities*, Clarendon Press, Oxford, 1990.

28. G. Rábai, M. Orbán and I. R. Epstein, *Acc. Chem. Res.* **23**, 258–263 (1990).

29. S. A. Giannos, S. M. Dinh and B. Berner, *J. Pharm. Sci.* **84**, 539–543 (1995).

30. G. P. Misra and R. A. Siegel, *J. Pharm. Sci.* **91**, 2003–2015 (2002).

31. G. P. Misra and R. A. Siegel, *J. Control. Release* **79**, 293–297 (2002).

32. J. Crank, *The Mathematics of Diffusion*, Clarendon Press, Oxford, 1975.

33. E. L. Cussler, *Diffusion*, Cambridge University Press, Cambridge, 1997.

34. G. Rábai and I. Hanazaki, *J. Phys. Chem.* **100**, 10615–10619 (1996).

35. M. Dolnik, T. S. Gardner, I. R. Epstein and J. J. Collins, *Phys. Rev. Lett.* **82**, 1582–1585 (1999).

36. A. J. Ryan, C. J. Crook, J. R. Howse, P. Topham, M. Geoghegan, S. J. Martin, A. J. Parnell, L. Ruiz-Pèrez and R. A. L. Jones, *J. Macromol. Sci. B Phys.* **44**, 1103–1121 (2005).

37. R. Yoshida, H. Ichijo, T. Hakuta and T. Yamaguchi, *Macromol. Rapid Commun.* **16**, 305–310 (1995).

38. J. R. Howse, P. Topham, C. J. Crook, A. J. Gleeson, W. Bras, R. A. L. Jones and A. J. Ryan, *Nano Lett.* **6**, 73–77 (2006).

39. A. J. Ryan, C. J. Crook, J. R. Howse, P. Topham, R. A. L. Jones, M. Geoghegan, A. J. Parnell, L. Ruiz-Pèrez, S. J. Martin, A. Cadby, A. Menelle, J. R. P. Webster, A. J. Gleeson and W. Bras, *Faraday Discuss* **128**, 55–74 (2005).

40. C. J. Crook, A. Smith, R. A. L. Jones and A. J. Ryan, *Phys. Chem. Chem. Phys.* **4**, 1367–1369 (2002).

41. P. Topham, J. R. Howse, C. J. Crook, S. P. Armes, R. A. L. Jones and A. J. Ryan, *Macromolecules* **40**, 4393–4395 (2007).

42. I. Varga, I. Szalai, R. Meszaros and T. Gilanyi, *J. Phys. Chem. B* **110**, 20297–20301 (2006).

43. R. Yoshida, M. Tanaka, S. Onodera, T. Yamaguchi and E. Kokofuta, *J. Phys. Chem. A* **104**, 7549–7555 (2000).

44. S. Sasaki, S. Koga, R. Yoshida and T. Yamaguchi, *Langmuir* **19**, 5595–5600 (2003).

45. R. Yoshida and T. Sakai, *Langmuir* **20**, 1036–1038 (2004).

46. R. Yoshida, T. Takahashi, T. Yamaguchi and H. Ichijo, *J. Am. Chem. Soc.* **118**, 5134–5135 (1996).

47. P. Labrot, P. DeKepper, J. Boissonade, I. Szalai and F. Gauffre, *J. Phys. Chem. B* **109**, 21476 (2005).

48. J. Boissonade, P. DeKepper, F. Gauffre and I. Szalai, *Chaos* **16**, 037110 (2006).

49. S. Villain, *Comportement mécanique de gels soumis à des réactions autocatalytiques.* Ph.D. Thesis, Université Paris 7 (2007).

50. J. Boissonade, *Phys. Rev. Lett.* **90**, 188302 (2003).

51. J. Boissonade, *Chaos* **15**, 023703 (2005).

52. R. A. Siegel and C. G. Pitt, *J. Control Release* **33**, 173–186 (1995).

53. R. A. Siegel, in: *Controlled Release: Challenges and Strategies*, edited by K. Park, American Chemical Society, Washington, DC, 1997, pp. 501–527.

54. J.-C. Leroux and R. A. Siegel, *Chaos* **9**, 267–275 (1999).

55. J.-C. Leroux and R. A. Siegel, in: *Intelligent Materials and Novel Concepts for Controlled Release Technologies*, edited by J. De Nuzzio, American Chemical Society, Washington, DC, 1999, pp. 98–112.

56. X. Zou and R. A. Siegel, *J. Chem. Phys.* **110**, 2267–2279 (1999).

57. R. A. Siegel, G. P. Misra and A. P. Dhanarajan, in: *Polymer Gels and Networks*, edited by Y. Osada and A. R. Khokhlov, Marcel Dekker, New York, 2002, pp. 357–372.

58. G. P. Misra and R. A. Siegel, *J. Control Release* **81**, 1–6 (2002).

59. A. P. Dhanarajan, G. P. Misra and R. A. Siegel, *J. Phys. Chem.* **106**, 8835–8838 (2002).

60. A. P. Dhanarajan, J. Urban and R. A. Siegel, in: *Nonlinear Dynamics in Polymeric Systems*, edited by P. J. Pojman and Q. Tran-Cong-Miyata, American Chemical Society, Washington, DC, 2003, pp. 44–57.

61. A. S. Bhalla, S. K. Mujumdar and R. A. Siegel, *Macromol. Symp.* **254**, 338–344 (2007).

62. A. Katchalsky and R. Spangler, *Quart. Rev. Biophys.* **2**, 127–175 (1968).

63. H.-S. Hahn, P. J. Ortoleva and J. Ross, *J. Theoret. Biol.* **41**, 503–521 (1973).

64. J.-F. Hervagault, M. C. Duban, J. P. Kernevez and D. Thomas, *Proc. Natl. Acad. Sci.* **80**, 5455–5459 (1983).

65. T. Teorell, *J. Gen. Physiol.* **42**, 831–845 (1959).

66. P. Meares and K. R. Page, *Proc. R. Soc. Lond. A.* **339**, 513–532 (1974).

67. R. Larter, *Chem. Revs.* **90**, 355–381 (1990).

68. P. E. Rapp, *J. Math. Biol.* **3**, 203–224 (1976).

69. T. Y.-C. Tsai, Y. S. Choi, W. Ma, J. R. Pomerening, C. Tang and J. E. Ferrell, *Science* **321**, 126–129 (2008).

70. K. Sekimoto, *Phys. Rev. Lett.* **70**, 4154–4157 (1993).

71. D. W. Urry, *J. Phys. Chem. B.* **101**, 11007–11028 (1997).

72. M. Shibayama, S. Mizutaini and S. Nomura, *Macromolecules* **29**, 2019–2024 (1996).

73. S. Sasaki and H. Maeda, *Phys. Rev. E.* **54**, 2761–2765 (1996).

74. S. Sasaki and F. J. M. Schipper, *J. Chem. Phys.* **115**, 4349–4354 (2001).

75. T. Tokuhiro and A. Tokuhiro, *Polymer* **49**, 525–533 (2008).

76. J. P. Baker and R. A. Siegel, *Macromol. Rapid Commun.* **17**, 409–415 (1996).

77. A. S. Bhalla, *Physicochemical Investigations of a Drug Delivery Oscillator*, Ph.D. Thesis, University of Minnesota, Minneapolis, MN, 2007.

78. A. Dhanarajan, *Mechanistic Studies and Development of a Hydrogel/Enzyme Drug Delivery Oscillator*, Ph.D. Thesis, University of Minnesota, Minneapolis, MN, 2004.

79. A. Dhanarajan and R. A. Siegel, *Macromol. Symp.* **227**, 105–114 (2005).

80. S. I. Kang and Y. H. Bae, *Macromol. Chem. Symp.* **14**, 145–155 (2001).

81. S. I. Kang and Y. H. Bae, *J. Control Release* **80**, 145–155 (2002).

STRUCTURE FORMATION AND NONLINEAR DYNAMICS IN POLYELECTROLYTE RESPONSIVE GELS

A.R. Khokhlov[*]
Chair of Physics of Polymers and Crystals, Physics Department, Moscow State University, Moscow 119991, Russia

I.Yu. Konotop, I.R. Nasimova, N.G. Rambidi
Chair of Physics of Polymers and Crystals, Physics Department, Moscow State University, Moscow 119991, Russia

Abstract. General principles of information processing by chemical light-sensitive reaction–diffusion media are discussed. New modes of image evolution in the process of its transformation by reaction–diffusion medium are proposed. New approach for the design of materials that are capable to exhibit oscillating swelling-deswelling behaviour based on complexes of catalyst of Belousov–Zhabotinsky (BZ) reaction with polymer gels is described.

Keywords: information processing, Belousov–Zhabotinsky reaction, mechano-chemical oscillation, polyelectrolyte responsive gels

1. Introduction

Various practical applications of chemical reaction–diffusion media based on nonlinear dynamic mechanisms such as pattern recognition, path planning, robot navigation, and many others were investigated during the last decade [1–15]. It is believed that nonlinear mechanisms perform an essential role in cardiac arrhythmia and fibrillation [16,17]. The design of new reaction-diffusion systems development of new approaches and techniques

---

[*] To whom correspondence should be addressed. e-mail: khokhlov@polly.phys.msu.ru

P. Borckmans et al. (eds.), *Chemomechanical Instabilities in Responsive Materials*,
© Springer Science + Business Media B.V. 2009

based on nonlinear media dynamics is one of the promising ways of investigation at the present time.

Between them practically important applications are based on the idea that complex responses of the medium to external prearranged physical stimuli could be equivalent to information processing operations [3–6]. In this case one can consider the reaction medium as continuous distributed information processing system operating in a highly parallel mode.

These approaches have been discussed in literature beginning from the late 1960s (see, for instance, [18]). Further development of image processing methods based on the unique properties of reaction–diffusion systems was achieved in pioneering studies performed by Lothar Kuhnert [1–3]. These important studies determined physical and chemical principles inherent in information processing by reaction diffusion media. Both, high parallelism and nonlinear dynamic mechanisms that increased greatly the medium behavioural complexity were in the basis of the elaboration of image processing methods. Detailed investigations of information processing capabilities in chemical light-sensitive reaction–diffusion media were performed in the beginning of nineties (see [4–6] and references therein). From those investigations it became clear that fundamental differences exist between:

- Image processing of positive and negative images of initial pictures
- Responses of the medium to the light excitations and information processing operations based on these responses and performed by the medium
- Information processing of black and white pictures, pictures having several levels of brightness, and half-tone pictures

There were several attempts during the last decade to develop numerical hardware independent techniques for image processing based on reaction–diffusion paradigm (see, for instance, [19–24]). They gave the possibility to enhance such important features of the picture under investigation as thin and thick contours, lines having different slope and corners of the image, its skeleton [23,24]. However, dynamic mechanisms inherent in real "hardware" chemical oscillators (Belousov–Zhabotinsky, Briggs–Rauscher and other media, see [25]) are sufficiently more sophisticated than theoretical descriptions used. As a result chemical oscillators demonstrate in experiment a variety of modes that could not be reproduced by numerical modelling. Some of them might be useful for image processing.

The basic goal of the investigations that were carried out in our group was to study the general principles and variety of modes inherent in chemical light-sensitive reaction–diffusion media that could be used for information (image) processing. The very important point was the employing modern polymeric materials as specialized matrices for the medium and immobilization of separate chemical components of the reaction–diffusion system [26].

New important regimes of image processing by reaction–diffusion media were found as a result of investigation based on improved modern technique. It enables one to explain qualitatively the whole set of the experimental data using simple Field–Körös–Noyes (FKN) model. And more, it became possible to make clear the adequacy between modes of reaction–diffusion medium and information processing operations performed by the medium.

The very attractive and principally new approach for the design of smart materials can be developed if the chemical reaction–diffusion media is combined with stimuli-responsive polymer gels. The introduction of components of Belousov-Zhabotinsky reaction to the gel media can induce the periodical volume changes as a result of propagation of chemical waves.

One of the pioneer studies considering such kind of systems was done by Yoshida et al. [27]. They reported the swelling-deswelling oscillations of the ionic N-isopropylacrylamide gel if the ruthenium tris(2,2′-bipyridine) units were covalently bonded to the polymer chains. Immersed in BZ reaction media, this $Ru(bpy)_3$ acts as a catalyst of BZ reaction in which periodic redox changes of catalyst moiety from $Ru(bpy)_3^{2+}$ to $Ru(bpy)_3^{3+}$ occurs. As it was shown, this chemical oscillation was converted into the mechanical oscillation of the polymer network. To introduce the $Ru(bpy)_3$ moieties to NIPAM gel, $Ru(bpy)_3$ with a double bond was especially synthesized.

Polymer gel is known to form complex systems with different types of organic compounds. Very stable complexes can be formed as a result of specific noncovalent interactions between polyelectrolyte gels and oppositely charged dyes [28–30]. The ordinary absorption of catalyst to the gel can significantly simplify the procedure of preparation of the catalyst containing gel media. New systems based on this approach were elaborated and investigated in our laboratory.

This paper has the following structure. In the second part, some methodological bases of image processing by reaction–diffusion media are discussed. Third part describes some results on the design of smart materials exhibiting mechano-chemical oscillation behaviour.

2. Distributed molecular information processing media

The methodological basis of image processing by reaction–diffusion media still remains vague enough till now. The process of the image evolution in the medium, not image processing, was considered virtually in the first study named "Image processing using light-sensitive chemical waves" [1–3] and in the following publications [4–6]. There were no attempts to understand how complete is the correspondence between modes of the image evolution in the medium and typical image processing operations used in practice. One of the most efficient ways to make clear this problem is to choose generally accepted and practically oriented information processing technique and to compare thoroughly basic principles, image processing operations and the variety of problems to be solved.

The important approach to consider this problem, that is the theoretical basis and technique of image processing named "mathematical morphology", was elaborated during the last decades.

Binary mathematical morphology operates with complex two-dimensional objects defined in discrete space. Object "A" is a set of pixel satisfied to the specific predetermined conditions:

In spite of numerical presentation of initial data the mathematical morphology operates images as a whole. Primitive operations of the mathematical morphology are dilation – $A \oplus B$ (that increases the image), and erosion $A\Theta B$ (that decreases it). Additional notion is introduced: together with object A; it is structural element B which determine the character of the shape changes of the object A at its border.

Two basic operations of the mathematical morphology are defined based on these elements: opening

$$A \circ B = (A\Theta B) \oplus B,$$

and closing

$$A \bullet B = (A \oplus B)\Theta B.$$

Detailed consideration shows that combined use of opening and closing enables to perform practically all basic image processing operations: contour enhancement, segmentation and so on.

Information processing features of mathematical morphology approach and image processing by Belousov-Zhabotinsky media are shown in Table 1. There are evident differences between these two approaches. First of all, one of these approaches is a numerical method realized by modern digital computers, whereas another is a material realization of reaction–diffusion media capabilities. Nevertheless the information processing origin of these methods is adequate. It is based on nonlinear mechanisms and wave character of information processing. Both of these approaches are biologically motivated. At the same time differences in information processing features of them are not a matter of principle. Therefore all operations inherent in mathematical morphology can be carried out by Belousov-Zhabotinsky media.

Processing of black and white images. There are two operations performed by Belousov-Zhabotinsky media that can be considered as initial elementary ones. They are "contour(+)" and "contour(−)" operations and represent contour enhancement of the input image following by expanding (contour(+)) or shrinking (contour(−)) of the contour figure revealed (Figure 1). The choice between these operations is determined by the negative or positive image form.

TABLE 1. Information processing features of mathematical morphology technique and reaction–diffusion processing.

Mathematical morphology	Chemical reaction–diffusion medium
Numerical method of image processing based on nonlinear transformation of images	Processing of images by a chemical medium based on nonlinear dynamic mechanisms
Operational data element in an image represented by a set of pixels	Operational data element is a single image
Consecutive pixel by pixel processing of the image by a numerical computer	Parallel processing of the image in all its points by a chemical reaction–diffusion medium
Multimode processing of images based on different types of structural elements	Only one circular structural element is virtually used
Dilation and erosion are initial image processing operations, two basic operations – opening and closing, could be reduced to initial ones	Contour(+) and contour(−) are initial image processing operations, opening and closing could be performed based on these operations
All practically actual black and white image processing operations could be performed	All practically actual black and white image processing operations could be performed
A variety of the half-tone image processing operations could be carried out	A variety of the half-tone image processing operations equivalent to operations performed by mathematical morphology could be carried out

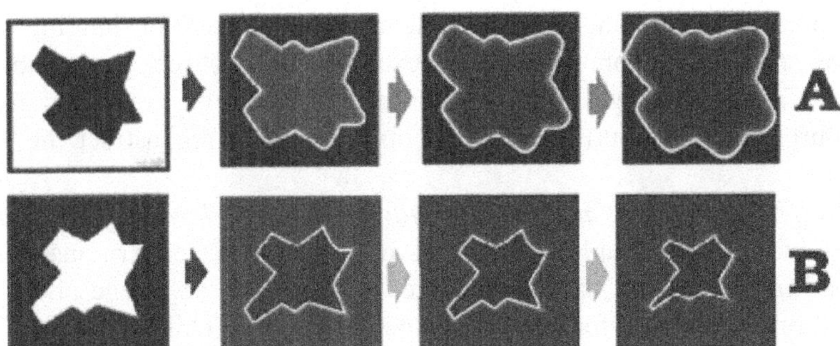

Figure 1. Contour(+) (A) and contour(−) (B) operations of a Belousov-Zhabotinsky medium [6].

It should be mentioned that considering image processing by Belousov-Zhabotinsky media positive and negative forms of the image should be distinguished.

To avoid uncertainties in the following discussion let us define the positive image of a picture as image corresponding to typical picture inherent in human surroundings. If the notion of "typical picture" is uncertain (suppose in the case of geometrical figures) let us define positive image as dark figure on the light background. Black and white and half-tone pictures will be used below. Images of these pictures can be considered as a set of optical density values D_i corresponding to each point of the picture ($0 < D_i < D_\infty$, where D_∞ is a maximum value of the optical density). The negative image of the picture was defined as a set of inverted density values ($D_i^N = D_\infty - D_i$).

These elementary operations are adequate to dilation and erosion operations of the mathematical morphology. Moreover, it is possible to reproduce also opening and closing operations, using contour(+) and contour(−) elementary operations performed by Belousov-Zhabotinsky medium. Therefore it is possible to conclude that reaction–diffusion medium can perform image processing operations typical for mathematical morphology.

All known image processing operation carried out by the mathematical morphology technique can be performed by Belousov-Zhabotinsky media. Most of them can be carried out using contour (+) and contour(−) operations, equivalent to dilation and erosion. A lot of examples were published elsewhere (see, for instance, [5–6]). They embrace: smoothing of immaterial features of the figure (enhancement its general shape) and segmentation of the figure, enhancement and removing of small features of the image, thinning, skeleton and Voronoi diagram calculation and so on.

Opening and closing operations, performed by Belousov-Zhabotinsky media enable to carry out more complex image processing operations. As examples, removing and deepening details of the image or thinning of the image are shown. More complicated example is defect repair and removing noises from an image (Figure 2). Here consecutive combined use of contour(+) and contour(−) gives the opportunity to reconstruct the initial image.

Half-tone images and images having several levels of brightness. Considerable and notable possibilities of the complex pattern analysis are opened in the case of half-tone images. In this case an image under consideration is first transformed into its negative form. This transformation is the continuous process when the negative image appears step by step beginning from the most dark (or the brightest, depending on positive or negative form of initial image) fields of the image. This capability of Belousov-Zhabotinsky media is one of the most remarkable medium feature. Namely,

Belousov-Zhabotinsky medium is the natural realization of the temporal sequence processor, that transforms complex spatial distribution of visual information into temporal sequence of its fragments. Because of this feature Belousov-Zhabotinsky media are capable to solve complex practically important problems. Let us give some examples.

The "hidden image" is a fragment of the picture having brightness very close to the brightness of the background. The example of this situation is shown in Figure 3. The difference in brightness between the eagle image and the background is 10 units of Photoshop HSB model. The evolution of the picture in the Belousov-Zhabotinsky medium enables to enhance the hidden image in spite of a very big difference of its brightness in comparison with the background.

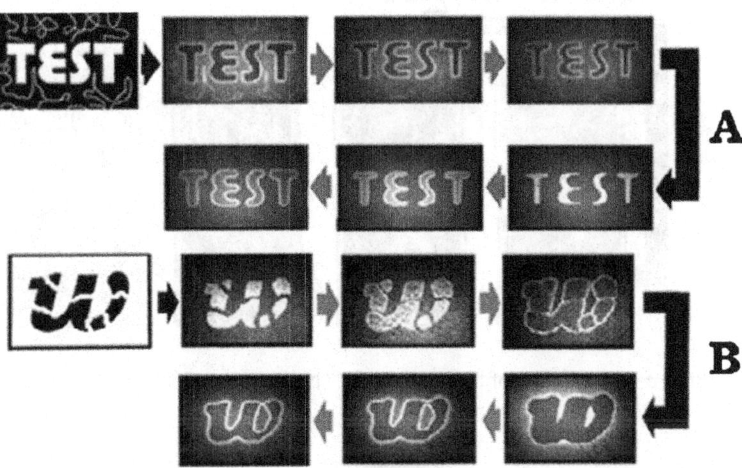

Figure 2. Removing noises (A) and defect repair (B) performed by Belousov-Zhabotinsky media [6].

Figure 3. Enhancement of the hidden image by a Belousov-Zhabotinsky medium [6].

The important part of the image analysis in medicine, material science and some other fields is to reveal consecutively fields of the image having increasing (or decreasing) brightness. Examples of this process performed by Belousov-Zhabotinsky medium are shown in Figure 4.

One of the useful tasks of the half-tone image processing is watershed operations. They give the opportunity to enhance the relief of the hill-like image. The example of the watershed operation performed by Belousov-Zhabotinsky medium is shown in Figure 5.

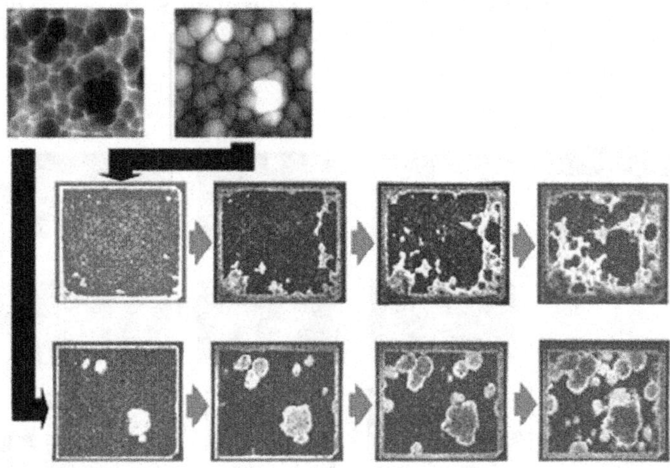

Figure 4. Processing of half-tone images by Belousov-Zhabotinsky media [6].

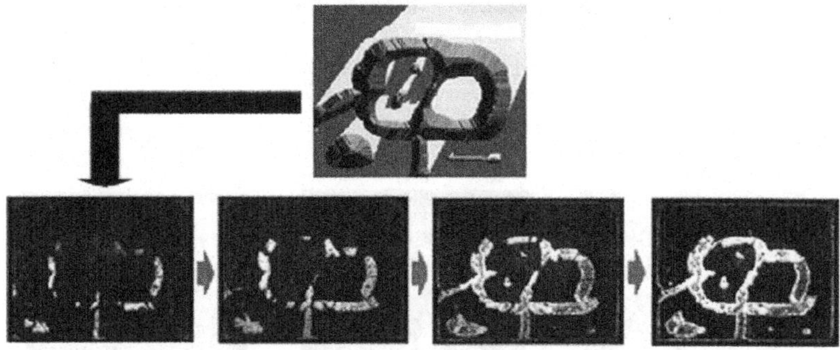

Figure 5. Watershed operation carried out by a Belousov-Zhabotinsky medium [6].

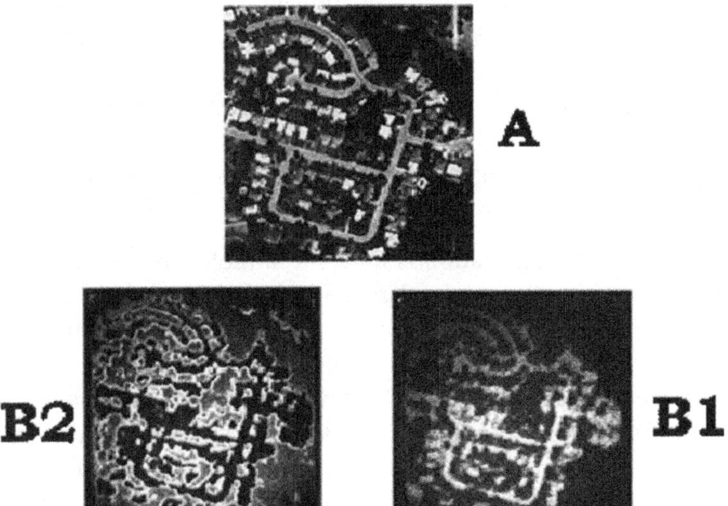

Figure 6. Urban road detection by mathematical morphology technique (B1) and Belousov-Zhabotinsky media (B2) using the same urban map (A) [6].

An attempt to use mathematical morphology technique for urban road detection was performed lately. The urban map and results of the preliminary detection of the road network by the mathematical morphology technique are shown in Figure 6. The detection of the road network by Belousov-Zhabotinsky type medium using the same urban map is shown in Figure 6. It is easy to see, that these two road networks are in a good correspondence.

What's next?

There are three mutually correlated hypostasis's of reaction–diffusion paradigm that define the understanding of its potentialities. They are:

- Theoretical reasons
- Operational opportunities to implement them
- The most important fields of application

Theoretical reasons. Unlike the digital von Neumann computer, the reaction–diffusion device is not a rigid structurally predetermined system. Its dynamics depends on the composition and control stimuli variations. It enables implementation of rather effective control of the dynamics. Even simple reaction–diffusion systems display high degree of parallelism, discrete and continuous dynamic mixed, and vertical flow of transmission and processing information. Even in a simple system the following features can be determined:

- The level of macro-micro type transformation of information (particularly, photochemical transformation of a light two-dimensional picture into pseudo two-dimensional distribution of reaction components)
- Dynamics at the molecular level implementing a definite information process
- The level of micro-macro information transformation, i.e. physico-chemical reading information

High degree of organization is inherent in molecular media of this type. Moreover, these media display gradualism, i.e. small changes of active medium composition, within defined limits, lead only to small quantitative, not qualitative changes of system dynamics.

It is essential that media of Belousov-Zhabotinsky type have also other characteristics necessary for displaying adaptive behaviour. Among these is the nature of interaction of the system with the environment, feedback organization, etc. Therefore it is possible to conclude that it could really be possible to create a device, capable of learning, having multilevel architecture and a high degree of behavioural complexity. In general the detailed comparison of the information features of the brain, the digital von Neumann computer and reaction–diffusion media leads to the conclusion that there is a sufficiently greater analogy between features of reaction–diffusion media and the brain, which is the system that solves problems of high computational complexity in a natural way.

Potentialities for the commercial use. Regretfully, experimental technique used now for material implementation of these theoretical reasons is in a primitive embryonic state. Therefore the laboratory-scale devices discussed above looks nowadays as beautiful toys, not as a basis for commercial devices. However the thorough analysis shows that reaction–diffusion media seem to be promising even now for fabrication of new effective information processing means.

Let us return to the example discussed above – contour enhancement by reaction–diffusion medium. The time of this primitive operation performed by the medium is about 1–5 s. It was shown above that the number of floating-point operations necessary to solve this problem by digital computer is about 5×106 if the resolution of the picture is 103x103. Average time of floating point operation (multiplication) performed by Pentium III processor (600 Mhz) is about 3×10^{-9} s. And the time of contour enhancement is $\sim 10^{-2}$ s. However this time increases greatly if the resolution of the picture is $104 \times 104 - \sim 1$ s, or $105 \times 105 - 100$ s. The computational complexity of the object in this example is not high. Nevertheless the computational performances of reaction–diffusion medium and digital computer can be comparable. Given a problem of high computational complexity the performance of the reaction–diffusion medium can be sufficiently increased even by orders-of-magnitude.

The very important advantages of reaction–diffusion means are simplicity and low commercial mass production costs. The material structure of reaction–diffusion devices is immeasurably simpler than VLSI. Therefore the technological basis for the fabrication of reaction–diffusion devices is simpler and cheaper. The reaction–diffusion media are impurity tolerant in comparison with semiconductor devices. It leads to additional opportunities to decrease processing capabilities and cost.

Fields of application. The progress of the human society faces more and more new challenges – simple, cheap, autonomous, massively deployed means capable to collect and store information, and to solve intellectual problems such as recognition, control and navigation. These systems should sense external stimuli and respond in real or near-real time. Important challenges embrace, for instance, the ocean self exploration, the nano satellite system deployment, military and dual-use applications such as mobile robots that work in dynamically changing environments.

The real time for these systems is often ~0.1 to 1.0 s. Moreover, the shelflife of such means can be rather short.

Chemical and biochemical reaction–diffusion media seem to be an attractive basis for fabrication of these devices. They are capable of natural style performing of intellectual operations. It's should be mention that information processing capabilities and problems that can be efficiently solved based on reaction–diffusion systems seem to be far from exhausted. A number of approaches could result in advanced powerful information processing means. Several possibilities should be mentioned which might be important for the future development reaction–diffusion information processing means. Promising theoretical and experimental investigations were performed during the last several years (see details in [5]) though most of them was not bound directly to information processing. However it should be reasonable to suppose that because of biological nature of the reaction–diffusion paradigm the biologically inspired principles would be indispensable to greatly broaden information processing capabilities. The most important between them seems to be the multi level organization of biological information processing. The understanding of the importance of multi level architecture was growing during the last years [5]. And it is necessary to take the next step – to realize, to choose, and to put together:

- Practical problems most adequate to non von Neumann computing
- To find multi level algorithms for solving these problems
- To elaborate experimental technique to implement these algorithms

The success of this approach depends greatly on the further development of reaction–diffusion "hardware".

3. Design of smart materials exhibiting mechano-chemical oscillation behaviour

As it was discussed in introduction, new approach for the design of smart materials can be developed if components of chemical reaction–diffusion media are incorporated to the stimuli-responsive polymer gels.

In recent years stimuli-responsive polymer gels are widely investigated due to their unique property undergo abrupt swelling-deswelling transition when the solvent quality changed (e.g. by changing the external temperature or adding poor solvent). This phenomenon is called polymer gel collapse. It was first predicted theoretically by Dusek [31] and observed experimentally by Tanaka [32]. Later it was realized that collapse is a consequence of the coil-globule transition of the network chains [32,33]. Character of the gel collapse depends essentially on the presence of charges in the network: for polyelectrolyte networks this is an abrupt, very cooperative first-order phase transition, while for neutral networks collapse is usually continuous [34]. This distinction in the behaviour is due to the additional exerting osmotic pressure of counter ions in polyelectrolyte gels [34,35]. This osmotic pressure can be formed only by mobile counter ions. Decrease of the concentration of mobile counter ions inside the gel and change of the ionization degree of the polyelectrolyte gel can easily affect its swelling degree, this is why gels containing components of BZ reaction can exhibit mechanical oscillation as a result of propagation of chemical waves.

One way of the immobilisation of a catalyst of BZ reaction to the stimuli-responsive polymer gels is catalyst absorption governed by electro-static interactions. This is very simple procedure that is not required any labour-consuming methods of system preparation. In this case, choice of catalyst for the reaction will not be restricted by the availability of catalyst-containing substances with a double bond or some other special chemical structure.

In our previous investigation the complex formation of polyelectrolyte gels and linear polymers with oppositely charged organic dyes was studied in detail to understand some general features of the polyelectrolyte gel/ oppositely charged substance complex formation [28–30].

The immersion of polyelectrolyte gel to the solution of organic dyes (see, for example, ref. [29]) leads to the effective dye absorption to the gel accompanied by the gel collapse. The absorption mechanism is as follows: due to the ion exchange reaction and hydrophobic interactions, the dye mole-cules are concentrated and aggregated in the gel phase. The aggregation of dye molecules in oppositely charged gel is more advantageous than in the solution since inside the gel the aggregate charges are neutralised by the counterions initially immobilised on the gel chains. In this case the losses in the translational entropy are smaller than in the solution. The stability of gel/dye complexes in water and salt-water solutions was investigated. It was shown that oppositely charged dyes generally form stable complexes with polyelectrolyte gels. It is interesting enough, that obtained complex systems exhibit all features of the individual compounds. For example, as it was shown [see ref. 29] for alizarin and catechol violet which are known to be chelating ligands for specific adsorption of aluminium ions, the instantaneous colour change of the gel phase is observed after the placement of the gel/dye complexes to the $Al_2(SO_4)_3$ solutions. Such kinds of spectra changes are characteristic for dye/Al^{3+} complexes.

On the base of these results, it can be assumed that complexes governed by electrostatic interactions seem to be very prospective system for the immobilisation of catalyst to the polymer matrix.

Choosing the proper system, it is necessary to take into account that gels used for this purpose, have to contain some charged constituents and exhibit strong enough elastic and mechanical properties.

In our studies few different polymer matrices were tested and compared according their absorption ability, mechanical strength and amplitude of mechano-chemical oscillation. Ferroin ($[Fe(o\text{-}phen)_3]SO_4$) was used as a catalyst of BZ reaction.

One of the systems that were investigated is polyacrylamide-silica gel. This kind of complex systems was already used for the incorporation of BZ reagents (see for example [36]). In these studies the effect of the mechanical deformations of an elastic excitable media on the wave propagation was investigated.

It was known that in water solution silica particles usually carry out negative charge on the surface (the principal mechanism by which glass and silica surfaces acquire a charge in contact with water is the dissociation of silanol groups). This is why silica particles immobilized into polymer gel should adsorb positively charged ferroin catalyst.

Another system used for catalyst immobilisation is anionic polyelectrolyte gels. In this case, negatively charged units are directly cross-linked to the gel network.

Let us now describe some obtained experimental results. Both polyacrylamide-silica gels and anionic gels were synthesized by free radical polymerisation of acrylamide in the presence of cross-linking agent. To obtain anionic gels, 10% to 20 mol. % of acrylic acid was added as a co-monomer to the reaction media. In the case of polyacrylamide-silica gels, polymerisation was carried out in the presence of different amount of Na_2SiO_3. As a result chemically cross-linked acrylamide gel with incorporated inside silica particles were formed simultaneously during the polymerisation process.

The absorption of positively charged ferroin catalyst into acrylamide-silica and acrylamide-co-acrylic acid gels was investigated as function of silica and acrylic acid concentration. It was found that ferroin catalyst was affectively absorbed by both systems but for the acrylamide-silica gels the absorbed ferroin concentration does not exceed 60% of the initial concentration in the solution.

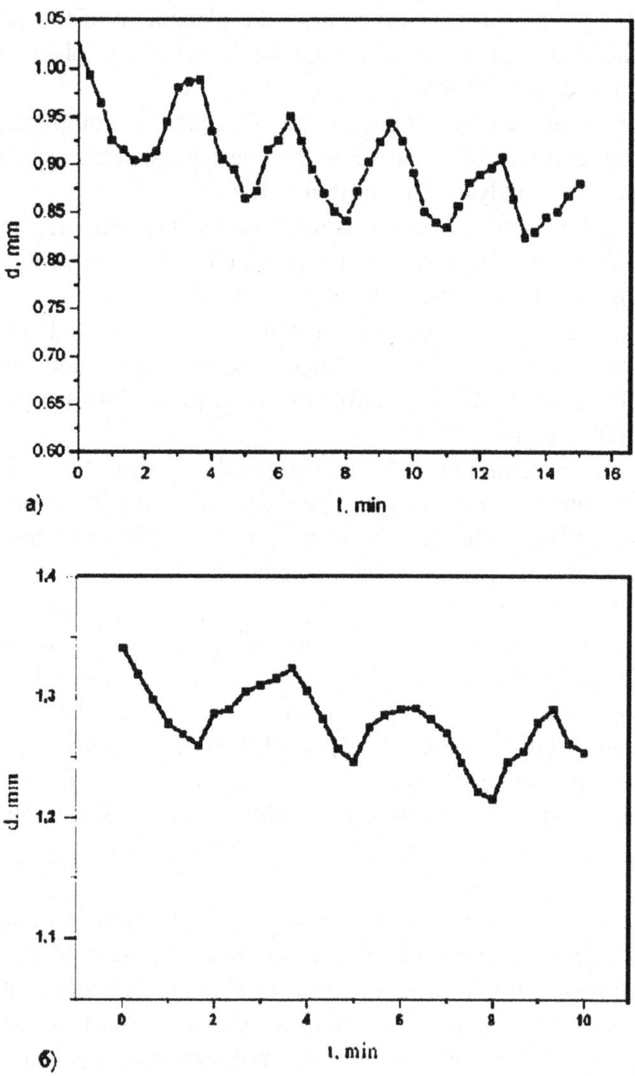

Figure 7. Variation of gel diameter with time. Acrylamide-silica/catalyst (a), acrylamide-co-acrylic acid/catalyst (b) complex systems [37].

Figure 7 represents the periodic changes of the diameter of the gel samples if they are immersed in the BZ reaction media (0.08 M $NaBrO_3$ solutions containing 0.6 M H_2SO_4 and 0.6 M malonic acid) for acrylamide-silica and acrylamide-co-acrylic acid gels/ferroin complexes.

Oxidation of ferroin to the ferrine results in the periodic gel shrinking. Possible mechanism of this chemically induced shrinking is the additional cross-linking of the gel sample when extra intercharge complexes are formed due to the overcharging of $Fe(phen)_3^{2+}$ to $Fe(phen)_3^{3+}$.

If to compare mechanical properties of the investigated systems it has to be mentioned that for the acrylamide-silica gels, increase of Na_2SiO_3 content in the reaction media leads to the decrease of the strength modulus of the gels (Figure 8). Less cross-linked acrylamide gel is formed in the presence of silica. At certain Na_2SiO_3 concentration the resulting gel becomes too fragile to be used as the mechano-chemical oscillating media. Unlike to the acrylamide-silica gels, introduction of charge groups to the acrylamide-co-acrylic acid gel almost does not affect mechanical properties, at least in the investigated range of acrylic acid concentrations. That is certainly one of the advantages of the acrylamide-co-acrylic acid gels in comparison with acrylamide-silica gels.

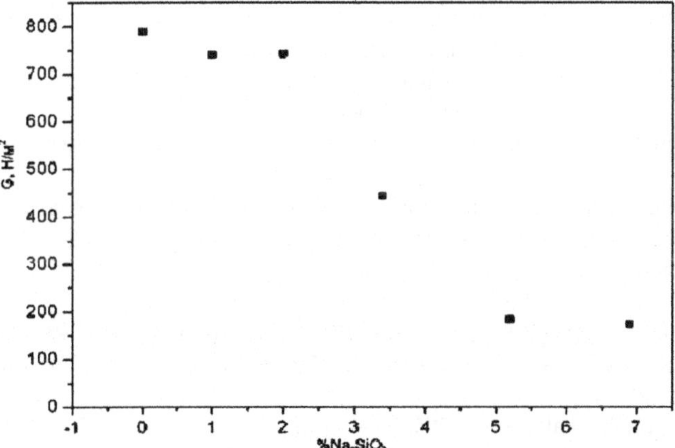

Figure 8. Elastic modulus of acrylamide-silica gel as function of Na_2SiO_3 concentration in polymerisation media [37].

In conclusion it has to be noted that principal possibility of application of electrostatic interactions for the immobilisation of catalyst to the polymer matrix and for the design of smart systems that undergo periodical volume changes without any external stimuli was shown. However, the amplitude of the mechanical oscillation that was observed for all system under investigation is not higher then 1% of the initial gel size. One possible way to increase it is use weakly-charged thermosensitive gels at temperatures close to the shrinking transition temperature as a gel matrix. In this case, any minor change of charge concentration in the system should significantly affect the swelling degree. Some preliminary experiments aimed to the development of such kind of systems were already started in our group.

References

1. L. Kuhnert, *Naturwissenschaften* **73**, 96 (1986).
2. L. Kuhnert, *Nature* **319**, 393 (1986).
3. L. Kuhnert, K.I. Agladze and V.I. Krinsky, *Nature* **337**, 244 (1989).
4. N.G. Rambidi and A.V. Maximychev, *Biosystems* **41**, 195 (1997).
5. N.G. Rambidi, *Int. J. Unconventional Comput.* **1**, 101 (2005).
6. N.G. Rambidi, S.G. Ulyakhin and A.S. Tsvetkov, *Unconventional Computing 2005: From Cellular Automata to Wetware* (Ch. Teucher and A. Adamatzky, Eds., Luniver Press, Beckington, 2005).
7. K.I. Agladze, N. Magome, R.R. Aliev, T. Yamaguchi and K. Yoshikawa, *Physica D* **106**, 247 (1997).
8. O. Steinbock, P. Kettunen and K. Showalter, *J. Phys. Chem.* **100**, 18970 (1996).
9. O. Steinbock, A. Toth and K. Showalter, *Science* **267**, 868 (1995).
10. N.G. Rambidi and D. Yakovenchuk, *Phys. Rev. E*, **63**, 026607 (2001).
11. A. Hjelmfelt, F.W. Schneider and J. Ross, *Science* **260**, 335 (1993).
12. J.-P. Laplante, M. Pemberton, A. Hjelmfelt and J. Ross, *J. Phys. Chem.* **99**, 10063 (1995).
13. I. Sendina-Nadal, A.P. Munuzuri, D. Vives, V. Perez-Munuzuri, J. Casademunt, L. Ramirez-Poiscina, J.M. Sancho and F. Sagues, *Phys. Rev. Lett.* **80**, 5437 (1998).
14. A. Adamatzky, B.P.J. De Lacy Costello, C. Melhuish and N. Ratcliffe, *J. Intell. Robotic Syst.* **37**, 233–249 (2003).
15. A. Adamatzky and B.P.J. De Lacy Costello, *Phys. Lett. A* **309**, 397–406 (2003).
16. J.M. Dadivenko, A.V. Pertsov, R. Salomonsz, W. Baxter and J. Jalife, *Nature* **355**, 349 (1992).
17. F.X. Witkowski et al., *Nature* **392**, 78 (1998).
18. W.K. Pratt, *Digital Image Processing* (Wiley, New York, 1978).
19. H. Meinhardt and M. Klinger, *J. Theor. Biol.* **126**, 63 (1987).
20. G. Turk, *Comput. Graph.* **25**, 289 (1991).
21. A. Witkin and M. Kass, *Comput. Graph.* **25**, 299 (1991).
22. C. B. Price, P. Wambacq and A. Oosterlinck, *IEEE Proc.* **137**, 136 (1990).
23. I.V. Nuidel and V.G. Yakhno, *Studia Biophysica* **132**, 137 (1989).
24. N.S. Belliustin, S.O. Kuznetsov, I.V. Nuidel and V.G. Yakhno, *Neurocomputing* **3**, 231 (1991).
25. R.J. Field and M. Burger (Eds.), *Oscillations and Traveling Waves in Chemical Systems* (Wiley-Interscience, New York, 1985).
26. N.G. Rambidi, T.O.-O. Kuular and E.E. Makhaeva, *Adv. Mat. Optics Electron.* **8**, 163–171 (1998).

27. R. Yoshida, E. Kokufuta and T. Yamaguchi, *Chaos* **9**, 260 (1999).
28. C.H. Jeon, E.E. Makhaeva and A.R. Khokhlov, *J. Polym. Sci. B Polym. Phys.* **37**, 1209 (1999).
29. I.R. Nasimova, E.E. Makhaeva and A.R. Khokhlov, *Polymer Science (Russian), Ser. A* **42**, 319–324 (2000).
30. E.G. Mamchits, I.R. Nasimova, E.E. Makhaeva and A.R. Khokhlov, *Polymer Science (Russian) Ser. A* **48**, 91 (2006).
31. K. Dusek and D. Patterson, *J. Polym. Sci., Polym. Phys. Ed.* **6**, 1209 (1968).
32. T. Tanaka, *Phys. Rev. Lett.* **40**, 820 (1978).
33. A.R. Khokhlov, *Polymer* **21**, 376 (1980).
34. T. Tanaka, D.J. Fillmore, S.-T. Sun, J. Nishio, G. Swislow and A. Shah, *Phys. Rev. Lett.* **45**, 1636 (1980).
35. V.V. Vasilevskaya and A.R. Khokhlov, in *Mathematical Methods for Polymer Studies* (I.M. Lifshitz, A.M. Molchanov, Eds., ONTI NCBI, Puschino (1982)), p. 45.
36. A.P. Muñuzuri, C. Innocenti, J.-M. Flesselles, J.-M. Gilli, K.I. Agladze and V.I. Krinsky, *Phys. Rev. E* **50**, R667 (1994).
37. I.Yu. Konotop, I.R. Nasimova, N.G. Rambidi and A.R. Khokhlov, Autooscillating systems on the base of polymer gels; *Polymer Science (Russian, Vysokomolekulyarnye Soedineniya) Ser. A.* **51**, 1 (2009).

NONLINEAR CHEMICAL DYNAMICS IN SYNTHETIC POLYMER SYSTEMS

John A. Pojman[*]
Department of Chemistry, Louisiana State University, Baton Rouge, LA 70803

Abstract. The application of the methods of nonlinear chemical dynamics to synthetic polymer systems is considered. We review the differences between polymers and inorganic systems that have been the subject of nonlinear dynamics. We consider two methods for approaching the problem – coupling polymers to other nonlinear systems and using inherent nonlinear behavior of polymers. We specifically focus on frontal polymerization.

Keywords: autocatalysis, feedback, frontal polymerization

1. Introduction

What is to be gained from applying the methods and concepts of nonlinear dynamics to polymer systems? Are there things that nonlinear dynamicists can learn, or that polymer scientists can make, that would not be possible without bringing these two apparently disparate fields into contact? First we briefly review some distinguishing characteristics of polymers. Next, we suggest three challenges that present themselves. We then examine sources of feedback in polymeric systems. Next, we propose several approaches to develop nonlinear dynamics with polymers. Finally, we give some examples of results that suggest that these approaches are likely to bear fruit.

We do not have the space to review polymers but refer the reader to several texts [1–3]. What we seek to provide here is a brief overview of the most important differences between polymeric systems and small molecule ones, review sources of feedback, approaches to nonlinear dynamics with polymers, including some specific examples.

We will not deal with biological systems, which certainly are polymeric systems. Biological systems are nonlinear dynamical systems but they are complex enough and so important that they warrant separate treatment. We refer the reader to Goldbeter's book [4] for an introduction to the topic.

---

[*] To whom correspondence should be addressed. e-mail: john@pojman.com

P. Borckmans et al. (eds.), *Chemomechanical Instabilities in Responsive Materials*,
© Springer Science + Business Media B.V. 2009

1.1. WHAT IS SPECIAL ABOUT POLYMERS?

The distinguishing feature of polymers is their high molecular weight. The simplest synthetic polymer consists from hundreds to even millions of a single unit, the monomer, that is connected end to end in a linear chain. However, a distribution of chain lengths always exists in a synthetic system. The molecular weight distribution can be quite broad, often spanning several orders of magnitude of molecular weight.

Linear polymers, are often *thermoplastic*, meaning they can flow at some temperature, which depends on the molecular weight, e.g., polystyrene. Polymers need not be simple chains but can be branched or networked. Crosslinked polymers can be gels that swell in a solvent or *thermosets*, which form rigid 3-dimensional networks when the monomers react, e.g., epoxy resins. This interconnectedness allows long-range coupling in the medium.

The physical properties of the reaction medium change dramatically during reaction. For example, the viscosity almost always increases orders of magnitude. These changes often will affect the kinetic parameters of the reaction and the transport coefficients of the medium.

Phase separation is ubiquitous with polymers. Miscibility between polymers is the exception.

1.2. CHALLENGES

In contemplating the possible payoffs from applying nonlinear dynamics to polymeric systems, one might ask

1. Are there new materials that can be made by deliberately exploiting the far-from-equilibrium behavior of processes in which polymers are generated?
2. Are there existing materials and/or processes that can be improved by applying the principles and methods of nonlinear dynamics?
3. Are there new nonlinear dynamical phenomena that arise because of the special properties of polymer systems?

2. Sources of feedback

In order to observe the types of nonequilibrium self-organization seen with inorganic systems such as the Belousov-Zhabotinsky reaction (see Epstein and Pojman for a discussion [5]), the polymer systems must exhibit feedback. Synthetic polymer systems can exhibit feedback through several mechanisms. The simplest is thermal autocatalysis, which occurs in any exothermic reaction. The reaction raises the temperature of the system, which increases the rate of reaction through the Arrhenius dependence of the rate constants. In a spatially distributed system, this mechanism allows propagation of thermal fronts. Free-radical polymerizations are highly exothermic.

Free-radical polymerizations of certain monomers exhibit autoacceleration at high conversion via an additional mechanism, the isothermal "gel effect" or "Norrish-Trommsdorff effect" [6–9]. These reactions occur by the creation of a radical that attacks an unsaturated monomer, converting it to a radical, which can add to another monomer, propagating the chain. The chain growth terminates when two radical chains ends encounter each other, forming a stable chemical bond. As the polymerization proceeds, the viscosity increases. The diffusion-limited termination reactions are thereby slowed down, leading to an increase in the overall polymerization rate. The increase in the polymerization rate induced by the increase in viscosity builds a positive feedback loop into the polymerizing system.

The reaction of dianhydrides with diamines can follow autocatalytic kinetics if performed in the proper solvent [10]. The reaction of the amine with the anhydride creates a carboxylic acid that catalyzes reaction of the amine with an anhydride. Amine-cured epoxy systems exhibit autocatalysis because the attack on the epoxy group is catalyzed by OH, and an OH is produced for every epoxy group that reacts [11–13]. The synthesis of polyaniline by oxidation of aniline has been shown to be autocatalytic, if performed electrochemically [14] or by the direct chemical oxidation [15,16]. Because polymerization reaction are organic reactions, more study should be made of autocatalysis in organic synthesis.

Some polymer hydrogels exhibit "phase transitions" as the pH and/or temperature are varied [17,18]. The gel can swell significantly as the conditions are changed and can also exhibit hysteresis [18,19].

Most polymers are not miscible. Introducing chemical reactions to an initially miscible polymer mixture often leads to phase separation [20]. Autocatalytic behavior driven by chemical reactions and concentration fluctuations in miscible polymer mixtures was recently found in photo-cross-linked polymer mixtures [21]. Concentration fluctuations increase as the reaction proceeds, leading to the condensation of photoreactive groups labeled on one of the polymer components. This condensation leads to an increase in the reaction yield that, in turn, accelerates the concentration fluctuations. A positive feedback can thus be built in the reacting mixture under appropriate conditions.

If two immiscible polymers are dissolved in a common solvent, which is then removed by evaporation, phase separation will occur. If the solvent is removed rapidly, non-equilibrium patterns may result [22].

The necking phenomenon observed upon stretching a polymer film at a constant temperature is a well-known consequence of a negative feedback loop driven by the interplay between the increase in temperature associated with the sample deformation and its glassification caused by the heat exchange with the environment [23]. Oscillatory behavior and period-doubling in the stress resulting from a constant strain rate have been experimentally observed.

Diffusion of small molecules, usually solvents, into glassy polymers exhibits "anomalous" or "non-Fickian" behavior [24]. As the solvent penetrates, the diffusion coefficient increases because the glass transition temperature is lowered. The solvent acts as a plasticizer, increasing the free volume and the mobility of the solvent. Thus we have an autocatalytic diffusion process. This can be relevant in Isothermal Frontal Polymerization, which we discuss below.

Dissolving of some polymers in aqueous media can proceed by a front [25]. Water dissolvable polymers are formed from esters, which create an acid upon hydrolysis that can catalyze further hydrolysis.

Finally, polymer melts and solutions are usually non-Newtonian fluids [26–28]. They often exhibit shear thinning, which means the viscosity decreases as the shear is increased. This can lead to unusual phenomena. For example, when a polymer melt is extruded through a die, transient oscillations can occur [29,30]. (Polymers can also exhibit shear thickening.)

An unusual phenomenon is the Weissenberg effect, or the climbing of polymeric liquids up rotating shafts [28]. A Newtonian fluid, on the other hand, is depressed by rotation because of centrifugal forces.

3. Approaches to nonlinear dynamics in polymeric systems

We propose three approaches to creating nonlinear dynamical systems with polymers:

- Couple polymers and polymer-forming reactions to other nonlinear systems (Type I)
- Create a dynamical system using the inherent nonlinearities in polymeric systems (Type II)
- Polymer systems are invariably characterized by polydispersity of the molecular weight distribution. One should be able to exploit the distribution of polymer lengths to amplify nonlinear effects in polymer systems, perhaps because of the molecular weight dependence of the diffusion coefficient. We know of no experimental work but there has been a theoretical work considering such an effect on ester interchange reactions [31].
- Investigate the effects of reaction-dependent diffusion coefficients on known instabilities.

3.1. TYPE I: COUPLING TO ANOTHER NONLINEAR SYSTEM

3.1.1. *Nonlinear chemical system*

Given the importance of the BZ reaction in nonlinear chemical dynamics, it is not surprising that polymers and polymerizations would be coupled to it. Váradi and Beck had shown that adding acrylonitrile to the BZ reaction could inhibition oscillations and a precipitate was produced that they assumed was polyacrylonitrile [32]. Pojman et al. studied the BZ reaction to which acrylonitrile was added and showed that, after an inhibition period, the polyacrylonitrile was produced periodically in phase with the oscillations (Figure 1) [33]. Given that radicals are produced periodically from the oxidation of malonic acid by ceric ion, it seemed reasonable to assume the periodic appearance of polymer was caused by periodic initiation. However, Washington et al. showed that periodic termination by bromine dioxide caused the periodic polymerization [34].

An exciting application of coupling to another nonlinear system was demonstrated by Yoshida et al. who created a self-oscillating gel by coupling a pH oscillating reaction with a polymeric gel that expands and contracts with changes in pH [35]. They have also used a gel in which the ruthenium catalyst of the BZ reaction is chemically incorporated into polymer [36,37]. Yashin and Balazs have also considered the coupling of the BZ system to gels [38,39].

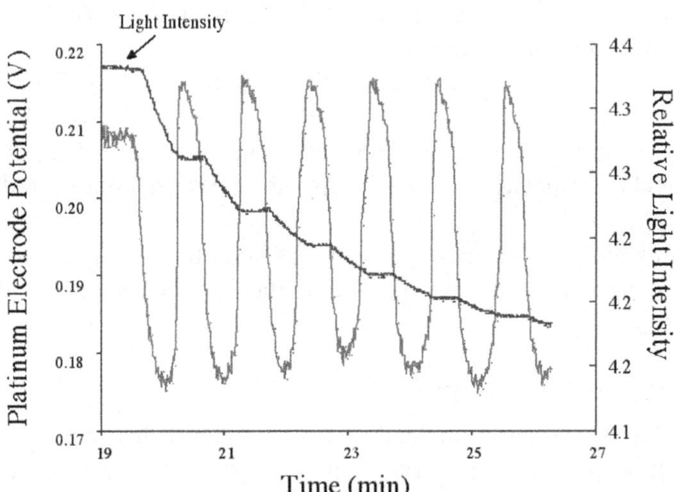

Figure 1. The evolution of a BZ reaction in which 1.0 mL acrylonitrile was present before the Ce(IV)/H_2SO_4 solution was added. [$NaBrO_3$]$_0$ = 0.077 M; [Malonic Acid]$_0$ = 0.10 M; [Ce(IV)]$_0$ = 0.0063 M; [H_2SO_4]$_0$ = 0.90 M. No oscillations occurred the first 15 min. (Adapted from [33].)

3.1.2. *Convective systems*

The nonlinear system need not be a chemical reaction. Kumacheva has used buoyancy-driven convection to generate patterns and then fixes them with polymerization [40,41]. Karthaus has used solvent dewetting to create micrometer sized domes of polymer on a solid substrate [42].

There is not always a clear distinction between Type I and Type II systems. In some cases the nonlinearities of the gel also play a role. Gauffre et al. and Labrot et al. have found that "if spatially bistable reaction systems are operated in size responsive chemosensitive gels, the size changes can provide a feedback which beyond plain reaction diffusion instabilities can be the source of new self-organizing phenomena, referred to as chemo-mechanical structures." [43–46]. Siegel and his colleagues utilized the hysteresis in a hydrogel's permeability to create autonomous chemomechanical oscillations in a hydrogel/enzyme system driven by glucose [47,48].

3.2. TYPE II: USING THE INHERENT NONLINEARITIES IN A POLYMER SYSTEM

3.2.1. *Oscillations in a CSTR*

With their combination of complex kinetics and thermal, convective and viscosity effects, polymerizing systems would seem to be fertile ground for generating oscillatory behavior. Despite the desire of most operators of industrial plants to avoid nonstationary behavior, this is indeed the case. Oscillations in temperature and extent of conversion have been reported in industrial-scale copolymerization [49].

Teymour and Ray reported both laboratory-scale CSTR experiments [50] and modeling studies [51] on vinyl acetate polymerization. The period of oscillation was long, about 200 min, which is typical for polymerization in a CSTR. Papavasiliou and Teymour reviewed nonlinear dynamics in CSTR polymerizations [52].

Emulsion polymerization as well has been found to produce oscillations in both the extent of conversion and the surface tension of the aqueous phase [53].

3.2.2. *Spatial pattern formation*

Typical phase separation leads to a two-phase disordered morphology. Multiphase polymeric materials with a variety of co-continuous structures can be prepared by controlling the kinetics of phase separation via spinodal decomposition using appropriate chemical reactions. By taking advantages of photo-crosslinking and photoisomerization of one polymer component in a binary miscible blend, Tran-Cong-Miyata and coworkers [54,55] have been able to prepare materials, known as semi-interpenetrating polymer networks, and polymers with co-continuous structures in the micrometer range.

4. Frontal polymerization

Frontal polymerization (FP) is a process of converting monomer into polymer via a localized reaction zone that propagates through the monomer. There are three modes of FP.

4.1. ISOTHERMAL FRONTAL POLYMERIZATION

Isothermal Frontal Polymerization (IFP), also called Interfacial Gel Polymerization, is a slow process in which polymerization occurs at a constant temperature and a localized reaction zone propagates because of the gel effect [56,57]. Figure 2 shows an image of a Gradient Refractive Index lens prepared by IFP.

Lewis et al. studied the mechanism of IFP with methyl methacrylate using Laser Line Deflection to determine front position and the front profile [58] and determined the factors that affect front propagation [59] Evstratova et al. confirmed that the process is indeed isothermal [60].

Figure 2. An image of a GRIN lens created by a radially-propagating front of methyl methacrylate polymerization from an annulus (1.5 cm) of poly(methyl methacrylate). Naphthalene was initially present in the monomer and accumulated as the front propagated inward.

Figure 3 presents a schematic of the mechanism. IFP proceeds as follows: Monomer and initiator dissolves into a polymer "seed", i.e., a high molecular weight piece of poly(methyl methacrylate). Because of the gel effect, the rate of polymerization is much faster in this viscous region than in the bulk solution. However, we must remember that the initiator is decomposing throughout the solution and so polymerization is occurring everywhere. It is also possible to add a polymeric inhibitor to extend the time of propagation [57]. Figure 4 shows a typical front's position as a function of time, for three different temperatures.

There are three distinct features of IFP. First, the total propagation distance is small. Secondly, the velocity is also very small and not constant. The propagation stops when the entire solution has polymerized. The front accelerates because polymerization is still occurring in the monomer-initiator solution away from the seed. The viscosity in the bulk solution is thus increasing slowly.

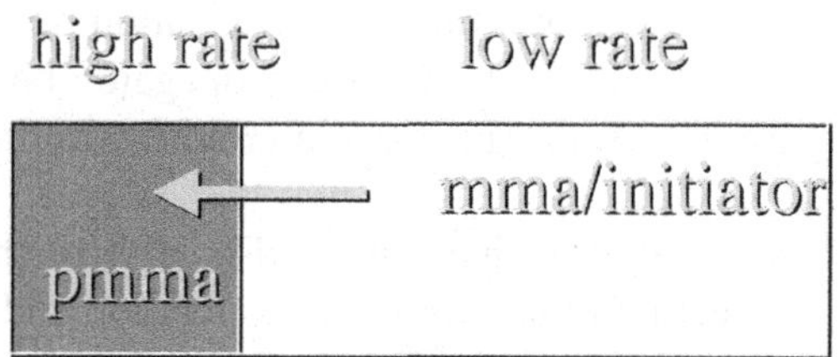

Figure 3. Mechanism for IFP. The polymerization rate is faster in the high viscosity region.

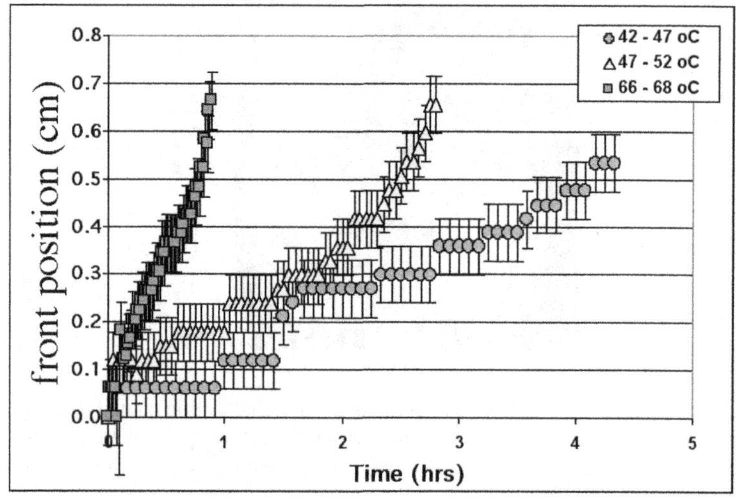

Figure 4. Propagation of IFP front as a function of time, for three different temperatures. (Adapted from [59].)

Photofrontal polymerization is process driven by the continuous influx of radiation. Typical systems involve a photoinitiator that absorbs the photons and dissociates into free radicals that initiate polymerization. If the dissociation products continue to absorb radiation, then the front position depends logarithmically on time [61]. If the initiator is photobleached, then the front position depends linearly on time [62–66].

4.2. THERMAL FRONTAL POLYMERIZATION

Frontal polymerization is a mode of converting monomer into polymer via a localized reaction zone that propagates, most often through the coupling of thermal diffusion and Arrhenius reaction kinetics. Frontal polymerization reactions were first discovered in Russia by Chechilo and Enikolopyan in 1972 [67]. They studied methyl methacrylate polymerization to determine the effect of initiator type and concentration on front velocity [68] and the effect of pressure [69]. The literature up to 1984 was reviewed by Davtyan et al. [70].

4.2.1. *Basic phenomena*

Frontal polymerization reactions are relatively easy to perform. In the simplest case, a test tube is filled with the reactants. The front is ignited by applying heat to one end of the tube with an electric heater. Fronts with free-radical polymerization propagate with velocities from 1 to 20 cm/min. The position of the front is obvious because of the difference in the optical properties of polymer and monomer (Figure 5).

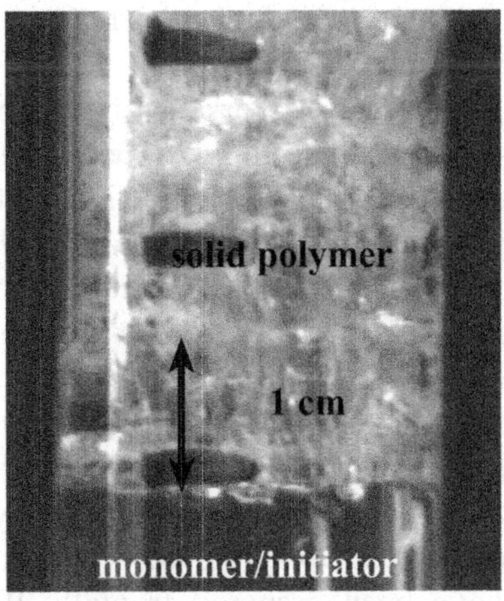

Figure 5. A descending case of frontal polymerization with triethylene glycol dimethacrylate and benzoyl peroxide as the initiator.

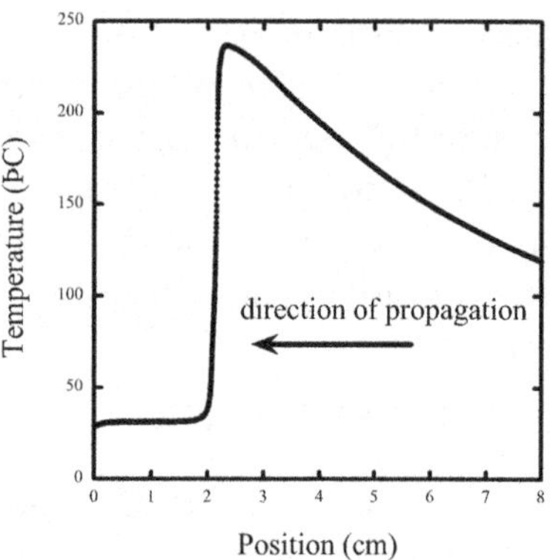

Figure 6. A typical temperature profile for a free-radical polymerization front.

The defining feature of frontal polymerization is the sharp temperature gradient present in the front. Figure 6 shows a typical temperature profile for a thiolene polymerization front [71]. Notice that the temperature jumps about 200°C over as little as a few millimeters, which corresponds to polymerization in a few seconds at that point. For multifunctional acrylates, such as trimethylolpropane triacrylate (TMPTA), maximum front temperature should exceed 400°C, if the reaction proceeded to 100% conversion. One factor that can limit conversion is the equilibrium dependence on temperature [72] and the other is the inherent low conversion obtained with multifunctional acrylates caused by the crosslinking.

4.2.2. *What systems can be performed frontally?*

The requirements for frontal polymerization are a system that does not react at the chosen initial temperature, but does react rapidly at an elevated temperature. The reaction must be exothermic.

An overwhelming majority of work has been on free-radical systems [73] with acrylates and methacrylates [74–76] because of the high reactivities of these monomers. Nason et al. studied the UV-ignited frontal polymerization of acrylates and methacrylates [77] Other free-radical systems can be used such as unsaturated polyester resins [78], and thiol-enes [71], Jiménez and Pojman studied frontal polymerization with polymerizable ionic liquid monomers [79].

Begishev et al. studied frontal anionic polymerization of ε-caprolactam [80,81] and epoxy chemistry has been used as well [82–88]. Mariani et al. demonstrated Frontal Ring-Opening Metathesis Polymerization [89] and Fiori et al. produced polyacrylate/poly(dicyclopentadiene) networks frontally [90]. Polyurethanes have been prepared frontally [91–93]. Frontal atom transfer radical polymerization (ATRP) has been achieved [94].

Photo-activated and induced epoxy systems have been cured frontally. Oxetanes and oxiranes via photoinduced cationic ring opening [95]. Mariani et al. developed an epoxy system in which UV light reacted with a cationic photoinitiator to start the frontal curing of an epoxy. The front propagated through the thermal-induced decomposition of benzoyl peroxide. The radicals produced reacted with the cationic photoinitiator to generate cations to initiate polymerization [88].

Solid monomers can be used if their melting point is sufficiently low [96,97].

Chen et al. reported on segmented polyurethane and polyurethane-nanosilica hybrid nanocomposites synthesized by frontal polymerization [93,98,99]. Chen et al. prepared epoxy resin/polyurethane hybrid networks [99] and urethane–acrylate copolymers [93,100].

A complete bibliography can be found at: http://www.pojman.com/FP_Bibliography.html.

4.2.3. *Applications*

1. Rapid curing of thick epoxy composites without an autoclave. White has shown it is possible to have a frontal curing of thick layers of a commercial epoxy prepreg with superior properties compared to homogeneous curing [84,87]. Chekanov et al. has shown that standard epoxy/amine systems can be cured an order of magnitude faster than batch methods while still achieving 90% of the mechanical properties [85].

2. Chekanov and Pojman demonstrated that functionally-gradient materials could be prepared with FP [101]. McCardle and Pojman patented the approach [102,103].

3. Special polymers: Steinbock and Washington prepared temperature-sensitive hydrogels [104]. Microporous polymers have been produced [105,106]. Bidali et al. demonstrated that frontal ATRP of a dimethacrylate resulted in a product with higher conversion and higher degradation temperatures [94].

4. Mariani has reported using FP to consolidate porous stone materials and wood [107–109].

5. HILTI Entwicklung Elektrowerkzeuge GmbH holds a patent on using FP for chemical anchors in which a "mortar" is injected into a hole surrounding a tie bar [110]. FP is initiated to rapidly cure and secure the tie bar.

4.2.4. *Free-radical polymerization kinetics*

A free-radical polymerization with a thermal initiator can be approximately represented by a three-step mechanism. First, an unstable compound, usually a peroxide or nitrile, decomposes to produce radicals:

$$I \rightarrow f 2R\bullet$$

where f is the efficiency, which depends on the initiator type and the solvent. A radical can then add to a monomer to initiate a growing polymer chain:

$$R\bullet + M \rightarrow P_1 \bullet$$
$$P_n\bullet + M \rightarrow P_{n+1}* \tag{P}$$

The propagation step (P) continues until a chain terminates by reacting with another chain (or with an initiator radical):

$$P_n\bullet + P_m\bullet \rightarrow P_n + P_m \ (\text{or } P_{n+m})$$

The major heat release in the polymerization reaction occurs in the propagation step. Frontal polymerization autocatalysis takes place through the initiator decomposition step because the initiator radical concentration is the main control for the total polymerization rate.

The rate of polymerization is given by:

$$\frac{d[M]}{dt} = \sqrt{\frac{fk_d[Initiator]}{k_t}} k_p [M]$$

This expression is only valid for low conversion at constant temperature because as the viscosity increases, the termination constant decreases. Typical values for the propagation constant of acrylates are on the order of $10^{-4} \ M^{-1}s^{-1}$ but termination constants are on the order of $10^{-7} \ M^{-1} \ s^{-1}$. An increase in viscosity decreases k_t more than k_p, which results in an overall increase in the rate of polymerization. If the viscosity is too high, then the rate of polymerization can be decreased [111].

Now we can understand an interesting phenomenon for frontal polymerization of acrylates. For a monoacrylate such as butyl acrylate, the front velocity is about 1 cm min^{-1}. Fronts with a diacrylate will be ten times faster. The propagation rate constant is about the same for monoacrylates and multifunctional acrylates but the termination constants are very different. At very low conversion, the polymerization occurs in a crosslinked gel in which termination can not occur. The polymerization rate is thus very high. Figure 7 shows the conversion versus time for both a diacrylate and a monoacrylate. Notice that the diacrylate polymerizes much faster.

This extreme gel effect has two other consequences. Conversion is often much lower for the multifunctional acrylates. Secondly, as the termination rate decreases, its contribution to the overall energy of activation is reduced. The steady-state assumption in the polymerization model gives an approximate relationship between the effective activation energy of the entire polymerization process and activation energy of the initiator decomposition reaction:

$$E_{\mathit{eff}} = E_p + (E_i/2) - (E_t/2)$$

where E_p is the activation energy of the propagation step, E_i is for the initiator decomposition and E_t is that for the termination step. Figure 7 shows the measured energy of activation for photopolymerization of a diacrylate and a triacrylate. The energy of activation for the photoinitiator step is zero so what was measured was $E_p - E_t/2$. Obviously, the energy activation is a strong function of conversion.

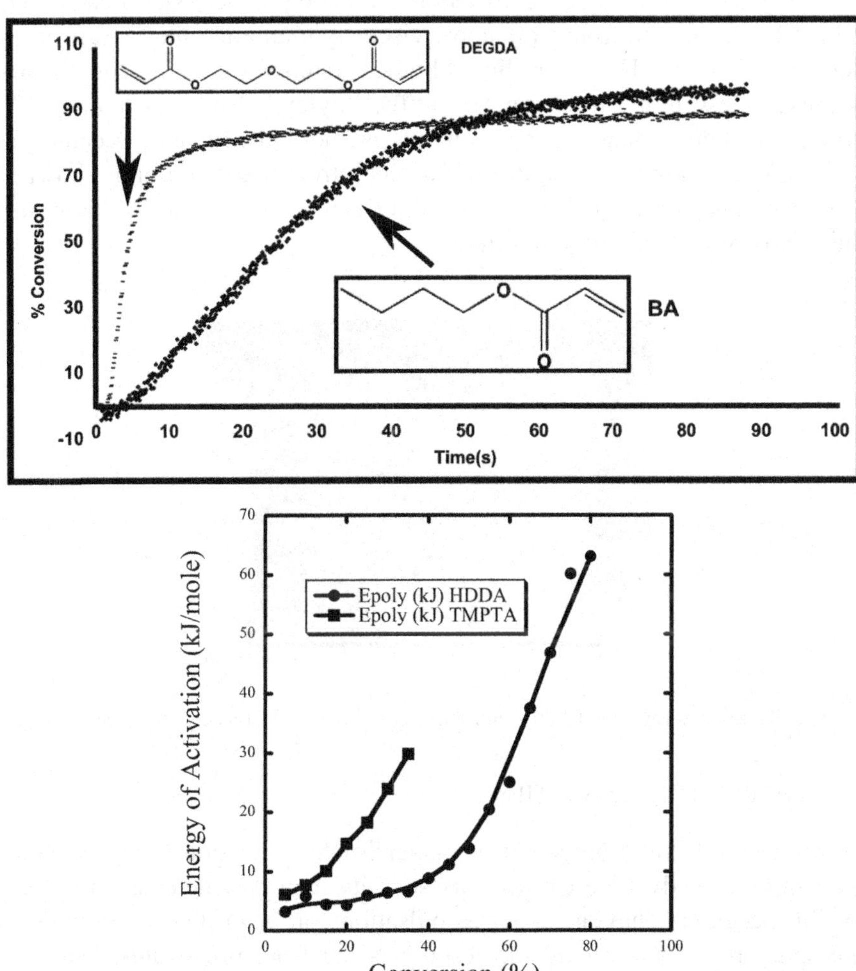

Figure 7. Top: Comparison of the conversion as a function of time for the photopolymerization of a diacrylate and a monoacrylate. (Image courtesy of Zulma Jiménez.) *Bottom*: Comparison of the energy of activation as a function of conversion for a 1,6 hexanedioldiacrylate (HDDA) and trimethylol propane triacrylate (TMPTA). (Adapted from [115].)

4.3. CONVECTIVE INSTABILITIES

Because of the large thermal and concentration gradients, polymerization fronts are highly susceptible to buoyancy-induced convection. Pojman et al. reviewed the work [112]. I wish to emphasize two points: First, FP systems demonstrated that chemical reactions must be taken into account when determining the stability conditions, that is, the front does not create density gradients and the fluid responds. The front velocity become another parameter in determining the critical conditions [113,114].

The second point is that convection can have practical significance. For FP producing a solid front from a low viscosity monomer, the front needs to descend the tube. However, liquid/liquid systems are more complicated because a descending front can exhibit the Rayleigh-Taylor instability. The product is hotter than the reactant but is more dense, and because the product is a liquid, fingering can occur. Such front degeneration is shown in Figure 8. The Rayleigh-Taylor instability can be overcome by increasing the viscosity with addition of a filler.

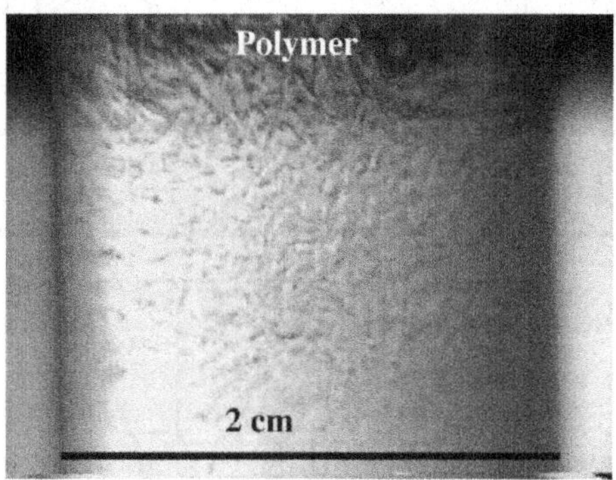

Figure 8. Rayleigh-Taylor instability with a descending front of butyl acrylate polymerization.

4.4. THERMAL INSTABILITIES

Fronts do not have to propagate as planar fronts. Analogously to oscillating reactions, a steady state can lose its stability as a parameter is varied and exhibit periodic behavior, either as pulsations or "spin modes" in which a hot spot propagates around the reactor as the front propagates, leaving a helical pattern.

The linear stability analysis of the longitudinally propagating fronts in the cylindrical adiabatic reactors with one overall reaction predicted that the expected frontal mode for the given reactive medium and diameter of reactor is governed by the Zeldovich number:

$$Z = \frac{T_m - T_o}{T_m} \frac{E_{eff}}{RT_m}$$

For FP, lowering the initial temperature (T_0), increasing the front temperature (T_m), increasing the energy of activation (E_{eff}) all increase the Zeldovich number. The planar mode is stable if $Z < Z_{cr} = 8.4$, and unstable if $Z > Z_{cr}$. The most commonly observed case with frontal polymerization is the spin mode in which a "hot spot" propagates around the front. A helical pattern is often observed in the sample (Figure 9). The first case was with the frontal polymerization of ε-caprolactam [80,81], and the next case was discovered by Pojman et al. in the methacrylic acid system in which the initial temperature was lowered [116].

At room temperature, multifunctional acrylates exhibit spin modes although monoacrylates do not. In fact, if an inert diluent, such as dimethyl sulfoxide (DMSO) is added, the spins modes are more apparent even though the front temperature is reduced [117]. We can understand this from Figure 7 in which the contribution to the energy activation from the polymerization kinetics depends on conversion and is always much higher than for a monoacrylate. Masere et al. found that changing the ratio of a monoacrylate to a diacrylate, keeping the front temperature constant, would cause a variety of spin modes. Changing the ratio of the acrylates changed the effective energy of activation for the front.

FP allows the study of spherically propagating fronts. Binici et al. developed a system that was a gel created by the base-catalyzed reaction of a trithiol with a triacrylate [118]. The system could still support a thermal front because of unreacted acrylate and the presence of a dissolved peroxide (Figure 10).

Figure 9. Helical patterns produced by "spin modes" in three different frontal polymerization systems. Tube diameters are approximately 1.5 cm.

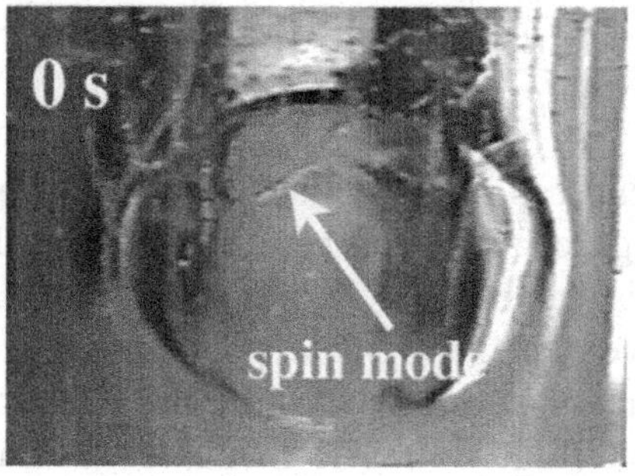

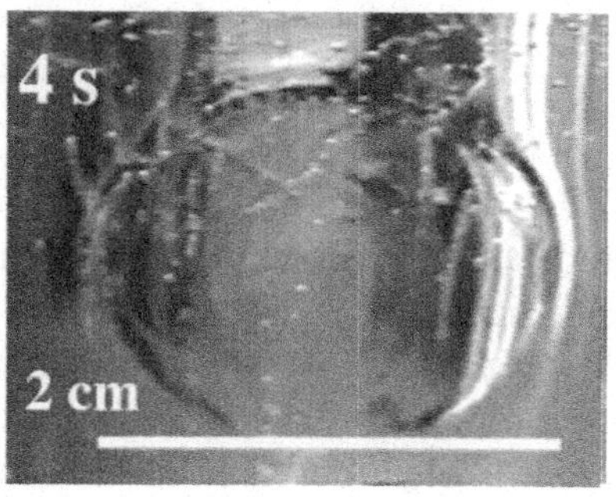

Figure 10. A spin mode on the surface of a spherically-expanding front of triacrylate polymerization.

5. Conclusions

Synthetic polymer systems can be created that exhibit a variety of interesting dynamical behavior. This can be achieved by coupling a polymerization to another dynamical system or by exploiting intrinsic nonlinearities of some polymerizations. Both methods have their virtues but not enough work has been done on identifying feedback mechanisms in polymerizations. Focusing on these feedback mechanisms and applying the experiences from inorganic systems, should lead to the development of new and useful systems.

References

1.   G. Odian, *Principles of Polymerization*, 4th Ed. (Wiley, New York, 2004).
2.   H. R. Allcock and F. W. Lampe, *Contemporary Polymer Chemistry* (Prentice-Hall, Englewood Cliffs, NJ, 1981).
3.   L. H. Sperling, *Introduction to Physical Polymer Science* (Wiley, New York, 1992).
4.   A. Goldbeter, *Biochemical Oscillations and Cellular Rhythms: The Molecular Bases of Periodic and Chaotic Behaviour* (Cambridge University Press, Cambridge, UK, 1996).
5.   I. R. Epstein and J. A. Pojman, *An Introduction to Nonlinear Chemical Dynamics: Oscillations, Waves, Patterns and Chaos* (Oxford University Press, New York, 1998).
6.   R. G. W. Norrish and R. R. Smith, *Nature* **150**, 336–337 (1942).
7.   E. Trommsdorff, H. Köhle and P. Lagally, *Makromol. Chem.* **1**, 169–198 (1948).
8.   G. O. O'Neal, M. B. Wusnidel and J. M. Torkelson, *Macromolecules* **31**, 4537–4545 (1998).
9.   G. O. O'Neal, M. B. Wusnidel and J. M. Torkelson, *Macromolecules* **29**, 7477–7490 (1996).
10.   R. L. Kaas, *J. Polym. Sci. Polym. Chem. Ed.* **19**, 2255–2267 (1981).
11.   J. Mijovic and J. Wijaya, *Macromolecules* **27**, 7589–7600 (1994).

12. I. P. Aspin, J. M. Barton, G. J. Buist, A. S. Deazle and I. Hammerton, *J. Mater. Chem.* **4**, 385–388 (1994).

13. J. P. Eloundou, M. Feve, D. Harran and J. P. Pascault, *Die Ang. Makrom. Chem.* **230**, 13–46 (1995).

14. S. Mu and J. Kan, *Electrochim. Acta* **41**, 1593–1599 (1996).

15. R. Mazeikiene and A. Malinauskas, *Synthetic Metals* **108**, 9–14 (2000).

16. C. Karunakaran and P. N. Palanisamy, *J. Mol. Catal. A Chem.* **172**, 9–17 (2001).

17. T. Tanaka, *Phys. Rev. Lett.* **40**, 820–823 (1978).

18. J. P. C. Addad, *Physical Properties of Polymer Gels* (Wiley, Chichester, U.K., 1996).

19. S. Hirotsu, Y. Hirokawa and T. Tanaka, *J. Chem. Phys.* **87**, 1392–1395 (1987).

20. Q. Tran-Cong and A. Harada, *Phys. Rev. Lett.* **76**, 1162–1165 (1996).

21. Q. Tran-Cong, A. Harada, K. Kataoka, T. Ohta and O. Urakawa, *Phys. Rev. E* **55**, R6340–R6343 (1997).

22. V. Edel, *Macromolecules* **28**, 6219–6228 (1995).

23. G. P. Andrianova, A. S. Kechkyan and V. A. Kargin, *J. Poly. Sci.* **9**, 1919–1933 (1971).

24. J. Crank, *Mathematics of Diffusion* (Clarendon Press, Oxford, 1975).

25. F. V. Burkersroda, L. Schedl and A. G. Opferich, *Biomaterials* **23**, 4221–4231 (2002).

26. E. T. Severs, *Rheology of Polymers* (Reinhold, New York, 1962).

27. J. D. Ferry, *Viscoelastic Properties of Polymers* (Wiley, New York, 1980).

28. R. K. Gupta, *Polymer and Composite Rheology* (Marcel Dekker, New York, 2000).

29. J. Meissner, *J. Appl. Polym. Sci.* **16**, 2877–2899 (1972).

30. R. B. Bird, R. C. Armstrong and O. Hassager, *Dynamics of Polymeric Liquids*, Vol. 1, 2nd Ed. (Wiley, New York, 1987).

31. J. A. Pojman, A. L. Garcia, D. K. Kondepudi and C. Van den Broeck, *J. Phys. Chem.* **95**, 5655–5660 (1991).

32. Z. Váradi and M. T. Beck, *J. Chem. Soc. Chem. Commun.* 30–31 (1973).

33. J. A. Pojman, D. C. Leard and W. West, *J. Am. Chem. Soc.* **114**, 8298–8299 (1992).

34. R. P. Washington, G. P. Misra, W. W. West and J. A. Pojman, *J. Am. Chem. Soc.* **121**, 7373–7380 (1999).

35. R. Yoshida, H. Ichijo, T. Hakuta and T. Yamaguchi, *Macromol. Rapid Commun.* **16**, 305–310 (1995).

36. R. Yoshida, S. Onodera, T. Yamaguchi and E. Kokufuta, *J. Phys. Chem. A* **103**, 8573–8578 (1999).

37. Y. Hara and R. Yoshida, *Langmuir* **21**, 9773–9776 (2005).

38. V. V. Yashin and A. C. Balazs, *Macromolecules* **39**, 2024–2026 (2006).

39. V. V. Yashin and A. C. Balazs, *Science* **314**, 798–800 (2006).

40. M. Li, S. Xu and E. Kumacheva, *Macromolecules* **33**, 4972–4978 (2000).

41. Z. Mitov and E. Kumacheva, *Phys. Rev. Lett.* **81**, 3427–3430 (1998).

42. O. Karthaus, K. Kaga, J. Sato, S. Kurimura, K. Okamoto and T. Imai, in *Nonlinear Dynamics in Polymeric Systems*, ACS Symposium Series No. 869, edited by J. A. Pojman and Q. Tran-Cong-Miyata (American Chemical Society, Washington, DC, 2003), pp. 199–211.

43. J. Boissonade, *Phys. Rev. Lett.* **90**, 188302 (2003).

44. F. Gauffre, V. Labrot, J. Boissonade and P. De Kepper, in *Nonlinear Dynamics in Polymeric Systems*, ACS Symposium Series No. 869, edited by J. A. Pojman and Q. Tran-Cong-Miyata (American Chemical Society, Washington, DC, 2003), pp. 80–93.

45. V. Labrot, P. De Kepper, J. Boissonade, I. Szalai and F. Gauffre, *J. Phys. Chem. B* **109**, 21476–21480 (2005).

46. J. Boissonade, *Chaos* **15**, 023703 (2005).

47. A. P. Dhanarajan, G. P. Misra and R. A. Siegel, *J. Phys. Chem. A* **106**, 8835–8838 (2002).

48. J.-C. Leroux and R. A. Siegel, *Chaos* **9**, 267–275 (1999).

49. T. R. Keane, in *Chemical Reaction Engineering: Proceedings of the Fifth European/ Second International Symposium on Chemical Reaction Engineering*, edited by J. M. H. Fortuin (Elsevier, Amsterdam, 1972), pp. A7-1–A7-9.

50. F. Teymour and W. H. Ray, *Chem. Eng. Sci.* **47**, 4121–4132 (1992).

51. F. Teymour and W. H. Ray, *Chem. Eng. Sci.* **47**, 4133–4140 (1992).

52. G. Papavasiliou and G. Teymour, in *Nonlinear Dynamics in Polymeric Systems*, ACS Symposium Series No. 869, edited by J. A. Pojman and Q. Tran-Cong-Miyata (American Chemical Society, Washington, DC, 2003), pp. 309–323.

53. F. J. Schork and W. H. Ray, *J. Appl. Poly. Sci. (Chem.)* **34**, 1259–1276 (1987).

54. A. Harada and Q. Tran-Cong, *Macromolecules* **30**, 1643–1650 (1997).

55. Q. Tran-Cong-Miyata, S. Nishigami, S. Yoshida, T. Ito, K. Ejiri and T. Norisuye, in *Nonlinear Dynamics in Polymeric Systems*, ACS Symposium Series No. 869, edited by J. A. Pojman and Q. Tran-Cong-Miyata (American Chemical Society, Washington, DC, 2003), pp. 276–291.

56. Y. Koike, Y. Takezawa and Y. Ohtsuka, *Appl. Opt.* **27**, 486–491 (1988).

57. B. R. Smirnov, S. S. Min'ko, I. A. Lusinov, A. A. Sidorenko, E. V. Stegno and V. V. Ivanov, *Vysokomol. Soedin., Ser. B* **35**, 161–162 (1993).

58. L. L. Lewis, C. A. DeBisschop, J. A. Pojman and V. A. Volpert, in *Nonlinear Dynamics in Polymeric Systems*, ACS Symposium Series No. 869, edited by J. A. Pojman and Q. Tran-Cong-Miyata (American Chemical Society, Washington, DC, 2003), pp. 169–185.

59. L. L. Lewis, C. S. DeBisschop, J. A. Pojman and V. A. Volpert, *J. Polym. Sci. A Polym. Chem.* **43**, 5774–5786 (2005).

60. S. I. Evstratova, D. Antrim, C. Fillingane and J. A. Pojman, *J. Polym. Sci. A Polym. Chem.* **44**, 3601–3608 (2006).

61. V. A. Briskman, in *Polymer Research in Microgravity: Polymerization and Processing*, ACS Symposium Series No. 793, edited by J. P. Downey and J. A. Pojman (American Chemical Society, Washington, DC, 2001), pp. 97–110.

62. V. Narayan and A. B. Scranton, *Trends Polym. Sci. (Camb.)* **5**, 415–419 (1997).

63. L. S. Coons, B. Rangarajan, D. Godshall and A. B. Scranton, in *Photopolymerization: Fundamentals and Applications*, ACS Symposium Series 673, edited by A. B. Scranton, C. N. Bowman and R. W. Peiffer (American Chemical Society, Washington, DC, 1997) pp. 203–218.

64. G. A. Miller, L. Gou, V. Narayanan and A. B. Scranton, *J. Poly. Sci. A Polym. Chem.* **40**, 793–808 (2002).

65. J. T. Cabral, S. D. Hudson, C. Harrison and J. F. Douglas, *Langmuir* **20**, 10020–10029 (2004).

66. J. A. Warren, J. T. Cabral and J. F. Douglas, *Phys. Rev. E* **72**, 021801 (2005).

67. N. M. Chechilo, R. J. Khvilivitskii and N. S. Enikolopyan, *Dokl. Akad. Nauk SSSR* **204**, 1180–1181 (1972).

68. N. M. Chechilo and N. S. Enikolopyan, *Dokl. Phys. Chem.* **221**, 392–394 (1975).

69. N. M. Chechilo and N. S. Enikolopyan, *Dokl. Phys. Chem.* **230**, 840–843 (1976).

70. S. P. Davtyan, P. V. Zhirkov and S. A. Vol'fson, *Russ. Chem. Rev.* **53**, 150–163 (1984).

71. J. A. Pojman, B. Varisli, A. Perryman, C. Edwards and C. Hoyle, *Macromolecules* **37**, 691–693 (2004).

72. J. A. Pojman, V. M. Ilyashenko and A. M. Khan, *J. Chem. Soc. Faraday Trans.* **92**, 2825–2837 (1996).

73. R. P. Washington and O. Steinbock, *Polym. News* **28**, 303–310 (2003).

74. J. A. Pojman, *J. Am. Chem. Soc.* **113**, 6284–6286 (1991).

75. J. A. Pojman, J. R. Willis, A. M. Khan and W. W. West, *J. Polym. Sci. A Polym. Chem.* **34**, 991–995 (1996).

76. P. M. Goldfeder, V. A. Volpert, V. M. Ilyashenko, A. M. Khan, J. A. Pojman and S. E. Solovyov, *J. Phys. Chem. B* **101**, 3474–3482 (1997).

77. C. Nason, T. Roper, C. Hoyle and J. A. Pojman, *Macromolecules* **38**, 5506–5512 (2005).

78. S. Vicini, A. Mariani, E. Princi, S. Bidali, S. Pincin, S. Fiori, E. Pedemonte, and A. Brunetti, *Polym. Adv. Technol.* **16**, 293–298, (2005).

79. Z. Jiménez and J. A. Pojman, *J. Polym. Sci. A Polym. Chem.* **45**, 2745–2754 (2007).

80. V. P. Begishev, V. A. Volpert, S. P. Davtyan and A. Y. Malkin, *Dokl. Akad. Nauk SSSR* **208**, 892 (1973).

81. V. P. Begishev, V. A. Volpert, S. P. Davtyan and A. Y. Malkin, *Dokl. Phys. Chem.* **279**, 1075–1077 (1985).

82. S. P. Davtyan, K. A. Arutyunyan, K. G. Shkadinskii, B. A. Rozenberg and N. S. Yenikolopyan, *Polymer Science U.S.S.R.* **19**, 3149–3154 (1978).

83. V. N. Korotkov, Y. A. Chekanov and B. A. Rozenberg, *Comp. Sci. & Tech.* **47**, 383–388 (1993).

84. S. R. White and C. Kim, *J. Reinforced Plastics and Comp.* **12**, 520–535 (1993).

85. Y. Chekanov, D. Arrington, G. Brust and J. A. Pojman, *J. Appl. Polym. Sci.* **66**, 1209–1216 (1997).

86. S. R. White and C. Kim, in *Thermochemical Modeling of an Economical Manufacturing Technique for Composite Structures*, 37th International SAMPE Symposium Exhibition (Long Beach, California, 1992), pp. 240–251.

87. C. Kim, H. Teng, C. L. Tucker and S. R. White, *J. Comp. Mater.* **29**, 1222–1253 (1995).

88. A. Mariani, S. Bidali, S. Fiori, M. Sangermano, G. Malucelli, R. Bongiovanni and A. Priola, *J. Poly. Sci. Part A: Polym. Chem.* **42**, 2066–2072 (2004).

89. A. Mariani, S. Fiori, Y. Chekanov and J. A. Pojman, *Macromolecules* **34**, 6539–6541 (2001).

90. S. Fiori, A. Mariani, L. Ricco and S. Russo, *e-Polymers* **29**, 1–10 (2002).

91. S. Fiori, A. Mariani, L. Ricco and S. Russo, *Macromolecules* **36**, 2674–2679 (2003).

92. A. Mariani, S. Bidali, S. Fiori, G. Malucelli and E. Sanna, *e-Polymers* **44**, 1–9 (2003).

93. S. H. Chen, J. Sui and L. Chen, *Colloid and Polymer Science* **283**, 932–936 (2005).

94. S. Bidali, S. Fiori, G. Malucelli and A. Mariani, *e-Polymers* **60**, 1–12 (2003).

95. J. V. Crivello, *J. Polym. Sci. Part A: Polym. Chem.* **45**, 4331–4340 (2007).

96. J. A. Pojman, I. P. Nagy and C. Salter, *J. Am. Chem. Soc.* **115**, 11044–11045 (1993).

97. D. I. Fortenberry and J. A. Pojman, *J. Polym. Sci. Part A: Polym. Chem.* **38**, 1129–1135 (2000).

98. S. Chen, J. Sui, L. Chen and J. A. Pojman, *J. Polym. Sci. Part A: Polym. Chem.* **43**, 1670–1680 (2005).

99. S. Chen, Y. Tian, L. Chen and T. Hu, *Chem. Mater.* **18**, 2159–2163 (2006).

100. T. Hu, S. Chen, Y. Tian, J. A. Pojman and L. Chen, *J. Poly. Sci. Part A: Polym. Chem.* **44**, 3018–3024 (2006).

101. Y. A. Chekanov and J. A. Pojman, *J. Appl. Polym. Sci.* **78**, 2398–2404 (2000).

102. J. A. Pojman and T. W. McCardle, Functionally gradient polymeric materials, U.S. Patent, 6,057,406 (2000).

103. J. A. Pojman and T. W. McCardle, Functionally gradient polymeric materials, U.S. Patent, 6,313,237 (2001).

104. R. P. Washington and O. Steinbock, *J. Am. Chem. Soc.* **123**, 7933–7934 (2001).

105. N. S. Pujari, A. R. Vishwakarma, M. K. Kelkar and S. Ponrathnam, *e-Polymers* **49**, 1–10 (2004).

106. N. S. Pujari, A. R. Vishwakarma, T. S. Pathak, A. M. Kotha and S. Ponrathnam, *Bulletin of Materials Science* **27**, 529–536 (2004).

107. S. Vicini, A. Mariani, E. Princi, S. Bidali, S. Pincin, S. Fiori, E. Pedemonte and A. Brunetti, *Polym. Adv. Technol.* **16**, 293–298 (2005).

108. A. Mariani, S. Bidali, P. Cappelletti, G. Caria and A. Colella, *e-Polymers* (2006).

109. A. Brunetti, A. Mariani and S. Bidali, *Nuovo Cimento- Societa Italiana di Fisica Sezione C* **30**, 407–416 (2007).

110. A. Pfeil, T. Burgel, M. Morbidelli and A. Rosell, Mortar composition, curable by frontal polymerization, and a method for fastening tie bars, U.S. Patent, 6,533,503 (2003).

111. H. Zhou, Z. Jiménez, J. A. Pojman, M. S. Paley and C. E. Hoyle, *J. Polym. Sci. A: Polym. Chem.* **46**, 3766–3773 (2008).

112. J. A. Pojman, S. Popwell, D. I. Fortenberry, V. A. Volpert and V. A. Volpert, in *Nonlinear Dynamics in Polymeric Systems*, ACS Symposium Series No. 869, edited by J. A. Pojman and Q. Tran-Cong-Miyata (American Chemical Society, Washington, DC, 2003), pp. 106–120.

113. G. Bowden, M. Garbey, V. M. Ilyashenko, J. A. Pojman, S. Solovyov, A. Taik and V. Volpert, *J. Phys. Chem. B* **101**, 678–686 (1997).

114. B. McCaughey, J. A. Pojman, C. Simmons and V. A. Volpert, *Chaos* **8**, 520–529 (1998).

115. K. N. Gray, *Photopolymerization Kinetics of Multifunctional Acrylates* (University of Southern Mississippi, Hattiesburg, MS, 1988).

116. J. A. Pojman, V. M. Ilyashenko and A. M. Khan, *Physica D* **84**, 260–268 (1995).

117. J. Masere and J. A. Pojman, *J. Chem. Soc. Faraday Trans.* **94**, 919–922 (1998).

118. B. Binici, D. I. Fortenberry, K. C. Leard, M. Molden, N. Olten, S. Popwell and J. A. Pojman, *J. Polym. Sci. A Polym. Chem.* **44**, 1387–1395 (2006).

INTERNAL STRESS AS A LINK BETWEEN MACROSCALE AND MESOSCALE MECHANICS

Ken Sekimoto (`ken.sekimoto@espci.fr`)
Matières et Systèmes Complexes, CNRS-UMR7057, Université Paris 7
and *Gulliver, CNRS-UMR7983, ESPCI, Paris, France*

Abstract. The internal (or residual) stress is among the key notions to describe the state of the systems far from equilibrium. Such stress is invisible on the macroscopic scale where the system is regarded as a black-box. Yet nonequilibrium macroscopic operations allow to create and observe the internal stress. We present in this lecture some examples of the internal stress and its operations. We describe the memory effect in some detail, the process in which the history of past operations is recalled through the relaxation of internal stress.

Keywords: internal stress, residual stress, momentum flux, non-equilibrium, memory effect, plasticity, yield

1. Introduction

In this lecture note, the *internal stress* is defined as the stress that is maintained within a system by itself, without the aid of external supports or constraints.[1] From outside, the internal stress is, therefore, invisible by purely mechanical means. But since the stress exists locally on smaller scales, it could be observed, for example, by stress sensitive optical probes. Also, once the internal mechanical balance is somehow broken, the internal stress becomes visible from outside. In this sense, the internal stress is a concept bridging between macroscale and mesoscale mechanics. The purpose of this lecture is to draw the reader's attention to this aspects of the internal stress.

We can find the internal stress everywhere around us: When we tighten the belt of our clothes, the internal stress keeps contact between the belt under tension and our body under compression. When we make an espresso coffee, the internal stress is between the high-pressure fluids inside the coffee-machine and the stretched metallic frame of the machine. If we make a cut midway into a red sweet pepper or a watermelon, they generally change

---

[1]The notion of the internal stress in this context therefore includes the *residual stress* on the one hand, since the latter has more strict definition than the former. Sometimes, however, the "internal stress" is used just as opposed to the stress on the surface of a material [1] on the other hand.

their shape, due to turgor pressure of plant cells which has accumulated
the internal stress. When a pressing machine in industry spreads a metallic
block, the framework of the machine and the block develop an internal stress,
i.e., under compression in the metallic block on the one hand, under tension
(and shear) in the machine's framework on the other hand. The *tensegrity* [2]
is a concept of self-standing architecture based on the internal stress. This
is formed by wires under tension, rods under compression and the joints
articulating those elements. The stable structure is realized by the internal
stress among these them. This idea of tensegrity is applied to analyze the
mechanical constitution in living cells [3]. The actin and microtubule filaments are thought to be joined by various active or passive binding proteins
or other macromolecules. In solid clusters atoms interacts with each other
by attractive and repulsive forces, making up a network of internal stress.
Even the perfect salt crystal (NaCl) may be regarded as a result of internal
stress due to electrostatic forces between like and unlike charge pairs. The
atmosphere is pulled by the earth through the gravitaty, while the air also
pushes the earth by the hydrostatic pressure. Any non-trivial self-sustaining
system, therefore, contains the internal stress.

If the internal stress is found everywhere, is this concept a "general
abstract nonsense" ?[2] Yes, perhaps. But this concept can provide with a
useful viewpoint. I will try to develop this hope in the following sections:
We first describe how the internal stress is maintained on the mesoscale
(Section 2). Then we describe a mesoscale *sensor* that uses internal stress to
realize the reliable observation under thermal fluctuations (Section 3). Next
we discuss briefly the generation of internal stress in the context of far-from
equilibrium process (Section 4). The last issue (Section 5) is the emergence
of the internal stress as a memory effect.

I wish that this essay serves for the analysis, modeling and designing of
various far from equilibrium phenomena such as glass dynamics, plasticity,
and active media.

2.   Mesoscale description of internal stress

2.1.   INCOMPATIBLE STRESS-FREE STATES OF CONSTITUENT
MODULES CAUSE THE INTERNAL STRESS

Under an internal stress, modules that constitute the system undergo some
deformations. Let us look back some of the examples mentioned above. When
a pressing machine presses down a metallic block, the latter is deformed

---

[2]The "general abstract nonsense" has been a word of criticism against the category theory
[4]. This theory turned out to provide a powerful tool, as well as a conceptual perspective, to
many field of mathematics.

from its natural form, and both the block and the machine develop the stress. Within a coffee maker, the water vapor is highly compressed with respect to the ambient pressure, and the container of the hot water/vapor opposes this pressure. The red sweet pepper before the cutting contains many cells whose shape is distorted with respect to their form under no constsraints. Inside a cell many biopolymeric filaments are deformed by other components of the cell. In clusters or in crystals, many pairs of atoms are not at the distance of the lowest energy for a particular pair-interaction.

Thus, on the mesoscale, the internal stress is caused by a quasi-static compromise among the constituent modules each of which insists on its own stress-free state[3] [5].

2.2.   INTERNAL STRESS IS THE CIRCULATION OF MOMENTUM FLUX

The state of a system having internal stress can be represented in terms of the momentum flux. We know that the total stress tensor σ with the minus sign carries the momentum. In (quasi-)static case of our interest the momentum flux density P, therefore, writes[4]

$$P = -\sigma. \tag{1}$$

The conservation law of momentum applied to the static state, $\nabla \cdot P = 0$, is nothing but the equation of static mechanical balance of the bulk without external force, $\nabla \cdot \sigma = 0$. If, for example, we focus on the flux of the x-component of momentum, $P \cdot \hat{x}$, the conservation law is $\nabla \cdot (P \cdot \hat{x}) = 0$, where $\hat{x}$ is the unit vector along x-axis. This is mathematically of the same form as the conservation of mass flux, Figure 1a shows an example of the permanent momentum flux.

2.3.   EXAMPLES IN SOFT MATERIALS

Surface buckling of swelling gels: Swelling of gel can take place by many different causes: it is sometimes enough to change the solvent [6] or the temperature [7]. The swelling often develops a buckling pattern on the free surface. While the surface region of the gel can swell freely in the direction normal to the surface, it is laterally constrained in order to fit with the undeformed portion of the gel inside the bulk. The surface region of the gel is therefore under lateral compression. The buckling of the gel surface [6] occurs when the resulting lateral compressible stress exceeds a threshold [8,9].

---

[3]If this static compromise becomes unstable, dynamical processes, such as an oscillation, explosion, chaos, can also occur.

[4]In general case we should add the momentum flux associated with the material flow.

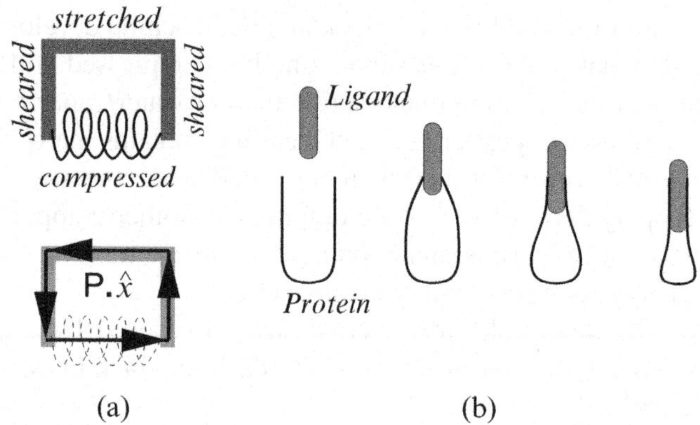

Figure 1. (a) (*Top*) An example of internal stress consisting of a spring and a framework. (*Bottom*) Schema of circulation of the flux of momentum in x direction, $P \cdot \hat{x}$. (b) "Induced fit" by *Protein* to a *Ligand*, realizing the arrival of signal to the conformational change of a receptor. Note that the ligand is under lateral stretching, while the bottom of the protein is under compression, as in (a).

A similar phenomena occurs in the peripheral growth of membrane such as some vegetable leaf [10]. See also [5]. Because of the freedom perpendicular to the membrane, the buckling pattern in this case occurs predominantly in this direction [11]. A complementary situation can also occur: If a gel is synthesized on a spherical substrate and pushed outward [12], the gel near the free outer surface develops a lateral tension, instead of compressive stress. This stress drives the lateral fracture of gel [13, 14], instead of buckling.

Water suction at the crack tip of developing gel fracture: The velocity of gel's fructure is known to change if the solvent is supplied at the crack tip of developing gel fracture [15]. The mode-I opening of the crack develops the negative pressure of the solvent within the gel. This negative pressure pulls the menisci of the solvent at the gel surface, which in turn compresses the gel toward inside. The internal stress establishes between the solvent under negative pressure and gel under compression. Recently [16] extrapolated the investigation of [15].

Drying gel: A similar situation occurs at the surface of drying gel or colloidal suspension. The surface layer of gel is under compression in the normal direction to the surface because it is pressed towards inside by the surface menisci of the solvent, on the one hand, and towards outside by the interior part of gel under osmotic pression.[5] A complementary situation occurs in

---

[5]The capillary pressure $p_{\mathrm{cap}} = \gamma(K_1 + K_2)$, where γ is the surface tension and $\{K_1, K_2\}$ are the principal curvatures of the meniscus, creates the permeation flow obeying the Darcy's

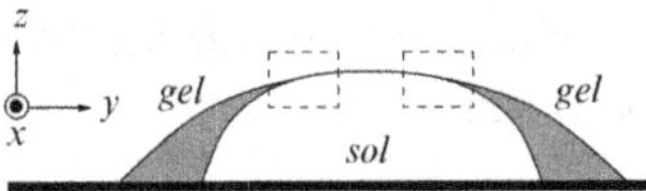

Figure 2.   Schematic sectional view of a drying droplet of colloidal suspension. The *shaded region* is in the gel phase, while the central part of the droplet is still in the sol phase. In the regions marked by dashed rectancles, the dry glassy surface layer is under compression in *x*-direction.

ionic gels: The counterions between the network of polyelectrolyte gel [7,17] push outward the free surface of the gel through the electrostatic double layer on the surface (Donnan effect) [18], while the network elasticity resists against the swelling.

Buckling of drying colloidal suspension: On continuing the discussion of the drying gel, an important phenomenon related to the drying is the glassification of the highly dried surface layer. This layer is very thin and of high elastic modulus [19]. When the interior elastic part exerts lateral compressible stresses on the glassy surface layer, it may cause surface buckling pattern, as is observed before the drying fracture [20], see (Figure 2).[6]

Permanent set of rubber crosslinking: Suppose we have a rubber network deformed by an external force. If we introduce further crosslinking into this deformed network, the network has virtually two networks whose stress free states are mutually incompatible. After removing the external force, the rubber returns to a new apparent "stress-free state." This state is called the *permanent set* [21].[7]

---

law, $\nabla p = -\zeta(1 - \phi_{\text{gel}})(\mathbf{u}_{\text{solv}} - \mathbf{v}_{\text{gel}})$, where ζ is a constant, ϕ_{gel} is the gel's volume fraction, and $\{\mathbf{u}_{\text{solv}}, \mathbf{v}_{\text{gel}}\}$ are the velocity of solvent and gel, respectively. The gel is pressed toward inside by the boundary condition, $\pi_{\text{gel}} = p_{\text{cap}}$, while it is pressed toward outside by the total mechanical balance in the gel, $\nabla \pi_{\text{gel}} = -\nabla p$, where π_{gel} is the osmotic pressure in the gel.

[6]We should note the glassification-induced update of the stress-free state. Without this, the vertical compression causes a lateral expansion as a secondary or Poisson effect. Given the normal compressive stress, $-\sigma_{zz} > 0$, the induced lateral compressive stress for a laterally constrained surface is $-\sigma_{xx} = -\sigma_{yy} = -\frac{\nu}{1-\nu}\sigma_{zz}$ for isotropic surface, or $-\sigma_{xx} = -\nu - \sigma_{zz}$ and $-\sigma_{yy} = 0$ when the y direction is not constrained [20], where ν is the Poisson ratio of the gel's osmotic elasticity.

[7]The purely entropic free energy of original rubber writes $-TS_1 = (k_B T \nu_0/2)(\lambda_x^2 + \lambda_y^2 + \lambda_z^2)$, where λ_x etc. are the elongation ratios along x-direction etc. with $\lambda_x \lambda_y \lambda_z = 1$. Then, the free energy, $-T(S_1 + S_2)$, after the second crosslinking introduced under deformation, λ_{x0} etc. with $\lambda_{x0} \lambda_{y0} \lambda_{z0} = 1$ is such that $-TS_2 = (k_B T \nu_1/2)[(\lambda_x/\lambda_{x0})^2 + (\lambda_y/\lambda_{y0})^2 + (\lambda_z/\lambda_{z0})^2]$. Then the permanent set has the deformation, λ_{xs} etc. obeys $\nu_0 \lambda_{xs}^2 + \nu_1(\lambda_{xs}/\lambda_{x0})^2 = C$ etc, where C is determined by the condition $\lambda_{xs} \lambda_{ys} \lambda_{zs} = 1$ [22]. Macroscopically, the new rubber has free-energy, $-TS_s = (k_B T \nu_c/2)[(\lambda_x/\lambda_{xs})^2 + (\lambda_y/\lambda_{ys})^2 + (\lambda_z/\lambda_{zs})^2]$, with $\nu_c^3 = (\nu_0 + \nu_1/\lambda_{x0}^2)(\nu_0 + \nu_1/\lambda_{y0}^2)(\nu_0 + \nu_1/\lambda_{z0}^2)$. See also [23].

We thus have seen that whether or not the internal stress is visible depends on the scale of description/observation.

3. Sensor working on the thermally fluctuating scale

The sensor of signal molecules (ligands) must fulfill apparently incompatible requirements: it must respond strongly enough to the arrival of the signal through the conformational change of the sensor itself, on the one hand, but the interaction between the ligand and the sensor must be mostly neutral to reflect the density of the ligands in the environment, on the other hand. The biological receptors solved this task using the the mechanism called the *induced fit* [24], which uses the internal stress [25]. Figure 1b explains how the attraction or tension between the ligand and the sensor deforms the latter. The analogy to the Figure 1a is evident. It is this deformation that represents the recognition of signal, satisfying the first requirement mentioned above. In energetics term, the gain of energy by binding interaction can be compensated by the cost of deformation of the sensor. This cancellation enables the bias-free arrival and departure of the ligand, satisfying the second requirement of the sensor.

We note that the reversible detection of signal described above has no contradiction with the theorem of Landauer and Bennett [26, 27], telling that we need at least $k_{\mathrm{B}}T \log 2$ to know the content of a single-bit memory. The latter process decreases the entropy by $\log 2$ of the "movable bit", or, the observer at the compensating external work of $k_{\mathrm{B}}T \log 2$. In the reversible detection, the system of ligand and detector is isolated from the external observer who gains no information.

4. Generation of the internal stress

In many cases, the states with internal stress are metastable. (The ionic crystal may be an exception.) To create these metastable states, we require far from equilibrium operations: The pressing machine turns a motor to raise the oil/air pressure in the piston. The coffee machine generates inner pressure by heating and boiling. Plant cells develop the pressure of vacuoles by pumps. Colloidal gel generates the stress through evaporation flux. The permanent set of rubber network is made by chemical reaction.

The generation of internal state as metastable state is closely related to the notion of the *plasticity*. It is because, during the generating process, some timescales of the system is momentarily diminished at the same time that the bias is put in favor of the (future) metastable state.[8]

---

[8]We may even compare with the positive discriminations to change the society's stable state, because excessive forcing is required momentarily in both case.

In order to create the internal scale of a certain spatial scale, we do not necessarily need the operations on that scale. Suppose that we bend strongly a metal rod. The inhomogeneity and anisotropy on the atomistic scale inside the bar can generate the metastable states of small scale, such as entangled dislocation network, a typical example of internal stress. After removing the external forcing, the rod once yielded will show the different elasticity as well as the different yield stress [28], as compared with those of the original state. The new metastable state reflects the preceding far from equilibrium operation.

5. Macroscale emergence of internal stress

We discussed above how an external macroscale forcing is "internalized" as the internal stress of mesoscale. Below we describe, to some detail, the opposite case: The internal state of mesoscale emerges as stress of macroscale through far from equilibrium processes.

5.1. RHEOLOGICAL MODEL OF RUBBER

We take a simple phenomenological constitutive equation for a linear segment of rubber:

$$\sigma_t = G_R\gamma_t + G_\infty \int_0^t \mathcal{G}(t - t')\dot{\gamma}_{t'}\,\mathrm{d}t', \tag{2}$$

where the tensile stress σ_t at time t consists of two terms. The first is a purely entropy-elastic stress, $G_R\gamma_t$, with γ_t being the elongation $(\lambda - 1)$ at the same time and G_R being the rubber elastic modulus. The second represents the rheological response of network chains, where $\mathcal{G}(\ddagger)$ is the relaxation kernel satisfying $\mathcal{G}(0) = 1$ and $\mathcal{G}(\infty) = 0$, and G_∞ is the lass elastic modulus. Mathematical structure of (2) is essentially the same as the constitutive equations of magnetic or dielectric materials. If $\mathcal{G}(z) = e^{-z}$, the model reduces to the well-known Maxwell model of rheology. We assume that $\gamma_t = 0$ for $t < 0$.

The following rewriting of (2) is physically appearing:

$$\sigma_t = G_R\gamma_t + G_\infty\left[\int_0^t \frac{\partial \mathcal{G}(t - t')}{\partial t'}(\gamma_t - \gamma_{t'})\mathrm{d}t' + \gamma_t\mathcal{G}(t)\right]. \tag{3}$$

If we interpret $G_\infty(\gamma_t - \gamma_{t'})$ as the force due to a Hooke spring created at the time t', the multiplicative factor, $\frac{\partial \mathcal{G}(t-t')}{\partial t'}\mathrm{d}t'$, is its survival probability until the present time t. The rheological response of network chains can, therefore, be viewed as the permanent set which is incessantly crosslinked but also uncrosslinked with the surviving probability $\mathcal{G}(()z)$ until z sec after the crosslinking.

5.2. INTERNAL STRESS IN THE RHEOLOGICAL MODEL

Next suppose that we stopped stretching at time t_0 and let loose the sample. Shortly after releasing the stress ($t = t_0^+ \equiv t_0 + \epsilon$ with small positive ϵ), the stretching $\gamma_{t_0^+}$ ($0 < \gamma_{t_0^+} < \gamma_{t_0}$) should obey the following equation:

$$0 = G_R \gamma_{t_0^+} + G_\infty \left[\int_0^{t_0^+} \mathcal{G}(t_0^+ - t)\dot{\gamma}_{t'} dt' + (\gamma_{t_0^+} - \gamma_{t_0}) \right]. \tag{4}$$

Since the crosslinks have finite lifetimes, the vanishing of the left hand side of (4) is the result of internal stress established as compromise among the temporary crosslinks or Hooke springs. Some springs are under tension but some others should be under compression. This internal stress can remain for very long time if the temperature is below the glass temperature.

5.3. MEMORY EFFECT OF RUBBER

Suppose we then fix again the sample's at the relaxed length $\gamma_{t_0^+}$. Initially $\sigma_{t_0^+} = 0$ by definition. But what will be the stress σ_t for $t > t_0^+$? We can show that σ_t for $t > t_0^+$ under fixed $\gamma_t = \gamma_{t_0^+}$ writes as follows:

$$\begin{aligned} \sigma_t &= G_R \gamma_t + G_\infty \left[\int_0^{t_0} \mathcal{G}(t - t')\dot{\gamma}_{t'} dt' + \mathcal{G}(t - t_0)(\gamma_{t_0^+} - \gamma_{t_0}) \right] \\ &= G_R[1 - \mathcal{G}(t - t_0)]\gamma_{t_0^+} + G_\infty \int_0^{t_0} [\mathcal{G}(t - t') - \mathcal{G}(t - t_0)\mathcal{G}(t_0 - t')]\dot{\gamma}_{t'} dt' \end{aligned}$$

$$\tag{5}$$

where we substituted $\mathcal{G}(t - t_0) \times$ (4) to go from the first line to the second. We note that $\mathcal{G}(t - t') - \mathcal{G}(t - t_0)\mathcal{G}(t_0 - t')$ in the last integral vanishes if $\mathcal{G}(z)$ is the Maxwell model, $\mathcal{G}(z) = e^{-z}$. Otherwise, a stress reappears before it returns finally to the rubber elasticity. This phenomenon was first found experimentally [29]. We call this reappearance of stress the memory effect. The Maxwell model, therefore, cannot explain the memory effect. Intuitively, if there is more than one characteristic times in the relaxation kernel $\mathcal{G}(z)$, the balance of internal stress on the mesoscale is transiently broken. And the uncompensated mesoscale stress appears as the macroscale stress. Due to the mathematical similarity of the constitutive equations, the memory effect in magnetic or dielectric systems is also understood with pertinent reinterpretation of the stress.

From information point of view, the *history* of operations in $0 < t < t_0$ can be read out from the memory effect in $t_0 < t$, up to the memory capacity (n) of the system, where n is the number of characteristic times in $\mathcal{G}(z)$, i.e. $\mathcal{G}(z) = \sum_{j=1}^{n} a_j e^{-z/\tau_j}$.

From energetics point of view, we can assess the energy stocked in the state of internal stress. We use an expression of σ_t similar to (3), which is of course equivalent to (5):

$$\sigma_t = G_R \gamma_{t_0^+} + G_\infty \left[\int_0^{t_0} \frac{\partial G(t_0 - t')}{\partial t'} [\gamma_{t_0^+} - \gamma_{t'}] dt' + G(t)\gamma_{t_0^+} \right]. \tag{6}$$

By an analogy to the potential energy of Hooke spring, the internal energy E_t^{int} contained by the mesoscale springs should be

$$E_t^{int} \equiv \frac{G_\infty}{2} \int_0^{t_0} [\gamma_{t_0^+} - \gamma_{t'}]^2 \frac{\partial G(t - t')}{\partial t'} dt', \tag{7}$$

where we have ignored $G(()t)$ term in (6). Experimentally, the corresponding quantity has been measured using the calorimetry [30]. They prepared glassy rubber samples with or without pre-stretching, and compared the exothermic heat upon slow warming. The pre-stretched sample, i.e. the sample containing the internal stress in the present context, released an excess heat. If this heat corresponds to the decrease of E_t^{int} in (7), the results implies that the mesoscale Hooke springs are *not* of entropic origin.

Acknowledgements

I thank my colleagues, Y. Tanaka, T. Ooshida, N. Suematsu, K. Kawasaki and Y. Miyamoto, J. Prost, F. Jülicher, H. Boukellal and A. Bernheim-Grosswasser, whose collaborations are cited above. I thank M. Ben-Amar for having shown unpublished paper, and A. Daerr for a critical comment. I appreciate the organizers of this School.

References

1. L. D. Landau and E. M. Lifshitz, *Mechanics (Course of Theoretical Physics, Volume 1, 3rd Ed.)*, (Reed Educational and Professional Publishing Ltd. Oxford, 2002).
2. B. Fuller, *SYNERGETICS – Explorations in the Geometry of Thinking, Volumes I & II, 2nd. Ed.* (Macmillan, New York, 1979).
3. D. E. Ingber, *Ann. Rev. of Physiol.* **59**, 575 (1997).
4. S. Mac Lane, *Categories for the Working Mathematician, 2nd. Ed.* (Springer, Berlin, 1998).
5. J. Dervaux and M. B. Amar, *Phys. Rev. Lett.* **101**, 068101 (2008).
6. T. Tanaka, S. Sun, Y. Hirokawa, S. Katayama, J. Kucera, Y. Hirose and T. Amiya, *Nature* **325**, 796 (1987).
7. M. Doi, http://www.ima.umn.edu/matter/fall/t1.html (2008).

8. K. Sekimoto and K. Kawasaki, *J. Phys. Soc. Jpn.* **56**, 2997 (1987).

9. N. Suematsu, K. Sekimoto and K. Kawasaki, *Phys. Rev. A* **41**, 5751 (1990).

10. B. Audoly and A. Boudaoud, *Phys. Rev. Lett.* **91**, 086105 (2003).

11. E. Cerda, K. Ravi-Chandar and L. Mahadevan, *Nature* **419**, 579 (2002).

12. F. Gerbal, P. Chaikin, Y. Rabin and J. Prost, *Biophysical J.* **79**, 2259 (2000).

13. K. Sekimoto, J. Prost, F. Jülicher, H. Boukellal and A. Bernheim-Grosswasser, *Eur. Phys. J. E* **13**, 247259 (2004).

14. J. van der Gucht, E. Paluch, J. Plastino and C. Sykes, *PNAS* **102**, 7847 (2005).

15. Y. Tanaka, K. Fukao, Y. Miyamoto, H. Nakazawa and K. Sekimoto: 1996, *J. Phys. Soc. Jpn.* **65**, 2349 (1996).

16. T. Baumberger, C. Caroli and D. Martina, *Nature Mater.* **5**, 552 (2006).

17. A. Khokhlov, these proceedings.

18. J. Ricka and T. Tanaka, *Macromolecules* **17**, 2916 (1984).

19. K. Huraux, PhD Thesis, Paris IV, (2008).

20. X. Ma, Y. Xia, E. Chen, Y. Mi, X. Wang and A. Shi, *Langmuir* **20**, 9520 (2004).

21. P. J. Flory, *Trans. Faraday Soc.* **56**, 722 (1960).

22. J. P. Berry, J. Scanlan and W. F. Watson, *Trans. Faraday Soc.* **52**, 1137 (1956).

23. S. Kaang, D. Gong and C. Nah, *J. Appl. Polym. Sci.* **65**, 917 (1997).

24. D. E. Koshland, *Sci. Am.* **229**, 52 (1973).

25. K. Sekimoto, *Physica D* **205**, 242 (2005).

26. R. Landauer, *Science* **272**, 1914 (1996).

27. C. H. Bennett, *Int. J. Theoret. Phys.* **21**, 905 (1982).

28. T. Ooshida and K. Sekimoto, *Phys. Rev. Lett.* **95**, 108301 (2005).

29. Y. Miyamoto, K. Fukao, H. Yamao and K. Sekimoto, *Phys. Rev. Lett.* **88**, 225504 (2002).

30. O. A. Hasan and M. C. Boyce, *Polymer* **34**, 5085 (1993).

ON SOME PASSIVE AND ACTIVE MOTION IN BIOLOGY

Chaouqi Misbah (`chaouqi.misbah@ujf-grenoble.fr`)
LSP, CNRS (UMR5588) and Université J. Fourier
BP 87 - 38402 Grenoble Cedex
France

Abstract. This contribution focuses on two main questions inspired by biology: (i) passive motion under flow, like advection of red blood cells in the circulatory system, and (ii) active motion generated by actin polymerization, as encountered in cells of the immune system and some micro-organisms (e.g. some bacteria and viruses). The first part is dedicated to the dynamics and rheology of vesicles (a simple model for red blood cells) under flow. Some results obtained on red blood cells are also presented and compared to vesicles. Vesicles and red blood cells under flow exhibit several interesting dynamics: tank-treading, tumbling, vacillating-breathing, and so on. These dynamics have a direct impact on rheology, as will be discussed both from the theoretical and experimental point of views. The second part addresses active motion. Some Bacteria (like Listeria) are known to transfect cells thanks to the polymerization on their surface of an actin gel. Monomeric actin proteins are recruited from the transfected cell when the bacteria gets in contact with the cell surface. It has been found that the bacteria propulsion into the cell occurs in the absence of molecular motors. Biomimetic experiments on beads and droplets have revealed that motion is a consequence of a spontaneous symmetry breaking that is accompanied with force generation. A simple basic model taking into account growth of actin and elasticity is sufficient to capture the essence of symmetry breaking and force generation, as will be presented in this contribution.

Keywords: vesicles, flow, red blood cells, blood rheology, actin-assisted motility

1. Introduction

In the realm of biology cell motion can be broadly classified into two important categories: (i) passive motion, and (ii) active motion. Passive motion refers to a situation where a cell is subject to the action of a force from outside. A typical example is the motion of blood cells (red blood cells, leucocytes, and platelets) in the blood circulatory system. The cells are simply advected by the flow. Active motion, in contrast, refers to the fact that motion may occur due to building of a force inside the cell itself. It must be kept in mind, however, that this motion is initially triggered by an external signal. A typical situation is cell motility of the immune system (leucocytes or white blood cells): due to a chemical signal (for example detection by membrane

P. Borckmans et al. (eds.), *Chemomechanical Instabilities in Responsive Materials*,
© Springer Science + Business Media B.V. 2009

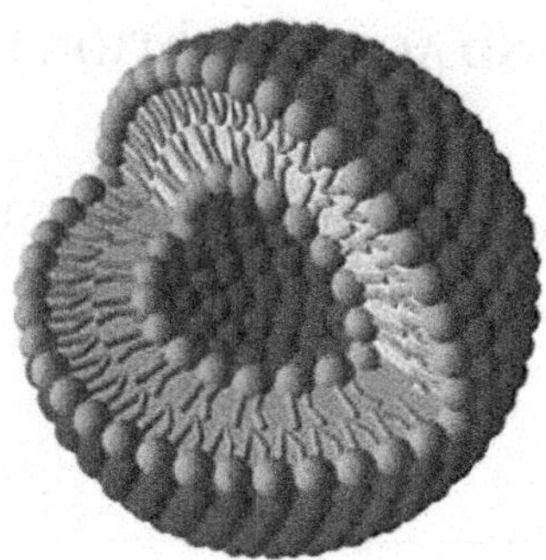

Figure 1.   A schematic view of a vesicle made of a bi-layer of phospholipid molecules. (From [18].)

receptors of trails of diverse molecules derived from microorganisms or damaged tissues) the cell builds internally a traction force that enables it to move forward. Active motion is also present within a variety of microorganisms (some viruses, bacteria and so on).

This contribution focuses on some aspects of these two motions. More precisely in the first part we shall study the dynamics of an individual biological red blood cell (RBC), or its biomimetic counterpart represented by a vesicle. A vesicle is a closed membrane made of a bilayer of phospholipid molecules (Figure 1). These molecules have a polar hydrophilic head which points towards the solvent, and hydrophobic tails which are directed inward in the bi-layer (Figure 1). Unlike vesicles, which are present in the cell cytoplasm which have a few *nm* of radius, the vesicles we refer too here are significantly bigger (their radius range typically between 10 and 100 μm), and thus sometimes called *giant vesicles*, or *giant liposomes*. Their large size allows for a direct observation under an optical (phase contrast) microscope. Both at room and physiological temperatures the bilayer is a two dimensional incompressible fluid (at lower temperature they are known to exhibit gel-like transition). We consider only their study in the fluid state. Human RBCs constitute one of the simplest cell in biology (Figure 2). Indeed the RBC is made, like a vesicle, of a phospholipid bi-layer, plus a protein network (spectrin), known also under the name of cytoskeleton (2). Its cytoplasm is devoid of a nucleus and organella; the internal content of a RBC is made of a hemoglobin solution, which is a simple newtonian fluid. It is thus hoped that vesicles may represent a simplistic staring point to understand viscoelastic properties, dynamics and rheology of bio-fluids, like blood.

We shall see that several dynamical and rheological features are common to RBC and vesicles. The study of vesicles offer a certain advantage. Indeed, unlike RBCs, they lend themselves to exploration of various parameters without affecting their structural properties. For example, their size can be varied

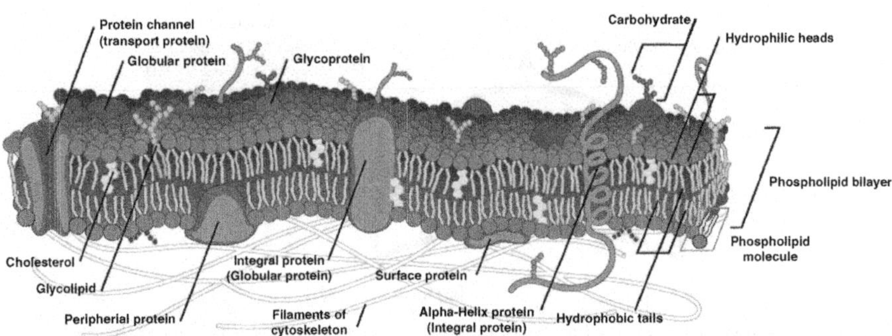

Figure 2. A schematic view of a red blood cell membrane. Besides the bi-layer of phospholipids, there is a network of proteins (called also cytoskeleton, shown in white) and other proteins (shown in blue and green). (Adapted from http://en.wikipedia.org/wiki/Cell_membrane.)

over a decade at least. Furthermore, they can be swollen or deflated by osmosis (and heating), so that one can get more or less deformable shapes. Finally, their internal content can be modified: one can change the internal solution in order to vary the internal viscosity, mechanical and viscoelastic properties, and so on... The ability to act on these parameters is essential with regard to scientific progress. For example, scanning a large parameter space offers the possibility to explore potential new phenomena. Another benefit is that this also allows to guide modeling (like checking some scaling relations which often requires exploration of parameter ranges over decades or more).

The second part will be dedicated to the active motion mediated by actin polymerization. Inside many biological cells there exists a reservoir of proteins, called actin monomers. When the membrane receptors detect a chemical signal (originating, for example, from surrounding pathogens) this elicits a series of chemical reactions inside the cell, which ultimately lead to polymerization of the actin monomers into a cross-linked gel. Polymerization occurs at the front of the cell, while depolymerization takes place at the rare (Figure 3), so that the polymerized volume remains about constant during the motility. Molecular motors (myosin molecules) intercalate between the actin filaments in the gel and are able to convert the ATP (Adenosine Tri-Phosphate) into mechanical work. This conversion is used by the cell to move itself forward in response to an infection or a tissue injury, and so on. It has been concluded quite recently that some organisms (e.g. the bacteria Listeria monocytogenes) move forward by using the actin gel assembly, in which no molecular motor has been found [45]. It has been further demonstrated that even simplified systems, and artificial beads or droplets (quoted with an enzyme in order to trigger polymerization) and put in a solution containing actin monomers, can be propelled to the sole effect of polymerization of the

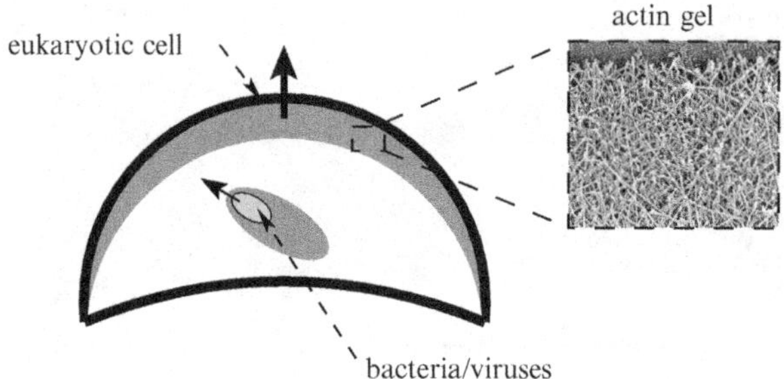

Figure 3. A schematic view of an actin network (in red) that develops inside the cell during active motion. Also shown is a schematic view of bacteria, or a virus, which has transfected a cell. The cross-linked gel is shown on the right. (From [41].)

actin gel (which is free of molecular motors) on their surface [3,16,29,44,45]. This has pointed to the fact that motion (i.e. force generation) is produced without the intervention of molecular motors.

Here we shall introduce a simple model based on growth and elastic stress that is built in the gel in order to show (i) that the growth of a symmetric gel is unstable against a spontaneous symmetry-breaking, and (ii) that this symmetry breaking results in a force generation that is capable of pushing the bead forward. It is also shown that the symmetry breaking leads to the formation of an actin comet, as is observed for the bacteria Listeria monocytogenes.

2. Basic model for passive motion

While this contribution is mainly devoted to non-equilibrium situations, we find it convenient to recall the main results of equilibrium vesicle shapes. This will also serve to introduce some preliminaries that are needed out of equilibrium.

2.1. A BRIEF SUMMARY ON EQUILIBRIUM SHAPES

We consider a vesicle (Figure 1), or a RBC, suspended in an aqueous solution. The membrane thickness (few nm) is small as compared to its radius (which is at least few μm), so that the membrane can be viewed as a 2D geometrical surface. The membrane is a 2D incompressible fluid. The only type of motion that is possible is bending. Think of a sheet of paper which is hard to stretch. The easiest (or soft) mode is bending. The bending is characterized by a mean curvature Helfrich [17] introduced the notion of curvature free energy for membranes. He suggested a model in which the cost in bending energy is given by

$$F = \frac{\kappa}{2} \int H^2 dA + \frac{\kappa_g}{2} \int K dA \tag{1}$$

H is the mean curvature, $H = (1/R_1 + 1/R_2)/2$, where $R_{1,2}$ are the principal radii of curvature, and K is the Gauss curvature, equal to $1/(R_1 R_2)$. The integrals are performed along the membrane area A, and κ and κ_g have the dimension of an energy (as does F), and represent the bending rigidity and the Gaussian rigidity, respectively. Expression (1) may be inferred from some general invariance considerations. Indeed, energy can not depend on reparametrization of the surface (for example change of coordinate systems). The bending of the membrane may be represented by a 2×2 matrix (for example, $\nabla_s \mathbf{n}$, where $\mathbf{n}$ is the unit vector normal to the membrane, and ∇_s is the gradient along the membrane) that measures the strength of local bending. A 2×2 matrix possesses two invariants, which are the trace and the determinant of $\nabla_s \mathbf{n}$. It is known in differential geometry that this matrix has $1/R_1$ and $1/R_2$ as eigenvalues. The trace is nothing but the mean curvature, and the determinant is the Gauss curvature. The energy should thus depend on these two quantities only. Furthermore, and because the membrane can be viewed as a two dimensional sheet, both concave and convex shapes having the same amplitude of their mean curvature should have the same energy (this is true to leading order where the internal structure of the membrane is ignored). Thus, the first plausible candidate in the free energy is[1] H^2. It is known that the second term in (1) is a topological invariant by virtue of a theorem of differential geometry (Gauss-Bonnet theorem). More precisely, for a given topology this term is a constant (independent of the shape; it has the same value for a sphere and an ellipse, etc...). That is to say, if one is not interested in a change of topology (like creation of two vesicles out of one, or transition from a closed membrane to a perforated membrane), then this contribution is a constant, and can be ignored. This is adopted here, since we do not consider topological changes.

Equilibrium shapes of vesicles then correspond to a minimum of $\kappa/2$ $\int H^2 dA$ subject to two constraints (i) fixed area (incompressible membrane), (ii) fixed volume (incompressible enclosed fluid). Mathematically this is dealt with by minimizing the quantity

$$E = \frac{\kappa}{2} \int H^2 dA + \zeta A + pV, \tag{2}$$

where ζ and p are Lagrange multipliers enforcing constant area A and constant volume V, respectively.

---

[1] Other candidates like $|H|$ could be considered. This singular behavior has no physical support.

Note that the curvature energy is scale invariant: changing lengths $\mathbf{r}$ by $\mu\mathbf{r}$ (where μ is a real constant) leaves the curvature energy invariant (since $H^2 \to H^2/\mu^2$ and $A \to A\mu^2$). This means that two vesicles having the same shape but having two different volume will cost the same curvature energy. This implies that the absolute scale is not selected by energy consideration. The relevant parameter is the reduced volume $v = [V/(4\pi/3)]/[A/4\pi]^{3/2}$. This definition measures the ratio of the actual volume over the volume of a sphere having the same area. Thus, a soccer ballon has $v = 1$. If the soccer ballon is deflated so that its actual volume is, say, $V = 0.7V_0$ (V_0 is the volume of the spherical ballon), then $v = 0.7$. v is a dimensionless number which is equal to one for a sphere, and $v < 1$ for any other shape. As a way of example, for a Human RBC $v \sim 0.6$.

It is obvious that for $v = 1$ the only possible shape is a sphere. For $v \neq 1$ there is an infinite number of different possible shape (a deflated ballon can be bent in a continuous manner by imposing infinitesimal deformations, so that an infinite family of shapes can be generated). Among this infinite manifold of possibilities, only special shapes are selected due the fact that they correspond to minimal curvature energy. The Helfrich model reveals a variety of equilibrium shapes, as shown on Figure 4. For Human RBCs, for example, where $v \sim 0.6$, the equilibrium shape obtained from the Helfrich model is the discocyte one shown on Figure 4. It is quite attractive to see that a simple model generates the shape of RBC's. The shapes shown on Figure 4 have been also observed experimentally on vesicles. The deflation of vesicles can be achieved by osmosis (addition of molecules in the suspending fluid) or by heating the solution (the thermal expansion of phospholipids is easier than that of the aqueous solution so that upon heating the surface of the vesicle increases faster than the enclosed volume, resulting thus in a decrease of the reduced volume v). The Helfrich model has given rise to numerous studies both theoretically and experimentally. The whole phase diagram[2] of equilibrium shapes is now fairly understood [36].

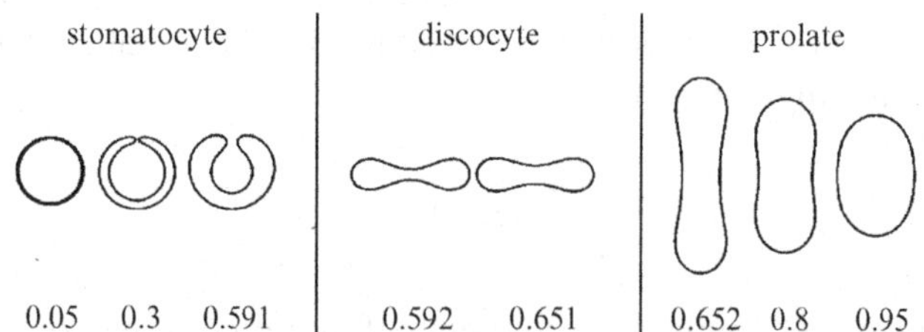

Figure 4.   Equilibrium shapes for different reduced volume. (From [36].)

---

[2]Note that the phase diagram is more complex than that shown on Figure 4. Indeed, there is at least a second new parameter which is the spontaneous curvature H_0 introduced in the

2.2. VESICLES UNDER NONEQUILIBRIUM CONDITIONS

If energy is injected into the system one can maintain the vesicle out-of equilibrium. Of particular interest is the motion under external flow, as occurs in
the blood circulatory system. It will be seen that even under a simple shear
flow the vesicle can exhibit several types of motions. An interesting fact is
that the type of motion directly impacts on rheology.

When a vesicle moves it has to advect the flow nearby. In addition, since
the membrane is fluid, and due to the fact that a shear flow has a rotational
velocity component, the membrane will rotate like a tank-tread. This induces a flow inside the vesicle. In most of experimental situations the viscous contribution prevails over inertia, so that the acceleration term in the
hydrodynamics equations can be neglected.

Let us first consider the case of a simple shear flow which will be written
as $\mathbf{U_0} = (\gamma y, 0, 0)$, where γ is the shear rate. The flow outside (and inside)
the vesicle is described, to a good approximation, by the Stokes equation (the
zero Reynolds number limit of the Navier-Stokes equations)

$$\eta \nabla^2 \mathbf{u} - \nabla p = 0, \quad \nabla.\mathbf{u} = 0, \tag{3}$$

where $\mathbf{u}$ and p are the velocity and the pressure fields respectively, and η
designates the viscosity. The fields referring to the interior of the vesicle
will be denoted with a bar (for example $\bar{p}$, $\bar{\mathbf{u}}$...). $\lambda = \bar{\eta}/\eta$ will designate the
viscosity contrast.

Some remarks are in order. Despite the fact that the Stokes equations
are linear the present problem is highly nontrivial due to the free boundary
character. Indeed, the shape of the vesicle is not known a priori and it has to be
solved for in a consistent manner. This triggers nonlinearities. In addition, the
problem is nonlocal in space. Indeed, any motion of the membrane at some
point will induce a flow that is felt somewhere else by distant point on the
membrane, since hydrodynamics are devoid of an intrinsic lengthscale (very
much like Coulomb interaction which is of long range). This is expressed
mathematically by the fact that the ith-component of the velocity of a given
point on the membrane (whose vector position at a given time is denoted
by $\mathbf{r}_m$) obeys a nonlinear integrodifferential equation given by [2] (actually

---

model as $\kappa/2 \int (H - H_0)^2 dA$, and which states that the natural curvature for an open membrane
would not be $H = 0$ but $H = H_0 \neq 0$, that is to say the membrane has a spontaneous curvature.
Other models are also introduced in the literature [36]).

the following equation is valid for any imposed external flow U_0 and not necessarily only a linear shear flow)

$$u_i(\mathbf{r}_m, t) = \frac{2U_0(\mathbf{r}_m)}{\lambda + 1} + 2\eta^{-1}(1 + \lambda)^{-1} \int dA\, G_{ij}(\mathbf{r}_m - \mathbf{r}'_m) F_j(\mathbf{r}_m, t)$$
$$+ 2\frac{\lambda - 1}{\lambda + 1} \int dA\, K_{ijl}(\mathbf{r}'_m - \mathbf{r}_m, t) u_j(\mathbf{r}'_m, t) n_l(\mathbf{r}'_m, t) \qquad (4)$$

where dA is the membrane area element (the integral is performed over the total area and $\mathbf{r}'_m$ sits on the membrane, and thus varies upon integration on the vesicle surface), n_i the ith component of the normal to the membrane. The second and third rank tensors, G_{ij} and K_{ijl} are given by

$$G_{ij}(\mathbf{r}) = \frac{1}{8\pi}\left[\frac{\delta_{ij}}{r} + \frac{r_i r_j}{r^3}\right], \qquad K_{ijl} = \frac{3r_i r_j r_l}{4\pi r^5} \qquad (5)$$

with $r = |\mathbf{r}|$. G_{ij} is also referred to as the Oseen tensor, or simply the Green's function associated with the Stokes flow. F_i is the ith component of the membrane force acting on the fluid. It is obtained from the (functional) derivative of the Helfrich energy (2) with respect to the membrane position. It can be split into a normal component F_n and a tangential one F_τ, $F_i = F_n n_i + F_\tau \tau_i$ (n_i and τ_i are the ith component of the unit normal and tangential vectors respectively). The normal component of the force [46] is given by (for a simple derivation in 2D, see [24]).

$$F_n = \kappa[2H(2H^2 - 2K) + 2\Delta_B H] - 2\zeta H \qquad (6)$$

Δ_B is the Laplace-Beltrami operator (or surface Laplacian), and $\zeta(\mathbf{r}_m, t)$ is a Lagrange multiplier which enforces local membrane incompressibility. Note that unlike the equilibrium situation where ζ is constant (Eq. (2)), here ζ depends on the given membrane point. Indeed, we must impose local membrane incompressibility, and not only a global one. The tangential [37] part of the force is given by

$$F_\tau = -g^{ij}\mathbf{R}_i \partial_i \zeta \qquad (7)$$

where g^{ij} are the elements of the inverse matrix of the metric $g_{ij} = \mathbf{R}_i.\mathbf{R}_j$ induced by the two tangential vectors $\mathbf{R}_i$. Repeated indices are to be summed over following Einstein's convention. Note that the Lagrange multiplier p associated with the volume constraint of the enclosed fluid is accounted for by $\nabla.\mathbf{u} = 0$. It is worthwhile to remark that at equilibrium the total force must vanish, and it follows from (7) that ζ is constant.

$\zeta(\mathbf{r}_m, t)$ is fixed from the surface projected divergence [31]

$$(\delta_{ij} - n_i n_j)\partial_i u_j = 0 \qquad (8)$$

ζ plays a similar role along the membrane (2D incompressible fluid) like the pressure field p which enforce local bulk fluid incompressibility.

Equation (4) is obtained by using the following conditions [2]: (i) continuity of the velocity of the fluid across the membrane. The continuity of the normal component follows from mass conservation. The continuity of the tangential velocity is a postulate based on the the non-slip condition. It is assumed also that the two monolayers forming the membrane constitute an entity, in that relative sliding is not allowed. This assumption is usually adopted. The effect of the monolayer sliding on membrane bending modes is discussed in [38]. (ii) The membrane velocity is equal to that of the adjacent fluid. This is valid as long as the the membrane is not permeable. (iii) The continuity of the total stress. More precisely, if σ_{ij} is the stress tensor of the fluid outside the vesicle, and $\bar{\sigma}_{ij}$ the corresponding quantity inside, we have

$$(\bar{\sigma}_{ij} - \sigma_{ij})n_i + F_j = 0 \tag{9}$$

This conditions expresses the fact that the resultant of forces due to hydrodynamics on both sides of the membrane are balanced by the membrane bending force F_j.

Brief Discussion on the methods   The linearity of the Stokes equations enables us to make use of the Green's function techniques. This would not have been possible if the full inertial term (including $\mathbf{u} \cdot \nabla\mathbf{u}$) were taken into account. It must be emphasized that thanks to the Green's function technique the velocity field inside and outside the vesicle has been integrated out. That is to say there is no need to solve for the field inside and outside in order to determine the evolution equation of a point of the membrane (the velocity of a membrane point is given by $u(\mathbf{r}_m, t)$ on the left hand side of Eq. (4)). Integrating out the velocity field in the bulk has been made at a certain price: nonlocality. In order to move a point on the membrane, there is a need to determine the shape and the velocity everywhere on the membrane (integral equation). What is gained by this technique is that there is no need to discretize the hydrodynamic equations in the bulk phases, which may prove to be a difficult problem, since the shape and the position of the boundary (the membrane) is not known a priori. Remeshing of the grids in the bulk phases, together with a mesh refinement close to the membrane, may prove necessary. This is why the Green's function technique is quite interesting, since it avoids these complications. However, if inertia are to be included (a situation which may prove essential if one is interested in flow in arteries and large veins in the circulatory system), or if the constitutive law is nonlinear (non newtonian fluids), then the Green's function techniques can not be used. An alternative approach is the phase-field method developed recently [4, 5].

If one is interested in the small Reynolds limit the integral formulation (sometimes also called the boundary integral formulation) is quite attractive and precise [2,6,7,24]. Note that if a wall is present within the system there is an additional integral (in addition to the second and third integral over the vesicle area whih appear one right hand side of Eq. (4)) which must be performed over the wall (see for example [9]). Actually a Green's function in the presence of a flat substrate is known, and an integration over a substrate can be circumvented (thanks to the technique of images, known in eletcro-dynamics) [34] if one makes use of that function. In the presence of two walls there is no explicit Green's function. The difficulty of obtaining an explicit form arises due to the presence of an infinite number of images (between two parallel mirrors, one would have an infinite number of images). Nevertheless, the free space Green's function (the Oseen tensor introduced above) can be used, but at the expense of additional surface integrals over the bounding walls (like in the case of a unique wall studied in [9] where the Oseen tensor was used).

Equation (4) constitutes the general evolution equation for a single vesicle in an unbouded flow. This equations has been exploited in 2D for vesicles (the 2D character saves computing time) in several circumstances [2, 7–9, 24], and in some cases in 3D [6, 26, 40]. Note also that other methods have been adopted as well in order to study vesicle dynamics. Of particular interest is the phase-field approach [5]. We shall focus below on some recent examples of dynamics before dealing with rheology.

2.3. VESICLES UNDER UNBOUNDED SIMPLE SHEAR FLOW

Under shear flow the vesicle may exhibit several interesting dynamics. Let us first introduce some preliminary notions. A simple shear flow, $\mathbf{U_0} = (\gamma y, 0, 0)$, is composed of a straining component along the angles $\pm\pi/4$ (see Figure 5) and a rigid rotation. For that purpose let us consider the strain rate $\nabla\mathbf{U_0}$ (which is a second rank tensor), which can be written as

$$\nabla\mathbf{U_0} = \begin{bmatrix} 0 & \gamma & 0 \\ 0 & 0 & 0 \\ 0 & 0 & 0 \end{bmatrix} = \begin{bmatrix} 0 & \frac{\gamma}{2} & 0 \\ \frac{\gamma}{2} & 0 & 0 \\ 0 & 0 & 0 \end{bmatrix} + \begin{bmatrix} 0 & \frac{\gamma}{2} & 0 \\ -\frac{\gamma}{2} & 0 & 0 \\ 0 & 0 & 0 \end{bmatrix} \tag{10}$$

If one considers only the first (symmetric) part one can write $\partial v_x/\partial y = \gamma/2$ and $\partial v_y/\partial x = \gamma/2$, which is solved by $v_x = \gamma y/2$ and $v_y = \gamma x/2$. Writing then $v_x = dx/dt$ and $v_y = dy/dt$ one easily finds the trajectory of a test particle to be given by $xdx - ydy = 0$, or $y^2 - x^2 = C$ (C is a constant of integration). This is a family of hyperboale (See Figure 5). This represents the

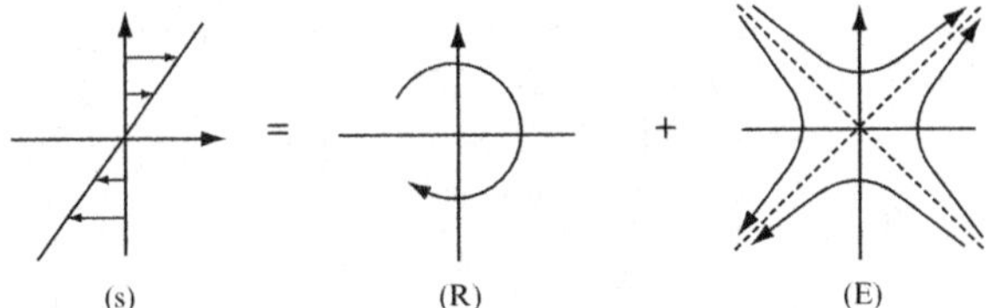

Figure 5.   A shear flow (S) can be split into a rotation (R) and an elongational part (E). (From [35].)

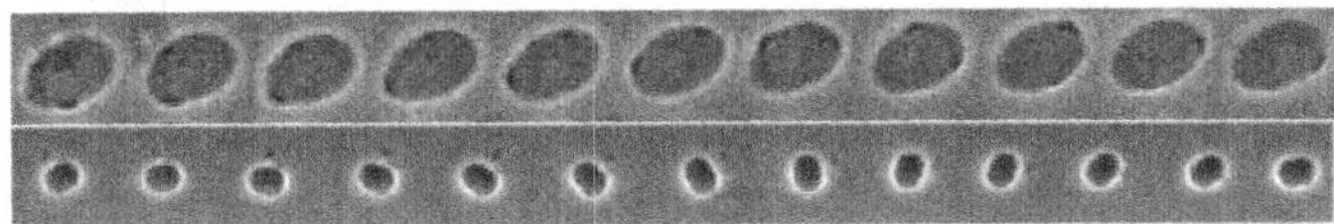

Figure 6.   A snapshot of a vesicle under a shear flow in the tank-treading regime (*upper panel*) and tumbling one (*lower panel*). In the TT regime the vesicle orients itself with a certain angle while its membrane moves like a tanktread, as shown by a dark defect bound to the membrane. (Courtesy of Thomas Podgorski.)

straining (or elongational) flow: straining along $\pi/4$ and compression along $-\pi/4$. A similar treatment of the second (antisymmetric) part of (10) yields $x^2 + y^2 = C$, which is the equation of a family of circles. This corresponds to a rigid body rotation. One can easily show that the rotation frequency is given by $-(\gamma/2)\hat{z}$. This means that if $\gamma > 0$ the rotation is clockwise.

Tank-treading   As seen above a simple shear flow has a rigid rotation component. This means that a solid body placed in a shear flow will undergo a rotation (or tumbling). However, due to the fact that the membrane is fluid the torque associated with the applied shear may be partially converted into a torque that causes the membrane to rotate while keeping the orientation of the vesicle fixed. This motion is known as tank-treading. Note that the orientation angle is always less than $\pi/4$ since the elongation is along $\pi/4$ but the torque due the applied flow, which causes a clockwise rotation, will always tend to decrease the orientation angle. Note also that a tank-treading motion is possible only if the (partial) conversion of the torque due to shear into the membrane is sufficient enough. The tank-treading motion can be occasionally visualized experimentally [30] thanks to the fact that sometimes a dust (or some defect) is bound to the membrane and in the course of time the defect rotates by following the membrane tank-treading, as seen on Figure 6 (top panel). Tank-treading of vesicles was studied numerically for vesicles by means of the boundary integral formulation [26].

Tumbling   The membrane tank-treading causes motion of the fluid inside and outside. Thus, if the internal fluid is too viscous (the ultimate limit is a

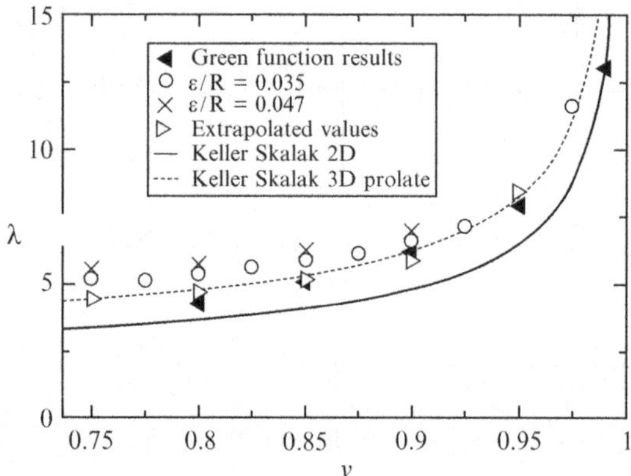

Figure 7.   The phase diagramme in the plane of the reduced volume v and the viscosity contrast λ. This is obtained by numerical simulations in 2D by using the boundary integral formulation (Eq. 4), and the phase field model (ϵ is the width of the diffuse interface in the phase field, and R is the vesicle radius), together with an analytical theory (assuming a fixed shape) due to Keller and Skalak. Shown also are the extrapolated values ($\epsilon \to 0$) obtained by the phase field model. (From [2].)

solid body) then the fluid dissipates too much energy and it becomes preferable for the fluid inside not to follow the membrane tank-treading. We will then have a tumbling bifurcation. Similarly, if the reduced volume v is small enough (deflated vesicle) then this means that the vesicle can be elongated sufficiently that the torque due to the applied shear becomes efficient and this promotes tumbling. We thus expect the bifurcation from tank-treading (*TT*) to tumbling (*TB*) to occur beyond a certain viscosity contrast which should decrease with the reduced volume. This task has been studied first numerically [2,5], and the results are summarized on Figure 7. Some remarks are in order: (i) The typical viscosity (i.e. for typical reduced volume close to that of human RBC) contrast at which tumbling takes place is not too high; it is of order $4-5$. (ii) Actually RBC undergoes tumbling even at a lower value, of about 2. This may be traced back to the fact that for vesicles the presence of the spectrin cytoskeleton is not accounted for. At present, it is still unclear how the cytoskeleton elasticity may affect the bifurcation towards tumbling.

Due to the complexity of the evolution equation, the phase diagramme (Figure 7) has been first obtained numerically [2,5] by using both the integral representation (Eq. 4) and the phase field model. More recently, an analytical theory has been presented [31]. For tractability the assumption of a quasi-spherical shape has been adopted. More precisely, the reduced volume v is supposed to be close to one. Alternatively, one can use the excess area Δ, defined by $A = 4\pi R_0^2 + \Delta$, A being the area of the vesicle and R_0 is the radius of a sphere having the same area. $\Delta = 0$ corresponds to a sphere; a quasi-spherical regime corresponds thus to small Δ. Δ and v are linked by the relation $\Delta = 4\pi[v^{-2/3} - 1]$. By assuming a shape preserving solution (that is the shape does not evolve) it is found [31] that the vesicle orientation angle is given by

$$\partial_t \psi = -\frac{\gamma}{2} + \gamma B \cos(2\psi), \quad B = \sqrt{\frac{2\pi}{15\Delta}\frac{60}{23\lambda + 32}} \tag{11}$$

For $B > 1/2$ there is a steady state solution given $\psi_0 \pm \frac{1}{2}\cos^{-1}(1/2B)$. It can easily be checked that the "$+$" solution is stable, while the "$-$" one is unstable. For $B < 1/2$ (or $\lambda > \lambda_c = -32/23 + 120/23\sqrt{2\pi/(15\Delta)}$) the tank-treading regime ceases to exist in favor of tumbling via a saddle-node bifurcation. Note also that the larger the excess area the smaller the critical viscosity contrast. For a sphere ($\Delta = 0$) λ_c diverges. For a phenomenological theory about tumbling, see [35].

Vacillating-breathing   Relaxing the assumption of a shape preserving solution (that is the shape is free to evolve) it has been found that the vesicle may undergo a new type of motion, that has been called vacillating-breathing (VB) [31]. In this motion the vesicle long axis undergoes oscillations about the flow direction, while its shape undergoes breathing (Figure 8). Let us define a dimensionless number C_a as

$$C_a = \frac{\eta \dot{\gamma} r_0^3}{\kappa} \equiv \tau \dot{\gamma} \tag{12}$$

where τ is a typical time scale for the relaxation of the vesicle towards its equilibrium when the flow is set to zero. In some sense C_a can also be viewed as a measure of how far from mechanical equilibrium the vesicle is in the course of its shear induced motion. C_a measures the competition between shear which tends to elongate the vesicle, and bending force that tend to maintain the shape close to equilibrium. A small C_a means a fast response to shear: the vesicle shape is slaved to shearing. On the shear time scale, the

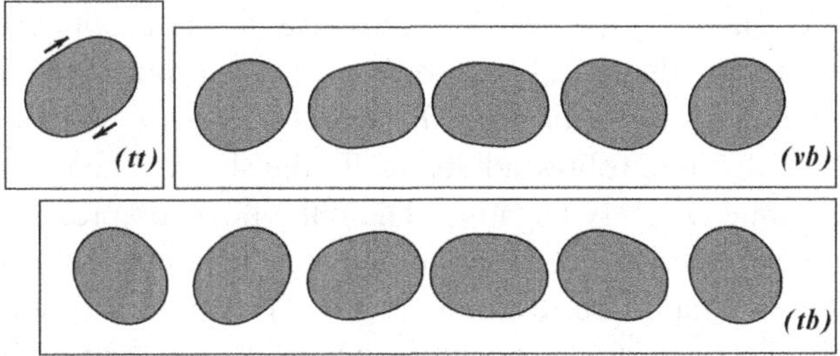

Figure 8.   A snapshot of the TT, VB and TB mode over one period. (From [10].)

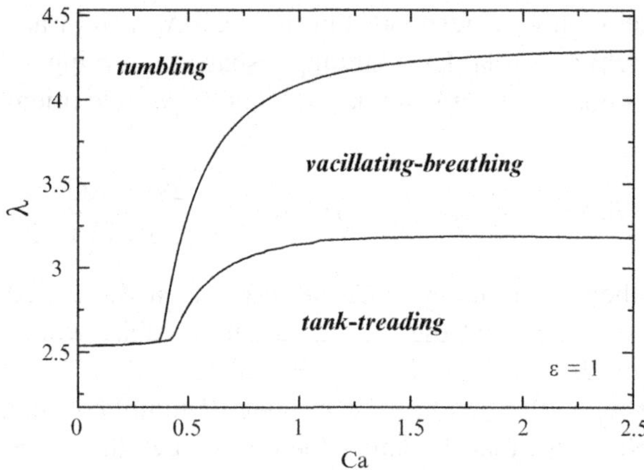

Figure 9. The phase diagramme for the *TT*, *VB* and *TB* modes. (Adapted from [10].)

vesicle behaves as a shape preserving entity. Contrariwise, a large C_a means that the shape exhibits a delay with respect to shearing. The phase diagram exhibiting the three types of motion (*TT*, *TB* and *VB*) is shown on Figure 9 [11, 28, 32]. At small C_a (quasi shape-preserving) one has a direct bifurcation from *TT* to *TB*. At $C_a \sim 1$, due to shape evolution, the *TB* is preceded by the new motion, namely *VB*. The *TT* towards the *VB* mode becomes a Hopf bifurcation, while the *VB* to *TB* mode is not a bifurcation, it is the continuation of the *VB* mode: upon increasing λ the *VB* excursion angle increases untill it reaches $\pm\pi/4$, in which case full rotation of the vesicles axes occurs: tumbling takes place.

The basic understanding of the *vb* mode is as follows. First we recall that a shear flow is a sum of a elongational part along $\pm\pi/4$ (which elongates the vesicle for $\psi > 0$ and compresses it for $\psi < 0$) and a rotational part, tending to make a clockwise *TB*. Due to the membrane fluidity the torque associated with the shear is partially transferred to *TT* of the membrane, so that (due to torque balance) the equilibrium angle for *TT* is $0 < \psi_0 < \pi/4$. Furthermore, an elongated vesicle tumbles more easily than a compressed one [5]. Suppose we are in the *tt* regime ($\psi_0 > 0$), but in the vicinity of *tb*, so $\psi_0 \simeq 0$. For small C_a the vesicle's response is fast as compared to shear, so that its shape is adiabatically slaved to shear (a quasi shape-preserving dynamics): a direct bifurcation from *TT* to *TB* occurs [5]. When $C_a \simeq 1$, the shape does not anymore follow adiabatically the shear. When tumbling starts to occur ψ becomes slightly negative. There the flow compresses the vesicle. Due to this, the applied torque is less efficient. The vesicle feels, so to speak, that its actual elongation corresponds to the *TT* regime and not to *TB*. The vesicle returns back to its *TT* position, where $\psi > 0$, and it feels now an elongation (which manifests itself on a time scale of the order of $1/\dot\gamma$). Due to elongation in this position, tumbling becomes again favorable, and the vesicle returns to $\psi < 0$, and so on. We may say that the vesicle *hesitates* or *vacillates* between *TB* and *TT*. The compromise is the *VB* mode.

Finally an other type of motion, namely spinning has been reported, and called by the authors spinning [27]. Spinning is a sort of kayaking, and is very similar in nature to the precession of a rigid ellipsoid under shear (when the ellipsoind is initially out of the shear plane), as discussed in the seminal paper by Jeffery [19]. We shall not dwell further on this mode.

Experimental studies on the three types of motion (*TT*, *TB* and *VB*) were considered in several papers. The *TT* of vesicles was studied in [12] and later in [22], and tumbling was extensively studied in [30] and [23]. The *VB* mode was briefly reported on in [23] (called there trembling) and in [30] (called there a transition region), but a systematic experimental analysis of this mode is lacking.

Dynamics of RBC   RBC are known to undergo tank-treading [15] and tumbling [25]. Normal RBC in the plasma undergo tumbling. Indeed, at physiological temperature the viscosity of hemo- globin (the internal solution) is of about $\lambda \sim 5 - 7$ times the viscosity of the plasma. If one takes vesicles as a model for RBC, one would also expect tumbling to take place at about $\lambda = 4$. Actually RBCs undergo a transition towards tumbling at about $\lambda \simeq 2$ (about a factor two as compared to the calculated values for vesicles). It is likely that the fact that the RBC have a cytoskeleton may explain this difference, but at present this discrepancy is not fully understood.

Other facts are observed with RBCs. For example, RBC dynamics depend quite significantly on the shear rate γ. The viscosity contrast can be modified by adding macromolecules (dextran molecules) in the suspended fluid; this increases the external viscosity, thus lowering the contrast λ. It is found that if the viscosity contrast is small enough (so that *TT* is expected), but if γ is small (say of about 1 *Hz*) then the RBC undergoes tumbling instead. Upon increasing γ, a transition from *TB* to *TT* is observed [1]. In addition, a somewhat similar motion to VB has been reported on [1]. This motion has been called *swinging*. It is not clear whether this is a new branch, or does it have something to do with the VB mode. More complex dynamics (quasi-periodic) are predicted [39] and observed [1].

3.   Rheology of vesicle and RBC suspensions: micro/macro link

Let us now turn to the rheology problem. We consider, from a theoretical point of view, a suspension of vesicles. Experimentally both systems have been analyzed. The concentration, or volume fraction ϕ of the suspension (that is the ratio of the volume of the vesicles over that of the full sample) is taken to be small enough so that we can neglect hydrodynamic interactions between vesicles. This has also the advantage of identifying some key ingredients before including interactions. Surprisingly enough, even in the very dilute regime the rheology has proven to be quite rich, and has been recently studied experimentally both for vesicle and RBC suspensions.

Ordinary fluids (simple fluids) and elastic solids are described by universal equations (Navier-Stokes and Lamé equations). The very basic idea in simple fluids is that the microscopic motion (molecular scale) is so rapid (of the order of, or less than, a nanosecond) so that at the scale of the macroscopic flow the micro/macro interaction can be neglected: the micro-modes are slaved to macro-modes. The fluid dynamics in that case is described by the macroscopic density and the velocity. Contrariwise, for a suspension of vesicles (and RBCs), and more generally for complex fluids like polymer solutions etc.., the dynamics of the suspended entities is slow and of comparable order to the macro-evolution, so that a separation of micro-scales (represented by the suspended entities) and the global scale of the flow is not legitimate. The constitutive law should carry information on the micro-scale despite the fact that the law of the composite fluid follows from an averaging procedure (the volume average, plus average over noise in the case of brownian particles – not considered here).

3.1. STRESS AVERAGE

As stated in Section 2.3 an analytical theory is possible for a small excess area from the sphere [11, 31]. This means that the velocity and the pressure field (at thus the stress tensor), and the evolution of the shape are known analytically [11,31]. Thus it is possible to make a volume average of the stress tensor, which contains a contribution from the fluid in the absence of vesicle and one stemming from vesicles. Once the volume averaging is extracted one can write the constitutive law [11]. Here we shall discuss a principal result regarding the behaviour of the effective viscosity. We shall also stress on how does the bifurcation from one mode (e.g. *TT*) to another one affect rheology. Our analysis shows a clear macroscopic signature of the underlying microscopic (individual vesicles) dynamics.

3.2. EFFECTIVE VISCOSITY

Let us put this study in the context of suspensions. The first theoretical treatment about the calculation of the effective viscosity of a suspension was given in [13, 14] who provided the famous expression of the effective viscosity of a dilute suspension of rigid spherical particles. The intrinsic viscosity is given by

$$[\eta] \equiv \frac{\eta_s - \eta}{\eta \phi} = \frac{5}{2} \tag{13}$$

where η_s is the suspension viscosity, η the viscosity of the ambient fluid, and ϕ is the volume fraction of the suspended particles. Later Taylor [42] provided the analogous expression for an emulsion (suspension of quasi-spherical droplets in an abient fluid having different viscosity)

$$[\eta] = \frac{5\lambda/2 + 1}{\lambda + 1} \tag{14}$$

which reduces to the Einstein relation for $\lambda \gg 1$. More recently an expression has been derived in the TT regime for quasi-spherical vesicles [31]

$$[\eta]_{TT} = \frac{5}{2} - \Delta \frac{23\lambda + 32}{16\pi} \tag{15}$$

(the subscript stands for tank-treading motion), where we recall Δ is the excess area relative to a sphere. For $\Delta = 0$ the Einstein viscosity is recovered, even though both the membrane and the enclosed solution are fluid (for a simple explanation, see [31]). For $\Delta \neq 0$, the case of interest, several noticeable differences with droplets are worth of mention: (i) the viscosity *decreases* with λ for vesicles, while the contrary is found for droplets. (ii) When λ is large expression (15) does not tend to $5/2$, as does (14). Indeed, as shown in [31] this expression is valid in the tank-treading regime only. For large enough λ the TT regime ceases to exist in favor of TB [31]. At low enough shear rate (or small C_a) there is a direct bifurcation from TT to TB (Figure 9). In the low shear rate regime, it is legitimate, in the TB regime, to assume a motion with a shape-preserving solution. Following the general expression given in [11] for the instantaneous viscosity, we have been able to compute analytically the effective viscosity in the TB regime. We give here the results, while technical computational details will be given elsewhere

$$[\eta]_{TB} = \frac{5}{2} + \sqrt{\frac{30}{\pi}} \left| \frac{\sqrt{\Delta - 4h^2}}{\sqrt{\Delta + 4h^2} + \sqrt{\Delta}} - h \right| \tag{16}$$

with $h = 60\sqrt{2\pi/15}/(23\lambda + 32)$. Note that the tumbling domain corresponds to $4h^2 < \Delta$ (the opposite limit is the domain of TT); see Eq. (11) and the discussion after that equation. Figure 10 shows the behavior of $[\eta]$ for vesicles in both the tt and TB regimes. At the bifurcation, one has a cusp singularity, with a linear behavior on the TT side, and a square root singularity on the

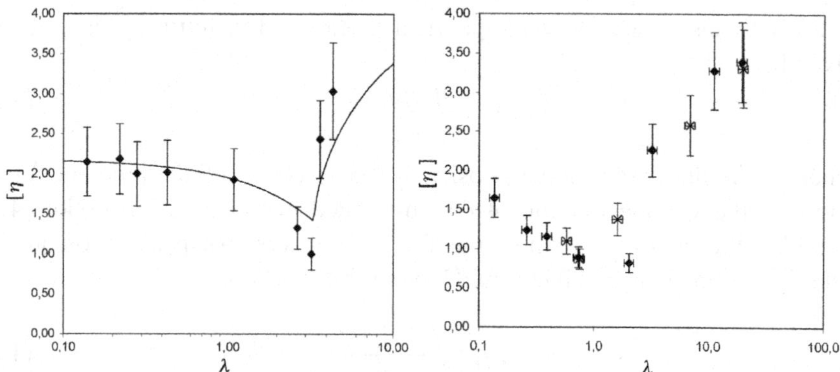

Figure 10. Left: The solid line represent the theoretical results (Eqs. (14,15), while the symbols are the experimental results for vesicle suspensions. Right: the experimental results for RBC suspensions. (From [43].)

TB side. The results obtained from experiments [43] are reported as well. In Figure 10 the behaviors of $[\eta]$ for RBC suspensions is shown. One sees that for both vesicle and RBC suspensions the same trend is obtained. For RBC suspensions we refrain from drawing the theoretical curve since the theory is developed for small Δ, while for RBC $\Delta \sim 4$. In addition, the cytoskeleton is not taken into account, and it will be essential to elucidate this task in future research. It is, however, quite surprising that the rheology of RBC remains qualitatively similar to that of vesicle suspensions.

4. Actin assisted motility

Hitherto we have considered the case of motion due to flow only. Cells of the immune system and many microorganisms (some viruses, and bacteria) use the actin growth machinery to propel themselves forward. When a cell receives an external signal the actin monomers (which are proteins) in the cell cytoplasm assemble into a cross-linked gel (see Figure 3). It has been believed for a long time that in addition to the actin polymerization, molecular motors (intercalated between the actin filaments) are necessary for cell motility. However, some bacteria, e.g. Listeria monocytogenes [45], move inside their host cells by using the actin polymerization machinery of the host, without the assistance of molecular motors. This discovery pointed to the important nontrivial fact that motors are not necessary to induce motion.

Actin polymerization and cross-linking is triggered by an enzyme (ActA or Wasp activates the so called Arp2/3 complex which nucleates new actin filaments on preexisting ones) on the external side of the bacterial membrane and leads initially to the growth of a symmetric gel around the bacterium.

Later, this actin shell undergoes a symmetry breaking and develops into a comet that propels the bacteria forward in their host cell.

Later, it has been demonstrated that an artificial bead (or a droplet) [3,44] which is coated with the ActA enzyme, and put in a solution containing actin monomers (and other necessary molecules, but not molecular motors), can show a similar phenomenon. More precisely, the gel grows initially in a symmetric fashion around the bead, until a thickness reaches a steady-state.[3] Then a spontaneous symmetry-breaking occurs leading to a comet formation and the motion of the bead.

5.   Model for symmetry breaking in actin gel

Our strategy is to understand if it is possible to capture the symmetry-breaking and the force generation in the most simple picture. The idea is to treat the gel as a continuum in the linear elasticity theory, and write simple kinetic relation expressing growth (or polymerization) [20]. We consider a bead (radius r_1) surrounded by a growing elastic actin gel (radius r_2) shown in Figure 11. The gel is stressed by a small molecular displacement in normal direction $u_r = L$ at the bead/gel interface. This choice is motivated by the microscopic picture, that for the addition of monomers, enzymes facilitate a molecular displacement in the gel. This displacement is the source of stress. Of course the history of the growth may induce some residual stress in the system, but we have found that including a prestress does not change the qualitative picture. At the bead as well as at the external gel surface no shear stress condition is applied, while the normal stress at the external

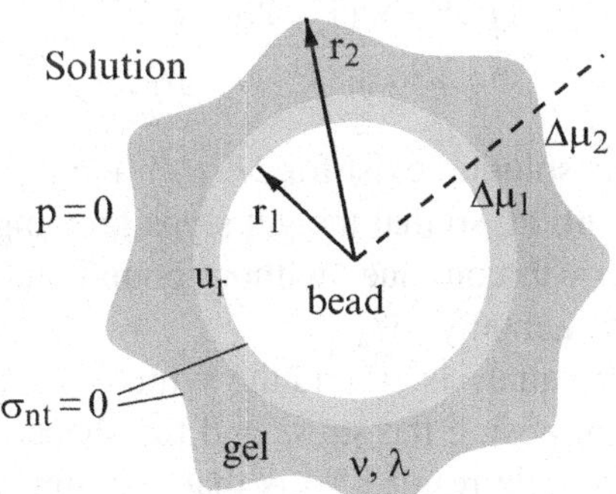

Figure 11.   Schematic view of a bead surrounded by an elastic gel with the Lamé coefficients v and λ. (From [20].)

---

[3]It may happen in some cases that no steady state thickness is attained, and the symmetry-breaking overrides the growing gel [33].

surface is set to zero (actually it can be set to the liquid pressure, but this is unimportant). Once the boundary conditions on the stress are known one can determine, in principle, the stress distribution in the gel. Next is to compute the cost in elastic energy per unit mass (or chemical potential) in order to insert a monomer on the bead. At the external gel surface one can also consider polymerization and depolymerization, but this does not affect the results from a qualitative point of view. Actually the internal surface is more active due to the presence of the enzyme. We then write a kinetic relation of the form

$$\partial_t r_2 = -M_1 \Delta\mu_1 \,, \tag{17}$$

where M_1 denotes a mobility and the difference in the chemical potential between a volume element in the gel and in solution at the internal (external) interface, respectively. Had we considered the external surface as well, we would then have included in the above equation a term like $-M_2\Delta\mu_2$, accounting for kinetics at the external surface. This is unimportant [20] for the main qualitative feature. Here we assume that the mobility is associated with the polymerization/depolymerization kinetics, which constitutes the prevailing dissipation mechanism. The chemical potential is composed of a contribution due to the gain in polymerization (denoted as $\Delta\mu_p < 0$) and an elastic part [20]

$$\Delta\mu_1 = \Delta\mu_p + \nu u_{ij} u_{ij} + \frac{\lambda}{2} u_{kk}^2 - \sigma_{nn}(1 + u_{kk}) \tag{18}$$

The stress problem can be solved analytically for a spherical geometry (symmetric growth) and it is found that the gel thickness is given by the following expression [20]

$$\bar{r}_2 = \left(2\frac{E\alpha - (1 - 2\sigma)\Delta\mu_p}{2E\alpha + (1 + \sigma)\Delta\mu_p} \right)^{(1/3)} r_1 \tag{19}$$

where $\alpha = L/r_1$. This solution exist for $2E\alpha/(1 + \sigma) \geq -\Delta\mu_p$: elasticity acts against monomer addition, so that the gel stops growing at that thickness. In the opposite limit growth continues without bound and both situations have been identified experimentally [33].

The linear stability analysis around the symmetric case can be performed analytically (by decomposing the stress and the shape evolution onto spherical harmonics). The basic result [20] is that a symmetric shape is unstable against symmetry breaking. The interesting fact is that the mode which corresponds to translation of the external surface with respect to the bead is the most unstable. In order to ascertain the subsequent evolution of the external boundary (i.e. in the fully nonlinear regime), a full numerical analysis must be performed. The result is the following. For all initial conditions used so

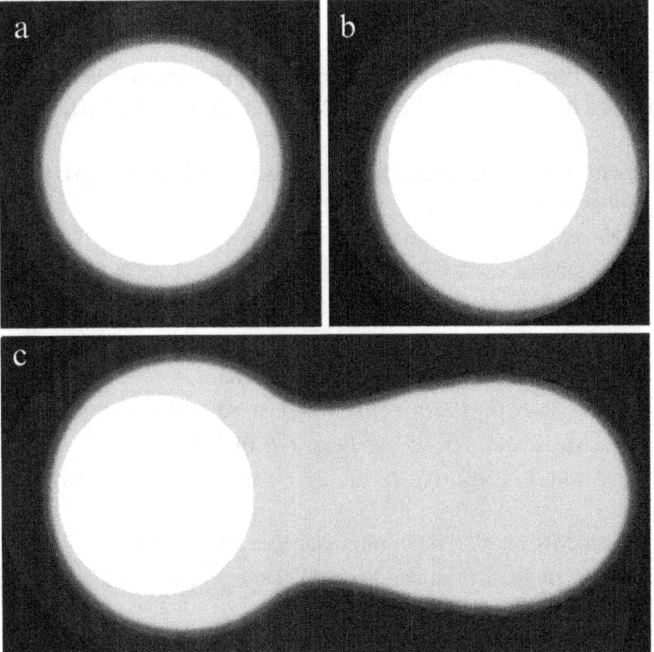

Figure 12. Symmetry-breaking of a circular gel. Shown is the evolution of the gel thickness, starting from a homogeneous thin gel (a) with random small amplitude perturbations. (b) shows the initial symmetry breaking, while (c) shows the subsequent evolution of the shape into a comet in the far nonlinear regime. (Adapted from [20, 21].)

far the ultimate stage is the formation of an actin comet, which is reminiscent of the comet developed by Listeria monocytogenes [20]. Figure 12 shows the result of numerical simulations.

This finding points to the fact that the comet formation is a robust feature; it results from simple physical prototypes. For a symmetric gel, the total normal stress integrated on the bead is zero. Symmetry-breaking leads to a net normal force, which is found to point in the opposite direction than the comet [21]. This means that this force is, in principle, able to push the bead forward. For the force to be communicated to the bead, a dissipation mechanism can be evoked. We may introduce a linear relationship between the force and velocity of the bead. This allows one to study the motion of the bead. This question is currently under investigation.

Acknowledgements

I would like to thank Gerrit Danker and Karin John for their help during the preparation of this contribution.

References

1. Abkarian, M., Faivre, M., and Viallat, A. (2007), *Phys. Rev. Lett.* **98**, 188302.

2. Beaucourt, J., Rioual, F., Séon, T., Biben, T., and Misbah, C. (2004), *Phys. Rev. E* **69**, 011906.

3. Bernheim-Groswasser, A., Wiesner, S., Golsteyn, R. M., Carlier, M.-F., and Sykes, C. (2002), *Nature* **417**, 308.

4. Biben, T., Kassner, K., and Misbah, C. (2005), *Phys. Rev. E* **72**, 041921.

5. Biben, T. and Misbah, C. (2003), *Phys. Rev. E* **67**, 031908.

6. Biben, T. and Misbah, C. (2008), *Three dimensional simulation of vesicles under flow, Preprint.*

7. Cantat, I. and Misbah, C. (1999a), *Phys. Rev. Lett.* **83**, 235.

8. Cantat, I. and Misbah, C. (1999b), *Phys. Rev. Lett.* **83**, 880.

9. Cantat, I. and Misbah, C. (2003), *Eur. Phys. J. E* **10**, 175.

10. Danker, G., Biben, T., Podgorski, T., Verdier, C., and Misbah, C. (2007), *Phys. Rev. E* **76**, 041905.

11. Danker, G. and Misbah, C. (2007), *Phys. Rev. Lett.* **98**, 088104.

12. de Haas, K. H., Blom, C., van den Ende, D., Duits, M. H. G., and Mellema, J. (1997), *Phys. Rev. E* **56**, 7132.

13. Einstein, A. (1906), *Annalen der Physik.* **19**, 289.

14. Einstein, A. (1911), *Annalen der Physik.* **34**, 591.

15. Fischer, T., Stohr-Lissen, M., and Schmid-Schonbein, H. (1978), *Science* **202**, 894.

16. Gerbal, F., Laurent, V., Ott, A., Carlier, M.-F., Chaikin, P., and Prost, J. (2000), *Eur. Biophys. J.* **29**, 134.

17. Helfrich, W. (1973), *Zeitschrift für Naturforschung Teil C* **28**, 693.

18. Jamet, D. and Misbah, C. (2007), *Phys. Rev. E* **76**, 051907.

19. Jeffery, G. (1922), *Proc. R. Soc. Lond.* **A101**, 161.

20. John, K., Peyla, P., Kassner, K., Prost, J., and Misbah, C. (2008a), *Phys. Rev. Lett.* **100**, 068101.

21. John, K., Caillerie, D., Peyla, P., Ismail, M., Raoult, A., Prost, J., and Misbah, C. in "Cell mechanics: From single scale-based models to multiscale modeling"(A. Chauviere, L. Preziosi, C. Verdier, Eds., CRC Press, Chapman and Hall, Florida), to appear (2009).

22. Kantsler, V. and Steinberg, V. (2005), *Phys. Rev. Lett.* **95**, 258101.

23. Kantsler, V. and Steinberg, V. (2006), *Phys. Rev. Lett.* **96**, 036001.

24. Kaoui, B., Ristow, G. H., Cantat, I., Misbah, C., and Zimmermann, W. (2008), *Phys. Rev. E* **77**, 021903.

25. Keller, S. and Skalak, R. (1982), *J. Fluid Mech.* **120**, 27.

26. Kraus, M., Wintz, W., Seifert, U., and Lipowsky, R. (1996), *Phys. Rev. Lett.* **77**, 3685.

27. Lebedev, V., Turitsyn, K., and Vergeles, S. (2008), *New J. Phys.* **10**, 043044.

28. Lebedev, V. V., Turitsyn, K. S., and Vergeles, S. S. (2007), *Phys. Rev. Lett.* **99**, 218101.

29. Loisel, T. P., Boujemaa, R., Pantaloni, D., and Carlier, M.-F. (1999), *Nature* **401**, 613.

30. Mader, M., Vitkova, V., Abkarian, M., Viallat, A., and Podgorski, T. (2006), *Eur. Phys. J. E* **19**, 389.

31. Misbah, C. (2006), *Phys. Rev. Lett.* **96**, 028104.

32. Noguchi, H. and Gompper, G. (2007), *Phys. Rev. Lett.* **98**, 128103.

33. Plastino, J., Lelidis, I., Prost, J., and Sykes, C. (2004), *Eur. Biophys. J.* **33**, 310.

34. Pozrikidis, C. (1992) *Boundary Integral and Singularity Methods for Linearized Viscous Flow*, Cambridge texts in applied maths 7 (Kindle Edition, 2001).

35. Rioual, F., Biben, T., and Misbah, C. (2004), *Phys. Rev. E* **69**, 061914.

36. Seifert, U. (1997), *Adv. Phys* **46**, 13.

37. Seifert, U. (1999), *Eur. Phys. J. B.* **8**, 405.

38. Seifert, U. and Langer, S. A. (1993), *Europhys. Lett.* **23**, 71.

39. Skotheim, J. M. and Secomb, T. W. (2007), *Phys. Rev. Lett.* **98**, 078301.

40. Sukumaran, S. and Seifert, U. (2001), *Phys. Rev. E* **64**, 011916.

41. Svitkina, T.M., Verkhovsky, A.B., McQuade, K.M. and Borisy, G.G. (1997), *J. Cell Biol.* **139**, 397.

42. Taylor, G. (1934), *Proc. R. Soc. Lond.* **A146**, 501.

43. Vitkova, V., Mader, M.-A., Polack, B., Misbah, C., and Podgorski, T. (2008), *Biophys. J. Lett.* **95**, L33.

44. Wiesner, S., Helfer, E., Didry, D., Ducouret, G., Lafuma, F., Carlier, M.-F., and Pantaloni, D. (2003), *J. Cell. Biol.* **160**, 387.

45. Yarar, D., To, W., Abo, A., and Welch, M. D. (1999), *Curr. Biol.* **9**, 555.

46. Zhong-can, O.-Y. and Helfrich, W. (1989), *Phys. Rev. A* **39**, 5280.

Made in the USA
Monee, IL
07 July 2026

56552848R00162